Topics in Spectral Geometry

GRADUATE STUDIES
IN MATHEMATICS **237**

Topics in
Spectral
Geometry

Michael Levitin
Dan Mangoubi
Iosif Polterovich

AMERICAN
MATHEMATICAL
SOCIETY

Providence, Rhode Island

EDITORIAL COMMITTEE

Matthew Baker
Marco Gualtieri
Gigliola Staffilani (Chair)
Jeff A. Viaclovsky
Rachel Ward

2020 *Mathematics Subject Classification.* Primary 35Pxx; Secondary 47A75, 58C40, 58J50, 58J53, 65N25.

For additional information and updates on this book, visit
www.ams.org/bookpages/gsm-237

Library of Congress Cataloging-in-Publication Data

Names: Levitin, Michael, 1963- author. | Mangoubi, Dan, 1974- author. | Polterovich, Iosif, 1974- author.
Title: Topics in spectral geometry / Michael Levitin, Dan Mangoubi, Iosif Polterovich.
Description: Providence, Rhode Island : American Mathematical Society, [2023] | Series: Graduate studies in mathematics, 1065-7339 ; volume 237 | Includes bibliographical references and index.
Identifiers: LCCN 2023030372 | ISBN 9781470475253 (hardcover) | ISBN 9781470475482 (paperback) | ISBN 9781470475499 (ebook)
Subjects: LCSH: Spectral geometry. | Eigenfunctions. | AMS: Operator theory – General theory of linear operators – Eigenvalue problems. | Global analysis, analysis on manifolds – Calculus on manifolds; nonlinear operators – Spectral theory; eigenvalue problems. | Global analysis, analysis on manifolds – Partial differential equations on manifolds; differential operators – Spectral problems; spectral geometry; scattering theory. | Global analysis, analysis on manifolds – Partial differential equations on manifolds; differential operators – Isospectrality. | Numerical analysis – Partial differential equations, boundary value problems – Eigenvalue problems.
Classification: LCC QA614.95 .L48 2023 | DDC 516/.07–dc23/eng/20230927
LC record available at https://lccn.loc.gov/2023030372

10 9 8 7 6 5 4 3 2 1   28 27 26 25 24 23

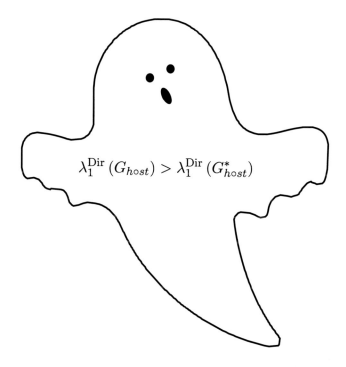

$$\lambda_1^{\text{Dir}}(G_{host}) > \lambda_1^{\text{Dir}}(G_{host}^*)$$

spectral *adj.* \ ˈspɛktrəl \

2. a. Having the character of a spectre or phantom; ghostly, unsubstantial, unreal.

5. a. Of or pertaining to, appearing or observed in, the spectrum. Also applied to a property or parameter which is being considered as a function of frequency or wave-length, or which pertains to a given frequency range or value within the spectrum.

From Oxford English Dictionary

Contents

Preface   ix

Introduction   xiii
   Overview   xiii
   Plan of the book   xiv
   Possible courses based on this book   xvi
   What is not in this book: Some further reading   xvii

Chapter 1. Strings, drums, and the Laplacian   1
   §1.1. Basic examples   1
   §1.2. The Laplacian on a Riemannian manifold   12

Chapter 2. The spectral theorems   25
   §2.1. Weak spectral theorems   25
   §2.2. Elliptic regularity and strong spectral theorems   40

Chapter 3. Variational principles and applications   55
   §3.1. Variational principles for Laplace eigenvalues   55
   §3.2. Consequences of variational principles   65
   §3.3. Weyl's law and Pólya's conjecture   82

Chapter 4. Nodal geometry of eigenfunctions   99
   §4.1. Courant's nodal domain theorem   99
   §4.2. Density of nodal sets   112
   §4.3. Yau's conjecture on the volume of nodal sets   117
   §4.4. Nodal sets on surfaces and eigenvalue multiplicity bounds 131

Chapter 5.   Eigenvalue inequalities   137
§5.1.   The Faber–Krahn inequality   137
§5.2.   Cheeger's inequality and its applications   149
§5.3.   Upper bounds for Laplace eigenvalues   158
§5.4.   Universal inequalities   173

Chapter 6.   Heat equation, spectral invariants, and isospectrality   183
§6.1.   Heat equation and spectral invariants   183
§6.2.   Isospectral manifolds and domains   193

Chapter 7.   The Steklov problem and the Dirichlet-to-Neumann map   213
§7.1.   The Steklov eigenvalue problem   213
§7.2.   The Dirichlet-to-Neumann map and the boundary Laplacian   229
§7.3.   Steklov spectra on domains with corners   237
§7.4.   The Dirichlet-to-Neumann map for the Helmholtz equation   253

Appendix A.   A short tutorial on numerical spectral geometry   265
§A.1.   Overview   265
§A.2.   Learning **FreeFEM** by example   273
§A.3.   List of downloadable scripts   284

Appendix B.   Background definitions and notation   287
§B.1.   Sets   287
§B.2.   Function spaces   288
§B.3.   Regularity of the boundary   290

Image credits   293

Bibliography   297

Index   321

Preface

Various distinct physical phenomena, such as wave propagation, heat diffusion, electron movement in quantum physics, oscillations of fluid in a container, can be modelled mathematically using the same differential operator — the Laplacian. Its spectral properties depend in a subtle way on the geometry of the underlying object, e.g., a Euclidean domain or a Riemannian manifold, on which the operator is defined. This dependence — or, rather, the interplay between the geometry and the spectrum — is the main subject of *spectral geometry*.

The roots of spectral geometry go back to the famous experiments of the physicist Ernst Chladni with vibrating plates in the late eighteenth century to early nineteenth century, as well as to the investigations of Lord Rayleigh on the theory of sound some decades later. The celebrated question of Mark Kac, "Can one hear the shape of a drum?", motivated a lot of research in the second half of the twentieth century and helped spectral geometry to emerge as a separate branch of geometric analysis.

Modern spectral geometry is a rapidly developing area of mathematics, with close connections to other fields, such as differential geometry, mathematical physics, number theory, dynamical systems, and numerical analysis. It is a vast subject, and by no means does this book pretend to be comprehensive. Our goal was to write a textbook that can be used for a graduate or an advanced undergraduate course, starting from the basics but at the same time covering some of the exciting recent developments in the area which can be explained without too many prerequisites. The authors have taught such courses over the past few years at different locations, in particular at the Université de Montréal and the Hebrew University of Jerusalem, and they have taught shorter courses at the Universities of Cardiff and Reading, as well as at several summer schools and instructional conferences; see, e.g., [**BouLev07**]. The present book is based in part on our lecture notes.

Acknowledgements. We gratefully acknowledge the influence of many earlier books on spectral geometry and related subjects, by Courant and Hilbert [**CouHil89**], Berger, Gauduchon, and Mazet [**BerGauMaz71**], Reed and Simon [**ReeSim75**], Bandle [**Ban80**], Chavel [**Cha84**], Bérard [**Bér86**], Davies [**Dav89**], [**Dav95**], Schoen and Yau [**SchYau94**], Rosenberg [**Ros97**], Henrot [**Hen06**], Helffer [**Hel13**], and Shubin [**Shu20**], to name just a few, as well as some of the more recent lecture notes by Laugesen [**Lau12**], Canzani [**Can13**], Buhovsky [**Buh16**], and Logunov and Malinnikova [**LogMal20**]. Of course, the standard disclaimer is that the choice of the topics in this book reflects the personal tastes and preferences of the authors.

Many people contributed to this book in different ways. It is a pleasure to thank our mentors Robert Brooks, Yakar Kannai, Victor Lidskii, Leonid Polterovich (to whom we are particularly thankful for encouraging this book project since its very early stages), and Dmitri Vassiliev, for introducing us to geometric spectral theory. Through the years, we have also been greatly influenced by collaborations and innumerable helpful discussions with Michiel van den Berg, E. Brian Davies, Lennie Friedlander, Nikolai Nadirashvili, Leonid Parnovski, and Mikhail Sodin, among others.

While preparing the manuscript we used the notes taken by Simon St-Amant at a course given by the third author, Iosif Polterovich, at the Université de Montréal. We are especially grateful to Matteo Capoferri, Philippe Charron, Stefano Decio, Emily Dryden, Alexandre Girouard, Asma Hassannezhad, Mikhail Karpukhin, Jean Lagacé, Antoine Métras, Nilima Nigam, and David Sher, who made a lot of insightful comments and suggestions on the preliminary draft of the book.

We thank Lyonell Boulton, Dorin Bucur, Simon Chandler-Wilde, Bruno Colbois, Dmitry Faifman, Bernard Helffer, Antoine Henrot, Dmitry Jakobson, Alexander Logunov, Eugenia Malinnikova, Marco Marletta, Egor Shelukhin, and Vukašin Stojisavljević for many useful discussions.

We would also like to thank the anonymous referees for multiple helpful suggestions.

All the remaining errors, omissions, and inaccuracies are of course entirely ours.

We are grateful to the staff at the American Mathematical Society, in particular to Sergei Gelfand for his encouragement and helpful advice throughout the preparation of the book and to Brian Bartling for his assistance on typesetting.

At various stages of work on this book, the first author, Michael Levitin, has been visiting the Centre de recherches mathématiques in Montréal. Its hospitality is gratefully acknowledged.

Image credits for all externally sourced illustrations are on page 293.

Last but not least, we would like to thank our families for their patience and support. Without them this project would have never been accomplished.

Introduction

Overview

The central theme of the book is spectral geometry of the Laplace operator on bounded Euclidean domains and compact Riemannian manifolds. Most of the time, we consider the classical Dirichlet or Neumann boundary conditions, except for the last chapter, where instead of the spectral parameter in the equation we look at the less explored Steklov problem with the spectral parameter in the boundary conditions.

The main topics discussed in the book can be summarised as follows:

- spectral theorems,
- eigenvalue inequalities,
- spectral asymptotics,
- nodal geometry,
- isospectrality and spectral invariants.

To cover these subjects we use a variety of techniques, such as variational principles, elliptic regularity, symmetrisation, conformal maps, harmonic analysis, and heat equation methods. Throughout the presentation we tried to keep a balance between the following principles:

- Focus on *phenomena*. For that reason, in many cases the proofs are given in the Euclidean setting, with indications on how the argument can be extended to the Riemannian case.
- Avoid *black boxes* as much as possible. While it is often unfeasible to present all the details, we at least tried to explain the main ideas behind the proofs.

- Keep *generality* reasonably wide to include most interesting examples. In particular, in the Euclidean setting we mostly consider Lipschitz boundaries, whilst on manifolds we deal with smooth Riemannian metrics.

The highlights of the book include:

- Spectral theorems and elliptic regularity. In particular, we discuss in detail both interior and boundary regularity of eigenfunctions.
- Weyl's law for the eigenvalue counting function.
- Friedlander–Filonov inequalities between Dirichlet and Neumann eigenvalues.
- Polya's conjecture for tiling domains and Berezin–Li–Yau–inequalities.
- Courant and Pleijel nodal domain theorems.
- Yau's conjecture on the size of the nodal sets.
- Isoperimetric inequalities for eigenvalues: Faber–Krahn, Cheeger, Szegő–Weinberger, and Hersch.
- Universal inequalities for eigenvalues.
- Heat trace asymptotics.
- Isospectrality and transplantation of eigenfunctions.
- Spectral geometry of the Steklov problem.

While many of these topics can be found in other books, having all these subjects under one cover makes this book quite different from the others. At times, our exposition of classical results contains some features which have not been emphasised previously. For example, we prove Courant's nodal domain theorem for Dirichlet eigenfunctions without any regularity assumptions on the boundary. Moreover, some of the material is based on recent research and therefore cannot be found in textbooks, such as the section on Yau's conjecture and essentially the entire chapter on the Steklov problem.

Plan of the book

The book is organised as follows.

In Chapter 1 we introduce our main hero, the Laplacian, and discuss several examples for which its eigenvalues and eigenfunctions can be calculated explicitly.

In Chapter 2 we lay the foundations for further material and explain the proofs of the weak and the strong spectral theorems for the Laplacian. This chapter includes mini-crash courses on the theory of selfadjoint unbounded linear operators, as well as on the Sobolev spaces and elliptic regularity. Our emphasis is on presenting the main tools and ideas, such as the Friedrichs extension, the a priori estimates, and Nirenberg's method of difference quotients, while referring the reader interested in full details to the existing literature.

Chapter 3 is concerned with the variational principles for eigenvalues and their applications. Apart from basic results such as domain monotonicity, Dirichlet–Neumann bracketing and Weyl's law, we prove the Friedlander–Filonov inequalities between Dirichlet and Neumann eigenvalues, the Berezin–Li–Yau inequalities, and Pólya's conjecture for tiling domains.

Chapter 4 focuses on the nodal geometry of eigenfunctions. We give a complete proof of Courant's nodal domain theorem, explaining some delicate issues arising for domains with nonsmooth boundary that have often been omitted in other sources. We also discuss Yau's conjecture on the volume of nodal sets, including recent breakthrough developments due to Logunov and Malinnikova. In particular, we give a sketch of the proof of a polynomial upper bound on the size of the nodal set. Some related topics, such as the density of the nodal set and the lower bound on the size of the nodal set in dimension two, are also presented. As an application of results on the local structure of the nodal set we prove multiplicity bounds for eigenvalues on surfaces.

In Chapter 5 we collect various geometric eigenvalue inequalities, such as the Faber–Krahn inequality, Cheeger's inequality, the Szegő–Weinberger inequality, as well as Hersch's inequality and other isoperimetric inequalities of surfaces. The latter is an actively developing subject and several recent advances are discussed in detail. This chapter also includes the universal inequalities, as well as related commutator identities.

The heat equation and results on heat kernel asymptotics are presented in Chapter 6. As an application, we prove Weyl's law on Riemannian manifolds. The spectral invariants arising from the heat asymptotics naturally lead us to the study of isospectrality. Some partial answers are given to the question "Can one hear the shape of a drum?", mentioned above. We present Milnor's example of flat isospectral tori which has fascinating connections to the theory of modular forms, and the celebrated Sunada construction of isospectral manifolds based on algebraic ideas. We also describe a rather elementary but ingenious transplantation technique that yields isospectral but not isometric planar domains. Some recent results on spectral rigidity are also discussed.

In the past decade, the study of the Steklov problem and of the Dirichlet-to-Neumann map became one of the most active directions in spectral geometry. This is the subject of Chapter 7. We define the Steklov spectrum and prove isoperimetric inequalities for Steklov eigenvalues. Using the connection between the Dirichlet-to-Neumann map and the boundary Laplacian, we obtain results on the asymptotics of the Steklov spectrum by means of the Hörmander–Pohozhaev identities and Weyl's law for the Laplacian on manifolds. We also provide a detailed exposition of recent results on the asymptotics of sloshing eigenvalues as well as Steklov eigenvalues on curvilinear polygons. Finally, we discuss the Dirichlet-to-Neumann map for the Helmholtz operator and use its properties to give another proof of the Friedlander–Filonov inequalities between Dirichlet and Neumann eigenvalues originally presented in Chapter 3.

Appendix A contains a short introduction to numerical spectral geometry, which provides the students with all the necessary tools for a quick numerical calculation of eigenvalues and eigenfunctions of planar domains.

In Appendix B we collect some standard background definitions and notation which we use throughout the book.

Possible courses based on this book

The book is to a large extent self-contained and is accessible to students and researchers with a basic knowledge of PDEs, functional analysis, and differential geometry. We do not *really* require prior knowledge of the theory of distributions and Sobolev spaces and explain the main notions we need. Throughout the book we often stay in the Euclidean setting and, where necessary, provide references for a reader unfamiliar with the fundamentals of Riemannian geometry. While graduate students in mathematics are the main targeted audience for the book, it could also be used, in parts, for teaching an advanced undergraduate course, as well as for both introductory and advanced mini-courses.

In our experience, essentially the whole book with the exception of the most advanced sections (§§2.2, 4.3, and 7.2–7.4) can be covered in a one-semester course. There are various ways to create shorter courses using the diagram of dependencies given on page xvii.

For example, one could teach the first three chapters only, or the first three chapters followed by one of the Chapters 4–7, with some minor additions and adjustments. Finally, the material of each of Chapters 1–3 can be taught as an introductory level mini-course, and each of the remaining chapters as a more advanced course.

Last, but not least, the book contains *many* exercises! The more difficult ones are provided with references and hints. A user-friendly tutorial

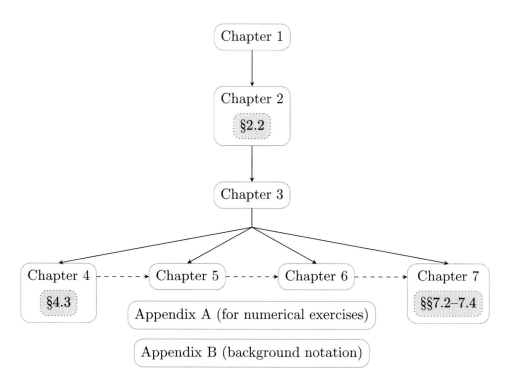

The diagram of chapter dependencies. The subsections in shaded boxes may be omitted for all but advanced courses. The dashed arrows indicate that while there is some dependency of material there, the corresponding chapters may be taught separately from each other.

on numerical spectral geometry presented in Appendix A could also help teachers who would like to introduce a computational component into their classes.

What is not in this book: Some further reading

As mentioned in the preface, the field of spectral geometry extends beyond the scope of this book. Below we discuss some interesting and important topics for further reading.

In order to keep the prerequisites to a minimum, we focused on results that can be presented without using pseudodifferential operators and micro-local analysis. As a consequence, apart from nodal geometry, we did not explore much the properties of eigenfunctions. We refer to [**Sog17**] and [**Zwo12**] for an exposition of results on asymptotic eigenfunction bounds, as well as questions arising in the fascinating area of *mathematical quantum chaos*, such as Shnirelman's quantum ergodicity theorem.

Throughout the book, we have almost exclusively dealt with the case of bounded domains and compact manifolds, for which the Laplace spectrum is discrete. A lot of interesting phenomena occur in other geometric set-ups. We refer to [**Bor16**] and [**DyaZwo19**] for recent developments of the spectral theory on infinite area hyperbolic spaces and the mathematical theory of resonances.

In this book, we focus on the Laplacian and the Dirichlet-to-Neumann map and do not touch other important operators. A modern exposition of the spectral theory of Schrödinger operators, with a particular focus on the celebrated Lieb–Thirring inequalities (closely linked to the Berezin–Li–Yau inequalities featured in Chapter 3), can be found in [**FraLapWei23**]. Many interesting geometric questions arise in the study of the spectrum of the Dirac operator, and we refer to [**BerGetVer04, Fri00, Gin09**] for further reading on this subject. Recent results on spectral geometry of potential operators, which are related to the Dirichlet-to-Neumann map, can be found in [**RuzSadSur20**]. A detailed introduction to the rich and actively developing theory of quantum graphs, which makes a cameo appearance in §7.3 of this book, can be found in [**BerKuc13**].

Chapter 1

Strings, drums, and the Laplacian

Pierre-Simon,
marquis de
Laplace
(1749–1827)

In this chapter, we introduce the Laplacian, both in the Euclidean space and on a Riemannian manifold, and consider the eigenvalue problems with the Dirichlet and Neumann boundary conditions. We discuss the related models of vibrating strings and drums and consider a few examples in which spectral problems can be explicitly solved.

Eugenio
Beltrami
(1835–1900)

1.1. Basic examples

1.1.1. The Laplace operator. In the Euclidean space $\mathbb{R}^d$ of dimension d with Cartesian coordinates $x = (x_1, \ldots, x_d)$, let

$$(1.1.1) \qquad \Delta f = \sum_{j=1}^{d} \frac{\partial^2 f}{\partial x_j^2},$$

where $f = f(x_1, \ldots, x_d)$ is a twice differentiable function.

Definition 1.1.1 (The Laplacian). The operator $-\Delta$ is called the *Laplace operator* (or the *Laplacian*) in $\mathbb{R}^d$.

Remark 1.1.2. There is no unique sign convention for Δ. In this book, we define Δ by (1.1.1), that is, in the analyst's sense; geometers often incorporate the minus sign into the definition of Δ. (The authors have argued long and hard about which notation to adopt.) Additionally, the term *Laplacian* may also be applied to the negative of our Laplacian.

One can rewrite (1.1.1) as

$$\Delta f = \operatorname{div} \nabla f,$$

where div denotes the divergence of a vector field and ∇ is the gradient of a scalar function; see §B.1. We will use this representation later on in order to define the Laplacian on a Riemannian manifold.

The Laplace operator appears in major partial differential equations arising in mathematical physics. Here are some examples; in all of them we set $\Delta := \Delta_x$; i.e., the operator acts only in the x variable.

- *Wave equation:*

$$\frac{\partial^2 U(t, x)}{\partial t^2} = \Delta U(t, x).$$

  Here $U(t, x)$ denotes the displacement from the equilibrium of the vibrating object at the point $x \in \mathbb{R}^d$ at time t.

- *Heat (or diffusion) equation:*

$$\frac{\partial U(t, x)}{\partial t} = \Delta U(t, x).$$

  Here $U(t, x)$ denotes the temperature of the object (or the density of the matter) at the point x at time t.

- *Laplace equation:*

$$\Delta U(x) = 0.$$

  The solutions of the Laplace equation are called *harmonic* functions. In hydrodynamics, the velocity potential $U(x)$ of an incompressible fluid flow is a solution of the Laplace equation.

- *Poisson equation:*

$$-\Delta U(x) = f(x).$$

  In electrostatics, $U(x)$ is interpreted as an electric potential corresponding to a given charge distribution f.

- *Schrödinger equation:*

$$\mathrm{i} \frac{\partial U(t, x)}{\partial t} = -\Delta U(t, x),$$

  where $\mathrm{i}^2 = -1$. In quantum mechanics, the solution $U(t, x)$ of this equation is called the wave function. Note that $U(t, x)$ is complex valued; the quantity $|U(t, x)|^2$ describes the probability density for a particle to be at the position x at time t.

Let us start with two simple real-life examples, which are also among the most relevant ones from the viewpoint of spectral geometry: the vibrating strings and drums.

1.1.2. Vibrating strings. Even if you never played a guitar yourself, you probably know that thicker guitar strings produce lower sounds and that pressing down on a string raises the pitch. These phenomena could be easily explained using a mathematical model of a vibrating string, given by the *one-dimensional wave equation*.

Consider a string of length l and uniform density ρ, fixed at both ends. Let $U : \mathbb{R}_+ \times [0, l] \to \mathbb{R}$ be a function, whose value $U(t, x)$ is equal to the deviation from the equilibrium of a transversally vibrating string at the point $x \in [0, l]$ at the time $t \in \mathbb{R}_+$ (transversal vibrations mean that each point of the string moves along the vertical line orthogonal to the equilibrium position). The function $U(t, x)$ satisfies the one-dimensional wave equation

$$(1.1.2) \qquad U_{tt} = a^2 \Delta U = a^2 U_{xx},$$

where the constant a can be expressed in terms of the tension τ of the string and the density ρ:

$$a = \sqrt{\tau/\rho}.$$

Since the string is attached at both ends, we impose the *Dirichlet* boundary conditions:

$$(1.1.3) \qquad U(t, 0) = U(t, l) = 0, \qquad t \in \mathbb{R}_+.$$

In order to find a solution of this equation we use the Fourier method.

Jean-Baptiste Joseph **Fourier**
(1768–1830)

The first step is to separate the variables and to look for a solution in the form

$$U(t, x) = T(t)X(x).$$

This is a so-called *standing wave*. From equation (1.1.2) we get

$$T''(t)X(x) = a^2 T(t)X''(x),$$

and, since $X(x)$ and $T(t)$ are not identically zero, we obtain

$$\frac{X''(x)}{X(x)} = \frac{T''(t)}{a^2 T(t)} = -\lambda,$$

where λ is some constant (the choice of the minus sign will become clear later). Indeed, the left-hand side of the equality does not depend on t, and the middle part is independent of x, so both are equal to a constant.

We now consider the equations for the functions $X(x)$ and $T(t)$ separately.

Taking into account (1.1.3), we obtain a Sturm–Liouville eigenvalue problem for the function $X(x)$ with Dirichlet boundary conditions:

$$(1.1.4) \qquad \begin{cases} -X''(x) = \lambda X(x), \\ X(0) = X(l) = 0. \end{cases}$$

Definition 1.1.3. A nontrivial solution $X(x)$ of the Sturm–Liouville problem (1.1.4) is called an *eigenfunction* corresponding to an *eigenvalue* λ.

■ **Exercise 1.1.4.** Show that the eigenvalues and eigenfunctions of the Sturm–Liouville problem (1.1.4) are given by

$$\lambda_m = \left(\frac{\pi m}{l}\right)^2, \qquad X_m(x) = \sin\left(\frac{\pi m}{l}x\right), \qquad m = 1, 2, \ldots.$$

■ **Exercise 1.1.5.** Show that for all natural numbers $k \neq m$,

$$\int_0^l X_k(x)X_m(x)\mathrm{d}x = 0.$$

Resolving a similar Sturm–Liouville problem for $T(t) = T_m(t)$ we obtain

$$T_m(t) = A_m \cos\left(\frac{a\pi m}{l}t\right) + B_m \sin\left(\frac{a\pi m}{l}t\right),$$

where A_m and B_m are arbitrary constants. Taking a superposition of the standing waves $U_m(t,x) = T_m(x)X_m(x)$, we get a formal solution of the wave equation (1.1.2):

$$(1.1.5) \quad U(t,x) = \sum_{m=1}^{\infty} \left(A_m \cos\left(\frac{a\pi m}{l}t\right) + B_m \sin\left(\frac{a\pi m}{l}t\right)\right) \sin\left(\frac{\pi m}{l}x\right).$$

■ **Exercise 1.1.6.** Show that the constants A_m and B_m, $m \in \mathbb{N}$, are uniquely determined by the initial conditions $u(0, x) = \varphi(x)$ (initial position), $u_t(0, x) = \psi(x)$ (initial velocity). Calculate A_m and B_m using the Fourier decompositions of the functions φ and ψ.

We are now in a position to address the questions about sounds emitted by a guitar string raised at the beginning of this section. As can be easily seen from (1.1.5), the natural frequencies of the string are given by

$$(1.1.6) \qquad \omega_m = a\sqrt{\lambda_m} = \frac{a\pi m}{l}, \qquad m \in \mathbb{N}.$$

The frequency ω_1 is called the *principal frequency*, or the *fundamental tone* of the string, and the higher frequencies are called *overtones*. It follows immediately from (1.1.6) that the frequencies decrease as the length l increases; in other words, shorter strings produce higher notes. This is precisely what we observe when pressing down on a guitar string (pressing down is essentially a way to change the length of the vibrating part of the string). Recall now that the constant a decreases as the density of a string increases. Therefore, the thicker the string is, the lower are the sounds that it emits. Similarly, the higher the tension of the string is, the higher the pitch is.

The eigenfunctions $X_m(x)$ describe the shape of the pure vibration modes. In particular, one may observe that for each $m = 1, 2, \ldots$, the eigenfunction $X_m(x)$ has precisely $m - 1$ zeros on the open interval $(0, l)$; see Figure 1.1. This fact has interesting higher-dimensional generalisations that we will discuss later.

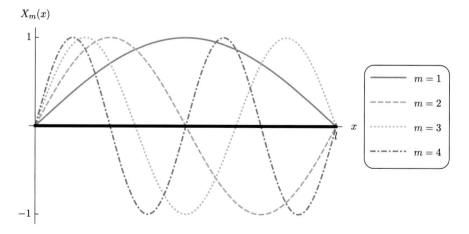

Figure 1.1. First four eigenfunctions of a fixed vibrating string.

■ **Exercise 1.1.7.** The vibrations of a free string of length l are modelled by the equation (1.1.2) with Neumann boundary conditions

$$(1.1.7) \qquad U_x(t,0) = U_x(t,l) = 0, \qquad t \in \mathbb{R}_+.$$

Find the eigenfrequencies of a free vibrating string and compare them with the (Dirichlet) eigenfrequencies given by (1.1.6).

Let us explain the physical meaning of the Neumann condition (1.1.7). As follows from the model leading to the wave equation (1.1.2), the tension force acting at the point x is equal to τU_x. *Free* vibration means that the endpoints of the string experience no tension, and therefore at these points U_x must vanish.

Example 1.1.8. Consider the vibrations of a string whose ends are neither fixed nor free but joined together in a circular loop. If the length of the string is 2π, we arrive, after the separation of variables, at the spectral problem

$$(1.1.8) \qquad \begin{cases} -X''(x) = \lambda X(x), \\ X(x) \text{ is } 2\pi\text{-periodic.} \end{cases}$$

Looking for the values of λ for which (1.1.8) has a nontrivial solution, we obtain

$$\lambda_0 = 0, \qquad X_0(x) = 1,$$

and also eigenvalues m^2, $m \in \mathbb{N}$, for each of which there are two linearly independent eigenfunctions $X_{m,1}(x) = \sin mx$ and $X_{m,2}(x) = \cos mx$.

1.1.3. Vibrating drums. Consider now a two-dimensional analogue of the problem discussed in the previous section. Imagine a drum with a membrane (drumhead) shaped as a bounded domain $\Omega \subset \mathbb{R}^2$. The function

$$U(t,x,y) : \mathbb{R}_+ \times \Omega \to \mathbb{R}$$

describing the vibration of the drumhead satisfies the wave equation

$$\begin{cases} U_{tt} - a^2 \Delta U = 0, \\ U|_{\partial\Omega} = 0, \end{cases}$$

where the constant a depends on the physical characteristics of the membrane. Again, searching for solutions in the form $U(t,x,y) = T(t)\,u(x,y)$, we get a familiar (ordinary) Sturm–Liouville equation for $T(t)$ and a Dirichlet eigenvalue problem for the function $u(x,y)$, that is the eigenvalue problem for the Laplacian in Ω,

$$(1.1.9) \qquad -\Delta u = \lambda u,$$

subject to the Dirichlet condition

(1.1.10)
$$u|_{\partial\Omega} = 0.$$

We say, as in Definition 1.1.3, that λ is an eigenvalue of the Dirichlet problem (1.1.9)–(1.1.10) if this problem has a nontrivial solution $u(x, y)$.

Unlike (1.1.4), the problem (1.1.9)–(1.1.10) usually cannot be explicitly solved. However, for certain geometries — for example, for a rectangle or for a disk — that could be done by using once again the separation of variables (in this case, the spatial variables x and y).

■ **Exercise 1.1.9.** Let $R_{a,b} = (0, a) \times (0, b)$ be a rectangle with sides a and b. Show that

(1.1.11)
$$\lambda^{D}_{k,m} = \pi^2 \left(\frac{k^2}{a^2} + \frac{m^2}{b^2} \right), \quad k, m = 1, 2, \ldots,$$

are the eigenvalues of the Dirichlet problem (1.1.9)–(1.1.10) on $R_{a,b}$, and the corresponding eigenfunctions are given by

(1.1.12)
$$u^{D}_{k,m}(x, y) = \sin \frac{k\pi}{a} x \, \sin \frac{m\pi}{b} y.$$

Prove that these functions form an orthogonal basis in $L^2(R_{a,b})$.

Remark 1.1.10. The separation of variables does not immediately imply that (1.1.11) and (1.1.12) provide *all* eigenvalues and eigenfunctions of the Dirichlet problem (1.1.9) on a rectangle. This has to be shown separately and, indeed, it follows from the fact that the set (1.1.12) forms a basis in $L^2(R_{a,b})$.

More generally, the fact that eigenfunctions of (1.1.9)–(1.1.10) in a bounded domain Ω can be chosen to form a basis in $L^2(\Omega)$ follows from the *spectral theorems*; see Chapter 2.

Definition 1.1.11. The *multiplicity* of an eigenvalue λ is the dimension of the corresponding eigenspace. If the dimension is equal to one, the eigenvalue is called *simple*.

■ **Exercise 1.1.12.** Show that if $\frac{a^2}{b^2}$ is irrational, then all the Dirichlet eigenvalues of a rectangle $R_{a,b}$ are simple.

Note that if $\frac{a^2}{b^2}$ is rational, then the multiplicities of the Dirichlet eigenvalues of $R_{a,b}$ can be arbitrarily large. This follows from number-theoretic results on the representation of integers as binary quadratic forms. In the case of a square, the precise answer could be found using the so-called sum of squares function, see [**HarWri08**, §16.9], and also Remark 1.2.14 below. For example, if $a = b = \pi$, one can check that the eigenvalue $\lambda = 5^{2k-1}$, $k \in \mathbb{N}$, has multiplicity $2k$.

Example 1.1.13. Since for an eigenvalue of multiplicity m we have an m-dimensional linear space of corresponding eigenfunctions, particular eigenfunctions may look quite unlike each other; see Figure 1.2.

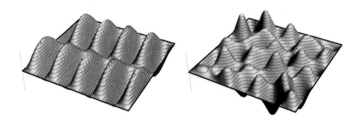

Figure 1.2. Two eigenfunctions corresponding to the same Dirichlet eigenvalue $85\pi^2$ of the unit square $[0,1]^2$: on the left, the eigenfunction $\sin(2\pi x)\sin(9\pi y)$, and on the right, the eigenfunction $\frac{1}{\sqrt{5}}(\sin(2\pi x)\sin(9\pi y) - \sin(9\pi x)\sin(2\pi y) - \sin(6\pi x)\sin(7\pi y) + 2\sin(7\pi x)\sin(6\pi y))$.

Along with the Dirichlet boundary condition $u|_{\partial\Omega} = 0$ corresponding to a membrane with a fixed boundary, one may consider the vibration of a free membrane. This problem gives rise to the Neumann boundary condition, which can be viewed as an appropriate generalisation of (1.1.7):

$$(1.1.13) \qquad\qquad \partial_n u = 0,$$

where from now on we set

$$\partial_n u := \langle (\nabla u)|_{\partial\Omega}, n \rangle$$

to denote the *normal derivative* of u. Here n is the exterior unit normal to the boundary $\partial\Omega$, and $\langle \cdot, \cdot \rangle$ stands for the standard vector inner product in $\mathbb{R}^d$ (or $\mathbb{C}^d$); see §B.1. It is clear that in order for the Neumann condition (1.1.13) to be well-defined, certain regularity of the boundary has to be assumed. For instance, if one assumes the boundary to be Lipschitz (i.e., locally representable as a graph of a Lipschitz function; see §B.3 for the definition), the normal derivative is well-defined at almost every point of the boundary. More general conditions under which the Neumann problem is well-defined will be discussed later.

■ **Exercise 1.1.14.** Show that

$$(1.1.14) \qquad \lambda_{k,m}^{\mathrm{N}} = \pi^2 \left(\frac{k^2}{a^2} + \frac{m^2}{b^2} \right), \qquad k, m \in \mathbb{N}_0 \ldots,$$

are the eigenvalues of the Neumann problem (1.1.9), (1.1.13) in the rectangle $R_{a,b}$, with the corresponding eigenfunctions

$$u_{k,m}^{\mathrm{N}}(x,y) = \cos\frac{k\pi}{a}x \, \cos\frac{m\pi}{b}y.$$

Note that the indices k, m of the Neumann eigenvalues may take the value zero, while in the Dirichlet case they start with one. In particular, the lowest Neumann eigenvalue is zero and the corresponding eigenfunction is a constant. In fact, this is true for any bounded domain Ω on which the Neumann problem is well-defined.

■ **Exercise 1.1.15.** Using the formula (1.1.11) for the eigenvalues of the Laplacian in an arbitrary rectangle with Dirichlet boundary conditions, find which rectangle minimises the first Dirichlet eigenvalue among all rectangles of fixed area. Similarly, using (1.1.14), find which rectangle of a fixed area maximises the first nonzero Neumann eigenvalue. What happens if we interchange minimisation and maximisation in these questions?

■ **Exercise 1.1.16.** Compute the Dirichlet and Neumann eigenvalues and eigenfunctions of a rectangular box in $\mathbb{R}^d$.

Example 1.1.17. Let us describe the eigenvalues and eigenfunctions of the Dirichlet and Neumann problems in the unit disk $\mathbb{D}$. Switching to polar coordinates (r, φ), using the standard expression

$$\Delta = \frac{\partial^2}{\partial r^2} + \frac{1}{r}\frac{\partial}{\partial r} + \frac{1}{r^2}\frac{\partial^2}{\partial\varphi^2}$$

for the Laplacian in planar polar coordinates, and looking for solutions of (1.1.9) in the form

$$u(r,\varphi) = \sum_{m=-\infty}^{+\infty} u_m(r)\mathrm{e}^{im\varphi},$$

we arrive at the equations

(1.1.15) $$u_m''(r) + \frac{1}{r}u_m'(r) + \left(\lambda - \frac{m^2}{r^2}\right)u_m(r) = 0$$

for unknown functions u_m.

The equations (1.1.15) are closely related to the *Bessel equation*

(1.1.16) $$y''(r) + \frac{1}{r}y'(r) + \left(1 - \frac{m^2}{r^2}\right)y(r) = 0.$$

For $m \in \mathbb{N}_0$, equation (1.1.16) possesses, up to a multiplicative constant, only one solution regular at $r = 0$. A specific choice of that constant corresponds to the solution defined via a power series

$$(1.1.17) \qquad J_m(r) = \left(\frac{r}{2}\right)^m \sum_{k=0}^{\infty} \frac{(-1)^k}{k!\Gamma(m+k+1)} \left(\frac{r}{2}\right)^{2k},$$

which is called the *Bessel function of the first kind* of order m. In fact, Bessel functions $J_\nu(r)$ can be defined in a similar manner for $\nu \in \mathbb{R}$ by taking $m = \nu$ in (1.1.17), see [**Wat95**, Chapter 3] for details, and it follows that $J_{-m}(r) = (-1)^m J_m(r)$ for $m \in \mathbb{N}$. We refer to [**Wat95**] for a complete treatment of the theory of Bessel functions and recall only some facts which we will use in the sequel. One can show that Bessel functions have infinitely many real zeros. Denote by $j_{m,k}$ the kth positive zero of the mth Bessel function $J_m(r)$ and by $j'_{m,k}$ the kth positive zero of the derivative $J'_m(r)$ (with the exception $j'_{0,1} = 0$ for the first zero of $J'_0(r)$; see [**DLMF22**, §10.21(i)]); cf. Figure 1.3.

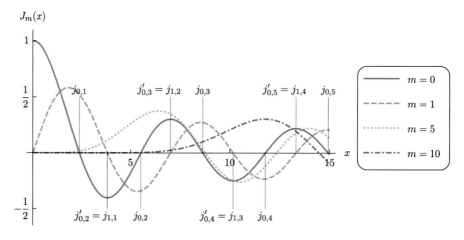

Figure 1.3. The graphs of some Bessel functions, with zeros of $J_0(x)$ and $J_1(x)$ marked. Note that $j_{1,k} = j'_{0,k+1}$ for $k \in \mathbb{N}$.

Returning now to the equations (1.1.15) and comparing to (1.1.16), one can easily see that the regular solutions of (1.1.15) are given, modulo a multiplicative constant, by $u_m(r) = J_m\left(\sqrt{\lambda} r\right)$.

Imposing the Dirichlet condition (1.1.10) now implies $u_m(1) = J_m\left(\sqrt{\lambda}\right) = 0$, and therefore the Dirichlet eigenvalues of the unit disk $\mathbb{D}$ are given by

$$j_{m,k}^2, \qquad m \in \mathbb{N}_0, \quad k \in \mathbb{N}.$$

For $m > 0$, the eigenvalues should be repeated with multiplicity two and the corresponding linearly independent eigenfunctions can be chosen either as

$$(1.1.18) \qquad J_m \left(j_{m,k} r \right) \sin m\varphi \quad \text{or} \quad J_m \left(j_{m,k} r \right) \cos m\varphi.$$

For $m = 0$ each eigenvalue is simple, with the corresponding eigenfunction $J_0 \left(j_{0,k} r \right)$ being radially symmetric. To ensure that we have found *all* the eigenfunctions we also need to prove that they form a basis in $L^2(\mathbb{D})$ as discussed in Remark 1.1.10; this is not entirely trivial and follows from the Sturm–Liouville theory; see [**CouHil89**, §V.5.5].

Similarly, imposing the Neumann condition (1.1.13) implies $u'_m(1)$ $= \sqrt{\lambda} J'_m \left(\sqrt{\lambda} \right) = 0$, and therefore the Neumann eigenvalues of the unit disk $\mathbb{D}$ are given by $j'^{\,2}_{m,k}$, $m \in \mathbb{N}_0$, $k \in \mathbb{N}$, where for $m > 0$ the eigenvalues should be repeated with multiplicity two. The eigenfunctions corresponding to $j'^{\,2}_{m,k}$ are given by either

$$(1.1.19) \qquad J_m \left(j'_{m,k} r \right) \sin m\varphi \quad \text{or} \quad J_m \left(j'_{m,k} r \right) \cos m\varphi$$

(as before we have only one eigenfunction for $m = 0$).

Finally, let us also note that the zeros of Bessel functions of different orders (respectively, of their derivatives) never coincide, and therefore there are no "accidental" multiplicities in the Dirichlet (respectively, Neumann) spectrum. In the Dirichlet case this follows from the proof of the celebrated Bourget hypothesis (1866) found by C. L. Siegel back in 1929; see [**Sie29**] and also [**Wat95**, pp. 484–485]. Essentially, Siegel proved a rather deep number-theoretic result: if $x \neq 0$ is an algebraic number, $J_m(x)$ is transcendental. At the same time, using relations between Bessel functions of different orders, one can show that if J_m and J_k share a common zero, it has to be an algebraic number. Therefore, the only possible common zero may be $x = 0$. The Neumann analogue of this result is also known; see [**HelSun16**].

◼ **Exercise 1.1.18.** Using integrals [**DLMF22**, formulas 10.22.37–10.22.38], check the orthogonality in $L^2(\mathbb{D})$ of the eigenfunctions (1.1.18) or (1.1.19) in either the Dirichlet or Neumann case. This is just an illustration of a much more general phenomenon which we will encounter later in Theorems 2.1.20, 2.1.36, and 2.2.21: the eigenfunctions of the Dirichlet or Neumann Laplacian can always be chosen to form an orthonormal basis in L^2.

Remark 1.1.19. In the same manner, the eigenvalues of the Dirichlet and Neumann Laplacians on circular sectors and annuli can be expressed in terms of zeros of some Bessel functions or their combinations, or of their derivatives. Similarly, the variables separate for ellipses, and the eigenvalues can be expressed in terms of zeros of some special functions; see [**GreNgu13**] and [**KutSig84**].

Remark 1.1.20. Apart from the Dirichlet and Neumann boundary conditions, there exist other types of selfadjoint boundary conditions, for example the Robin ones or Zaremba (mixed) ones, which we discuss later in §3.1.3. The Robin conditions arise, for example, when the boundary is neither free nor fixed, but attached by a spring or some elastic material. Dirichlet, Neumann, and Robin conditions also have other physical interpretations, notably in terms of the heat equation; see [**Str08**] for further details.

1.2. The Laplacian on a Riemannian manifold

1.2.1. The Laplace–Beltrami operator. In this section we use various basic notions from Riemannian geometry which can be found in standard textbooks. In particular, lecture notes [**Bur98**] contain a concise and clear exposition of essentially everything that is needed.

Consider a smooth closed (that is, compact without boundary) manifold M of dimension $\dim M = d$ endowed with the Riemannian metric $g = \{g_{ij}\}$, $i, j = 1, \ldots, d$.

For any differentiable function f on M one can define the gradient ∇f; it is a vector field such that for any $p \in M$ and for any vector $\xi \in T_pM$ the following identity holds:

$$(1.2.1) \qquad \langle \nabla f, \xi \rangle_g = df_p(\xi) =: \xi f,$$

where $\langle \cdot, \cdot \rangle_g$ is a scalar product on T_pM defined by the Riemannian metric; we will usually omit the subscript g. We say that ξf is the *directional derivative* of the function f in the direction of the vector ξ at the point p. It is easy to check that for the Euclidean space (1.2.1) yields the usual definition of the gradient.

Let us now introduce the *divergence* $\operatorname{div} X$ of a vector field X on a Riemannian manifold. Let dV_g be the volume density on (M, g). In local coordinates $x_1, \ldots, x_d$ it takes the form

$$dV_g = \sqrt{\det g}\, dx_1 dx_2 \ldots dx_d.$$

We will sometimes write this as

$$dV = dV_g = \sqrt{\det g}\, dx$$

for brevity. Given a smooth vector field X, one can define div X as a smooth function on M satisfying the identity

$$(1.2.2) \qquad \int_M f \,\mathrm{div}\, X \, dV_g = -\int_M \langle \nabla f, X \rangle \, dV_g$$

for all $f \in C^1(M)$. To verify that the divergence exists, we note that using a partition of unity it suffices to check (1.2.2) for functions f supported in a coordinate chart, which is done below. We refer to [**Ros97**, §1.2.3] for a discussion concerning this approach.

Let us calculate the gradient and the divergence in local coordinates $(x_1, \ldots, x_d)$. The corresponding basis in the tangent bundle TM is given by $\left(\frac{\partial}{\partial x_1}, \ldots, \frac{\partial}{\partial x_d} \right)$ satisfying

$$(1.2.3) \qquad \left\langle \frac{\partial}{\partial x_i}, \frac{\partial}{\partial x_j} \right\rangle = g_{ij}.$$

The gradient ∇f in this basis is given by

$$(1.2.4) \qquad \nabla f = \sum_{j=1}^{d} c^j(x) \frac{\partial}{\partial x_j}$$

for some coefficients $c^j(x)$. Applying formula (1.2.1) we get

$$\sum_{j=1}^{d} c^j(x) g_{ji} = \left\langle \sum_{j=1}^{d} c^j \frac{\partial}{\partial x_j}, \frac{\partial}{\partial x_i} \right\rangle = df\left(\frac{\partial}{\partial x_i} \right) = \frac{\partial f}{\partial x_i}.$$

Applying the inverse matrix $\{g^{ij}\}$ and substituting the values of c^j into (1.2.4) we obtain

$$(1.2.5) \qquad \nabla f = \sum_{i,j=1}^{d} g^{ij} \frac{\partial f}{\partial x_i} \frac{\partial}{\partial x_j}.$$

Let us now calculate the divergence. Let f be a differentiable function compactly supported in a coordinate chart. Applying formula (1.2.2) to a vector field $X = \left(a^1(x), \ldots, a^d(x) \right)$ and substituting (1.2.5) in the right-hand side we obtain

$$\int_M f \operatorname{div} X \sqrt{\det g}\, \mathrm{d}x_1 \ldots \mathrm{d}x_d$$

(1.2.6)
$$= -\int_M \left\langle \sum_{i,j=1}^d g^{ij} \frac{\partial f}{\partial x_i} \frac{\partial}{\partial x_j}, \sum_{i=1}^d a^i \frac{\partial}{\partial x_i} \right\rangle \sqrt{\det g}\, \mathrm{d}x_1 \ldots \mathrm{d}x_d$$

$$= -\int_M \sum_{i=1}^d \frac{\partial f}{\partial x_i} a^i \sqrt{\det g}\, \mathrm{d}x_1 \ldots \mathrm{d}x_d$$

$$= \int_M f \sum_{i=1}^d \frac{\partial}{\partial x_i} \left(a^i \sqrt{\det g} \right) \mathrm{d}x_1 \ldots \mathrm{d}x_d.$$

The second equality follows from (1.2.3), and the last equality is a result of the integration by parts. Since formula (1.2.6) holds for any such function f, comparing its left- and right-hand sides we get

(1.2.7)
$$\operatorname{div} X = \frac{1}{\sqrt{\det g}} \sum_{i=1}^d \frac{\partial}{\partial x_i} \left(a^i \sqrt{\det g} \right).$$

Recall that for a vector field X in the Euclidean space $\mathbb{R}^d$,

(1.2.8)
$$\operatorname{div} X = \sum_{i=1}^d \frac{\partial a^i}{\partial x_i}.$$

It is easy to check that (1.2.7) agrees with (1.2.8) in this case.

Remark 1.2.1 (Definitions of the divergence). There are several equivalent ways to define the divergence. Note that the right-hand side of (1.2.8) can be represented as the trace of the operator $\xi \mapsto \xi X := (\xi a^1, \ldots, \xi a^d)$ acting on vector fields. On a Riemannian manifold, the analogue of the directional derivative ξX is the covariant derivative $\nabla_\xi X$, where ∇ denotes the Levi–Civita connection. Thus, a standard way to define the divergence in Riemannian geometry is

(1.2.9)
$$\operatorname{div} X = \operatorname{trace} [\xi \mapsto \nabla_\xi X];$$

see, for example, [**Bur98**, §2.2] or [**Cha84**, §I.1].

On an orientable manifold one can also define the divergence in a coordinate-free way using differential forms; see [**BerGauMaz71**, §II.G.I]. Let $\omega_g = \sqrt{\det g}\, \mathrm{d}x_1 \mathrm{d}x_2 \ldots \mathrm{d}x_d = \mathrm{d}V_g$ be the volume form corresponding to the Riemannian metric g on M. One can show (see, for instance, [**Pet06**,

Corollary 46]) that the Lie derivative of ω_g in the direction of a vector field X is given by

$$(1.2.10) \qquad \mathcal{L}_X(\omega_g) = (\operatorname{div} X)\omega_g.$$

This formula explains the meaning of the term divergence; it measures the rate of expansion of the volume element as it flows along the vector field X.

■ **Exercise 1.2.2.** Show that formulas (1.2.9) and (1.2.10) yield the same expression (1.2.7) for the divergence in local coordinates. See [**Cha84**, §I.1] and [**Ros97**, §1.2.3] for a solution.

Let us now state the main definition of this subsection.

Definition 1.2.3. The operator $-\Delta := -\operatorname{div} \nabla$ defined on smooth functions is called the *Laplacian* (or the *Laplace–Beltrami operator*) on the manifold (M, g). We will sometimes write it as $-\Delta_g = -\Delta_M$ to distinguish a particular manifold or metric.

Combining the formulas (1.2.5) and (1.2.7) we obtain the following expression for the Laplacian:

$$(1.2.11) \qquad -\Delta f = -\frac{1}{\sqrt{\det g}} \sum_{i,j=1}^{d} \frac{\partial}{\partial x_i}\left(g^{ij}\sqrt{\det g}\,\frac{\partial f}{\partial x_j}\right).$$

Example 1.2.4. Let $g_{ij} = \delta_{ij}$, where δ_{ij} is the Kronecker symbol. Then the metric is flat and the Laplacian takes the form

$$-\Delta f = -\operatorname{div}\left(\frac{\partial f}{\partial x_1}, \ldots, \frac{\partial f}{\partial x_n}\right) = -\sum_{i=1}^{d} \frac{\partial^2 f}{\partial x_i^2},$$

and we recover the usual definition (1.1.1) of the Laplace operator in the Euclidean space.

■ **Exercise 1.2.5.** Recall that given two Riemannian manifolds (M, g) and (N, h), a diffeomorphism $F : (M, g) \to (N, h)$ is called an *isometry* if it preserves the Riemannian metric, i.e., $F^*h = g$, where F^*h denotes the pullback metric; see, for example [**BerGauMaz71**, Definition A.2]. Using the invariance properties of the divergence and the gradient, show that the Laplace operator commutes with isometries: $-\Delta_g(u \circ F) = (-\Delta_h u) \circ F$ for any function $u \in C^\infty(N)$.

■ **Exercise 1.2.6.** Given $u, v \in C^\infty(M)$, show that

$$\Delta(uv) = v\Delta u + 2\langle \nabla u, \nabla v \rangle_g + u\Delta v.$$

Example 1.2.7. Suppose that the Riemannian metric in local coordinates (x, y) on a surface is given by $ds^2 = h(x, y)(dx^2 + dy^2)$, where $h(x, y) > 0$. Such coordinates are called *isothermal* and they locally exist on any surface; see [**Spi88**, Addendum 1, Chapter 9]. Show that the Laplacian in isothermal coordinates has the form

$$-\Delta = -\frac{1}{h(x, y)}\left(\frac{\partial^2}{\partial x^2} + \frac{\partial^2}{\partial y^2}\right).$$

Remark 1.2.8 (Manifolds with boundary). In what follows, we will also consider compact Riemannian manifolds M with boundary $\partial M \neq \emptyset$. Note that in contrast to domains, which are open sets, by definition $\partial M \subset M$. Somewhat abusing notation, when talking about differential expressions or function spaces on a Riemannian manifold M with boundary, we always have in mind the *interior* of M, that is, $M \setminus \partial M$, without indicating this explicitly. Let us also mention that the definition of the divergence given above has to be adjusted accordingly in case of a manifold with boundary; the equality (1.2.2) should hold for all $f \in C_0^1(M)$, where by our convention $C_0^1(M) := C_0^1(M \setminus \partial M)$.

1.2.2. The Laplacian on a flat torus. Consider a two-dimensional flat square torus $\mathbb{T}_a^2 = \mathbb{R}^2/(a\mathbb{Z})^2$. Separating variables, and using Example 1.1.8, we can find its eigenfunctions using complex notation; they are of the form $e^{\frac{2\pi i \langle x, m \rangle}{a}}$, where $x = (x_1, x_2) \in \mathbb{T}_a^2$, and $m = (m_1, m_2) \in \mathbb{Z}^2$ is a vector with integer coordinates. The eigenvalues are given by $\lambda_{m_1, m_2} = \frac{4\pi^2}{a^2}(m_1^2 + m_2^2)$. In particular, we have a constant eigenfunction coming from the vector $m = (0, 0)$ and corresponding to the eigenvalue zero. The first nonzero eigenvalue $\lambda_1 = \frac{4\pi^2}{a^2}$ is of multiplicity four and comes from the eigenfunctions with $m = (\pm 1, 0)$ and $m = (0, \pm 1)$. The corresponding eigenfunctions may be chosen to be real as

$$\cos\frac{2\pi x_1}{a}, \quad \sin\frac{2\pi x_1}{a}, \quad \cos\frac{2\pi x_2}{a}, \quad \sin\frac{2\pi x_2}{a}.$$

☐ **Numerical Exercise 1.2.9.** Show that the multiplicity of an eigenvalue $\lambda \in \mathbb{N}$ of the torus $\mathbb{T}_{2\pi}^2$ is equal to the sum of squares function

(1.2.12) $$r_2(\lambda) := \#\left\{(m_1, m_2) \in \mathbb{Z}^2 : \lambda = m_1^2 + m_2^2\right\};$$

cf. Exercise 1.1.12. Use this to compile a table of all the distinct eigenvalues of $\mathbb{T}^2_{2\pi}$ less than 2,500 together with their multiplicities.

■ **Exercise 1.2.10.** Calculate the eigenvalues of the Laplacian on a flat rectangular d-dimensional torus

$$\mathbb{T}^d_{(a_1,\ldots,a_d)} = \mathbb{R}/(a_1\mathbb{Z}) \times \cdots \times \mathbb{R}/(a_d\mathbb{Z}),$$

using separation of variables and the spectrum of the Laplacian on a circle from Example 1.1.8.

■ **Exercise 1.2.11.** Find the eigenvalues and eigenfunctions of an arbitrary flat d-dimensional torus $\mathbb{T}^d_\Gamma = \mathbb{R}^d/\Gamma$, where Γ is an arbitrary lattice in $\mathbb{R}^d$. (You can find the answer in [**Cha84**, §II.2], [**BerGauMaz71**, §III.B.1], and [**Can13**, §5.2].)

A flat torus is a rare example of a manifold for which the eigenvalues and eigenfunctions can be calculated explicitly. However, even in this case, some basic questions regarding the properties of eigenvalues turn out to be very difficult.

Let us introduce, for a closed manifold M, the *counting function* of the eigenvalues of the Laplace–Beltrami operator on M,

$$\mathcal{N}_M(\lambda) = \mathcal{N}(\lambda) := \#\{j : \lambda_j(M) \le \lambda\}.$$

Each eigenvalue is counted with its multiplicity. The behaviour of the function $\mathcal{N}_M(\lambda)$ for large values of λ describes the asymptotic distribution of eigenvalues as $\lambda \to +\infty$. Understanding the properties of the counting function is one of the fundamental questions in spectral geometry.

Let us estimate $\mathcal{N}(\lambda) := \mathcal{N}_{\mathbb{T}^2_a}(\lambda)$ for a flat square torus. Each eigenvalue

$$\lambda_{m_1,m_2} = \frac{4\pi^2}{a^2}\left(m_1^2 + m_2^2\right)$$

corresponds to a point with integer coordinates (m_1, m_2) on the plane, and we are counting the number

$$\mathcal{G}(\rho) := \#\left\{(m_1, m_2) \in \mathbb{Z}^2 : m_1^2 + m_2^2 \le \rho^2\right\}$$

of such points inside a circle of radius $\rho := \frac{a\sqrt{\lambda}}{2\pi}$; we have

(1.2.13) $$\mathcal{N}_{\mathbb{T}^2_a}(\lambda) = \mathcal{G}\left(\frac{a\sqrt{\lambda}}{2\pi}\right).$$

Clearly, an approximate number of integer points inside the circle is given by the area of the circle. Therefore, in this case

$$(1.2.14) \qquad \mathcal{N}(\lambda) = \mathcal{G}\left(\frac{a\sqrt{\lambda}}{2\pi}\right) = \frac{a^2\lambda}{4\pi} + R(\lambda) = \frac{\mathrm{Area}(\mathbb{T}_a^2)\lambda}{4\pi} + R(\lambda),$$

where $R(\lambda) = o(\lambda)$ as $\lambda \to \infty$. Note the appearance of area in this asymptotic formula — as we will see later, this is not a coincidence. The asymptotic formula (1.2.14) for the counting function of the torus is known as *Weyl's law*; see §3.3.1.

What more can be said about the size of the remainder $R(\lambda)$?

Lemma 1.2.12. *The remainder in Weyl's law (1.2.14) on a square torus satisfies the estimate*

$$R(\lambda) = O(\sqrt{\lambda}) \qquad as\ \lambda \to +\infty.$$

Proof. For simplicity, set $a = 2\pi$; the result would follow for an arbitrary a by rescaling; see Exercise 2.1.42. Let us identify each unit square with integer coordinates in the plane with its left bottom corner (m, n). Then if $m^2 + n^2 < \lambda$, the whole square (corresponding to that corner) is contained inside the disk of radius $\sqrt{\lambda} + \sqrt{2}$; see Figure 1.4.

Therefore, $\mathcal{N}(\lambda) < \pi(\sqrt{\lambda} + \sqrt{2})^2$. Similarly, if the square has a nontrivial intersection with the open disk of radius $\sqrt{\lambda} - \sqrt{2}$, then $m^2 + n^2 < \lambda$. Note that the union of such squares fully covers the disk of radius $\sqrt{\lambda} - \sqrt{2}$, and therefore $\mathcal{N}(\lambda) > \pi(\sqrt{\lambda} - \sqrt{2})^2$. Combining the two bounds on $\mathcal{N}(\lambda)$ we get

$$|\mathcal{N}(\lambda) - \pi\lambda| \le 2\pi\sqrt{2\lambda} + 2\pi,$$

which implies the statement of the lemma.   □

This result was known to C. F. Gauss, and the problem of counting the number $\mathcal{G}(\rho)$ of integer points inside a disk of radius ρ is called *Gauss's circle problem*. However, the estimate given by Lemma 1.2.12 is quite far from the optimal one.

Conjecture 1.2.13. *For any $\varepsilon > 0$, we have $R(\lambda) = O(\lambda^{1/4+\varepsilon})$ as $\lambda \to +\infty$.*

This conjecture is due to G. H. Hardy (1916) and has remained wide open for more than a century. It is one of the most famous open problems in analytic number theory. It is known that without ε in the exponent the conjecture is false — this follows from a quite nontrivial lower bound due to Hardy and E. Landau. It was shown by G. Voronoi (1903), W. Sierpiński (1906), and J. G. van der Corput (1923) that the upper bound holds with the

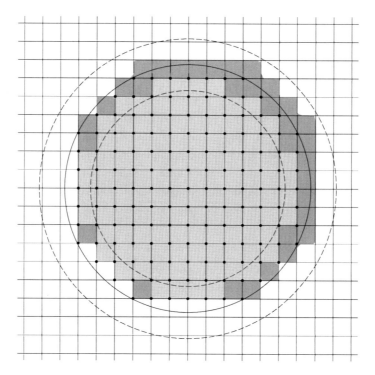

Figure 1.4. Estimating the number of integer points in a disk. The radii of the three concentric circles are $\sqrt{\lambda} - \sqrt{2}$, $\sqrt{\lambda}$, and $\sqrt{\lambda} + \sqrt{2}$.

exponent $\frac{1}{3}$. At present, the best upper bound for $R(\lambda)$ is due to J. Bourgain and N. Watt [**BouWat17**] with the exponent approximately equal to 0.3137.

Remark 1.2.14. There is a surprising link between the eigenvalue counting function for a flat square torus and the Bessel functions which appear in the spectral problems in the disk; see Example 1.1.17. Consider, once more, the torus $\mathbb{T}^2_{2\pi}$. Its eigenvalue counting function $\mathcal{N}_{\mathbb{T}^2_{2\pi}}(\lambda)$ coincides with the disk lattice point counting function $\mathcal{G}(\lambda)$ by (1.2.13). Consider, for an integer $m \geq 0$, the sum of squares function defined by (1.2.12). Then

$$\mathcal{G}(\rho) = \sum_{m=0}^{\lfloor \rho^2 \rfloor} r_2(m),$$

where $\lfloor \cdot \rfloor$ denotes the integer part. The function $\mathcal{G}(\rho)$ experiences a jump whenever ρ^2 is an integer with $r_2(\rho^2) > 0$. The identity due to Hardy [**Har15**] (in some form suggested by S. Ramanujan) is then

$$\mathcal{G}(\rho) - \frac{r_2(\rho^2)}{2} = \pi\rho^2 + \rho \sum_{n=1}^{\infty} \frac{r_2(n)}{\sqrt{n}} J_1\left(2\pi\rho\sqrt{n}\right),$$

thus bringing the Bessel function J_1 into play; see also [**BerDKZ18**] for some historical remarks and generalisations involving other Bessel functions.

1.2.3. The Laplace operator on spheres.

This section is based on the material that can be found in [**BerGauMaz71**, §III.C.1], [**Shu01**, §III.22], [**Cha84**, §II.4], and [**AxlBouWad01**, Chapter 5].

Let $(\xi_1, \ldots, \xi_d)$ be local coordinates on the unit sphere $\mathbb{S}^d \subset \mathbb{R}^{d+1}$ centred at the origin. Consider the corresponding spherical coordinates $(r, \xi_1, \ldots, \xi_d)$ defined in some open cone in $\mathbb{R}^{d+1}$, where $r > 0$ is the radial variable. The standard Euclidean coordinates can be expressed as $x_i = r\varphi_i(\xi_1, \ldots, \xi_d)$, $i = 1, \ldots, d+1$, where φ_i, $i = 1, \ldots, d+1$, are smooth functions parametrising the unit sphere. Given a function $f \in C^\infty(\mathbb{R}^{d+1})$, we obtain using the chain rule

$$
\begin{aligned}
\frac{\partial f}{\partial r} &= \sum_{i=1}^{d+1} \frac{\partial f}{\partial x_i}\frac{\partial x_i}{\partial r} = \sum_{i=1}^{d+1} \varphi_i \frac{\partial f}{\partial x_i}, \\
\frac{\partial f}{\partial \xi_j} &= \sum_{i=1}^{d+1} \frac{\partial f}{\partial x_i}\frac{\partial x_i}{\partial \xi_j} = r\sum_{i=1}^{d+1} \frac{\partial \varphi_i}{\partial \xi_j}\frac{\partial f}{\partial x_i}, \qquad j = 1, \ldots, d.
\end{aligned}
$$

(1.2.15)

Consider the basis $\left(\frac{\partial}{\partial r}, \frac{\partial}{\partial \xi_1}, \ldots, \frac{\partial}{\partial \xi_d}\right)$ in the tangent space $T_x\mathbb{R}^d$. Then formulas (1.2.15) imply

$$
\left\langle \frac{\partial}{\partial r}, \frac{\partial}{\partial r} \right\rangle_{g_{\mathbb{R}^{d+1}}} = \sum_{k=1}^{d+1} \varphi_k^2 = 1,
$$

$$
\left\langle \frac{\partial}{\partial r}, \frac{\partial}{\partial \xi_j} \right\rangle_{g_{\mathbb{R}^{d+1}}} = r\sum_{k=1}^{d+1} \varphi_k \frac{\partial \varphi_k}{\partial \xi_j} = 0,
$$

$$
\left\langle \frac{\partial}{\partial \xi_i}, \frac{\partial}{\partial \xi_j} \right\rangle_{g_{\mathbb{R}^{d+1}}} = r^2 \sum_{k=1}^{d+1} \frac{\partial \varphi_k}{\partial \xi_i}\frac{\partial \varphi_k}{\partial \xi_j} = r^2 \left\langle \frac{\partial}{\partial \xi_i}, \frac{\partial}{\partial \xi_j} \right\rangle_{g_{\mathbb{S}^d}},
$$

where $g_{\mathbb{S}^d}$ denotes the standard round metric on the sphere $\mathbb{S}^d$, that is, the metric induced by the Euclidean metric $g_{\mathbb{R}^{d+1}}$. Note that the last equality on the first line is simply the equation of the unit sphere; differentiating it with respect to ξ_j we obtain the last equality on the second line.

In view of the formulas above, the Euclidean metric in spherical coordinates $(r, \xi_1, \ldots, \xi_d)$ is given by

$$
g_{\mathbb{R}^{d+1}} = \begin{pmatrix} 1 & 0 \\ 0 & r^2\, g_{\mathbb{S}^d} \end{pmatrix}.
$$

Therefore, applying formula (1.2.11) for the Laplace operator we obtain

$$(1.2.16) \qquad \Delta_{g_{\mathbb{R}^{d+1}}} = \frac{\partial^2}{\partial r^2} + \frac{d}{r}\frac{\partial}{\partial r} + \frac{1}{r^2}\Delta_{g_{\mathbb{S}^d}}.$$

Let $\mathcal{P}_m$ be the space of homogeneous polynomials in $\mathbb{R}^{d+1}$ of degree m. By definition, $P \in \mathcal{P}_m$ if and only if $P = r^m \cdot P|_{\mathbb{S}^d}$. In particular,

$$(1.2.17) \qquad \frac{\partial P}{\partial r} = mr^{m-1} \cdot P|_{\mathbb{S}^d}, \qquad \frac{\partial^2 P}{\partial r^2} = m(m-1)r^{m-2} \cdot P|_{\mathbb{S}^d}.$$

We will denote by

$$\widetilde{\mathcal{P}}_m := \mathcal{P}_m|_{\mathbb{S}^d} = \{P|_{\mathbb{S}^d} : P \in \mathcal{P}_m\}$$

the restriction of $\mathcal{P}_m$ to the sphere $\mathbb{S}^d$.

Let

$$\mathcal{H}_m := \left\{ P \in \mathcal{P}_m : \Delta_{g_{\mathbb{R}^{d+1}}} P = 0 \right\}$$

be the space of all harmonic homogeneous polynomials of degree m, and let

$$\widetilde{\mathcal{H}}_m := \mathcal{H}_m|_{\mathbb{S}^d} = \{P|_{\mathbb{S}^d} : P \in \mathcal{H}_m\}$$

be the space of their restrictions to the sphere $\mathbb{S}^d$. It is easy to check that the spaces $\mathcal{H}_m$ and $\widetilde{\mathcal{H}}_m$ are isomorphic; indeed, the restriction map $\mathcal{H}_m \to \widetilde{\mathcal{H}}_m$ has an inverse given by

$$(1.2.18) \qquad \widetilde{P} \mapsto r^m \widetilde{P}.$$

Moreover, applying the left- and the right-hand sides of (1.2.16) to $r^m \widetilde{P}$ and taking into account (1.2.17) we obtain

$$0 = r^{m-2}\left(-\Delta_{g_{\mathbb{S}^d}}\widetilde{P} - m(d+m-1)\widetilde{P} \right),$$

which immediately implies that $\widetilde{P}$ is an eigenfunction of the Laplacian on the sphere with the eigenvalue $m(d+m-1)$. In other words, we have proved the following.

Proposition 1.2.15. Any element of the space $\widetilde{\mathcal{H}}_m$ is an eigenfunction of the Laplacian on the sphere corresponding to the eigenvalue $\lambda = m(d+m-1)$.

The space $\widetilde{\mathcal{H}}_m$ of such eigenfunctions is called the space of *spherical harmonics* of degree m. Let us now calculate the multiplicities of the eigenvalues $m(d+m-1)$, $m \in \mathbb{N}_0$, and show that there are no other eigenvalues of the Laplacian on the sphere.

Theorem 1.2.16. The eigenvalues of the Laplace operator on the standard sphere $\mathbb{S}^d$ are given by $m(d+m-1)$, $m \in \mathbb{N}_0$, and the corresponding eigenspaces coincide with $\tilde{\mathcal{H}}_m$. The multiplicity of the eigenvalue $\lambda = m(d+m-1)$ is equal to

$$(1.2.19) \qquad \kappa_{d,m} := \dim \tilde{\mathcal{H}}_m = \binom{d+m}{d} - \binom{d+m-2}{d}.$$

In order to prove this theorem we use the following proposition.

Proposition 1.2.17. For any $m \geq 0$, the following decomposition of $\mathcal{P}_m$ into a direct sum holds:

$$\mathcal{P}_m = \mathcal{H}_m \oplus r^2 \mathcal{P}_{m-2}.$$

Here and further on we assume that $\mathcal{P}_m = \{0\}$ if $m < 0$.

Proof. We prove the statement by induction in m. For $m = 0, 1$ the result is trivially true. Assume that it is true for all $l < m$ and let us show that it holds for $l = m$. First, let us show that

$$(1.2.20) \qquad \mathcal{H}_m \cap r^2 \mathcal{P}_{m-2} = \{0\}.$$

Indeed, suppose there exists $P \in \mathcal{H}_m \cap r^2 \mathcal{P}_{m-2}$. Consider its restriction on the sphere $\tilde{P} \in \tilde{\mathcal{H}}_m \cap \tilde{\mathcal{P}}_{m-2}$. Note that $\tilde{\mathcal{P}}_m$ is isomorphic to $\mathcal{P}_m$, with the inverse to the restriction map given by the same formula as (1.2.18).

As we have already shown, the space $\tilde{\mathcal{H}}_m$ is contained in the eigenspace of the Laplacian corresponding to the eigenvalue $\lambda = m(m+d-1)$. At the same time, by the induction hypothesis, the space $\tilde{\mathcal{P}}_{m-2}$ could be represented as a direct sum of certain spaces $\tilde{\mathcal{H}}_j$, and for all of them $j < m$. Using integration by parts it is easy to show that Laplace eigenfunctions corresponding to distinct eigenvalues are orthogonal in $L^2(\mathbb{S}^d)$. Therefore, we conclude that $\tilde{P} \equiv 0$. Since $P = r^m \tilde{P}$ by (1.2.18), we obtain $P \equiv 0$, which implies (1.2.20).

We have thus shown that $\mathcal{P}_m \supset \mathcal{H}_m \oplus r^2 \mathcal{P}_{m-2}$, and therefore

$$(1.2.21) \qquad \dim \mathcal{H}_m \leq \dim \mathcal{P}_m - \dim \mathcal{P}_{m-2}.$$

At the same time, consider the Laplace operator as a map $\Delta : \mathcal{P}_m \to \mathcal{P}_{m-2}$. Its kernel is precisely $\mathcal{H}_m$, and therefore

$$(1.2.22) \qquad \dim \mathcal{H}_m \geq \dim \mathcal{P}_m - \dim \mathcal{P}_{m-2}.$$

Combining (1.2.21) and (1.2.22) we conclude that

$$(1.2.23) \qquad \dim \mathcal{H}_m = \dim \mathcal{P}_m - \dim \mathcal{P}_{m-2}.$$

It then follows that the map $\Delta : \mathcal{P}_m \to \mathcal{P}_{m-2}$ is surjective, and by the dimension count $\mathcal{P}_m = \mathcal{H}_m \oplus r^2 \mathcal{P}_{m-2}$, which completes the proof of the proposition. $\qquad \square$

Proof of Theorem 1.2.16. Let us show first that

(1.2.24) $$L^2(\mathbb{S}^d) = \bigoplus_{m=1}^{\infty} \tilde{\mathcal{H}}_m.$$

Indeed, applying inductively (1.2.20) and taking restriction to the sphere, we get

$$\bigoplus_{m=1}^{\infty} \tilde{\mathcal{H}}_m = \bigoplus_{m=1}^{\infty} \tilde{\mathcal{P}}_m.$$

Note that the direct sum on the right is isomorphic to the space of all polynomials in $\mathbb{R}^{d+1}$ restricted to $\mathbb{S}^d$. Formula (1.2.24) then holds since polynomials are dense in $L^2(\mathbb{R}^{d+1})$. Hence, the first assertion of Theorem 1.2.16 follows from Proposition 1.2.15.

It remains to note that (1.2.19) follows from (1.2.23), taking into account the isomorphism $\mathcal{H}_m \cong \tilde{\mathcal{H}}_m$ and Lemma 1.2.18 below. $\square$

Lemma 1.2.18. The dimension of the space $\mathcal{P}_m$ of homogeneous polynomials of order m in $\mathbb{R}^{d+1}$ is given by

(1.2.25) $$\dim \mathcal{P}_m = \binom{d+m}{d} = \frac{(m+d)(m+d-1)\cdots(m+1)}{d!}.$$

Proof. The basis in $\mathcal{P}_m$ is given by monomials $x_1^{m_1}\ldots x_{d+1}^{m_{d+1}}$, such that $m_1 + \cdots + m_{d+1} = m$. Therefore, the dimension of $\mathcal{P}_m$ is the number of ordered partitions of m into a sum of $d+1$ nonnegative integers. Finding it is equivalent to finding the number of sequences of zeros and ones of length $d+m$ with exactly d zeros (summing up the ones between the neighbouring zeros we get precisely the required partition of m), which is clearly given by (1.2.25). $\square$

■ **Exercise 1.2.19.** Show that the coordinate functions $x_1, \ldots, x_{d+1}$ restricted to the sphere $\mathbb{S}^d$ form a basis of the first eigenspace on $\mathbb{S}^d$.

■ **Exercise 1.2.20.** Show that the eigenvalue counting function of the Laplacian on the sphere $\mathbb{S}^d$ satisfies the asymptotics

(1.2.26) $$\mathcal{N}_{\mathbb{S}^d}(\lambda) = \frac{2}{d!}\lambda^{\frac{d}{2}} + O\left(\lambda^{\frac{d-1}{2}}\right),$$

and the power in the remainder estimate cannot be improved. *Hint:* Find the asymptotic behaviour of multiplicities. A complete solution to this exercise can be found in [**Shu01**, §III.22].

■ **Exercise 1.2.21.** Using formula (1.2.16) and separation of variables, find eigenvalues and eigenfunctions of the Dirichlet and Neumann Laplacians for Euclidean balls in $\mathbb{R}^d$. In particular, show that for the d-dimensional unit ball $\mathbb{B}^d$, the Dirichlet eigenvalues are

$$\lambda^{\mathrm{D}}_{m,k}(\mathbb{B}^d) = \left(j_{m+\frac{d}{2}-1,k} \right)^2, \qquad m \in \mathbb{N}_0, \quad k \in \mathbb{N},$$

with multiplicity $\kappa_{d-1,m}$ given by (1.2.19), where $j_{m+\frac{d}{2}-1,k}$ is the kth positive zero of the Bessel function $J_{m+\frac{d}{2}-1}(x)$; see Example 1.1.17. Show also that the Neumann eigenvalues are

$$\lambda^{\mathrm{N}}_{m,k}(\mathbb{B}^d) = \left(p'_{d,m,k} \right)^2, \qquad m \in \mathbb{N}_0, \quad k \in \mathbb{N},$$

with the same multiplicity $\kappa_{d-1,m}$, where $p'_{d,m,k}$ is the kth positive zero of the derivative $U'_{d,m}(x)$ of the *ultraspherical Bessel function*

$$U_{d,m}(x) := x^{1-\frac{d}{2}} J_{m+\frac{d}{2}-1}(x),$$

with the exception $p'_{d,0,1} := 0$. For $d = 2$, compare your results with those given in Example 1.1.17.

Chapter 2

The spectral theorems

Johann Peter
Gustav
Lejeune
Dirichlet
(1805–1859)

In this chapter, we present the weak and strong spectral theorems for the Dirichlet and Neumann Laplacians, as well as for the Laplace–Beltrami operator on a Riemannian manifold. We present the fundamentals of the theory of Sobolev spaces and define the notion of weak solutions. We also recall some basic facts about selfadjoint unbounded linear operators and introduce the Friedrichs extension. In order to prove local and global regularity of eigenfunctions we give a brief overview of the theory of elliptic regularity, based on a priori estimates and Nirenberg's method of difference quotients.

Carl Gottfried
Neumann
(1832–1925)

2.1. Weak spectral theorems

2.1.1. Spectral theorems: An overview and the roadmap. Generally speaking, a spectral theorem is a result stating that subject to certain conditions an operator can be in some sense *diagonalised*. More specifically, in application to the eigenvalue problem (1.1.9) for the Laplacian in a bounded domain $\Omega \subset \mathbb{R}^d$ with Dirichlet (1.1.10) boundary conditions, it says that the eigenvalues form a discrete sequence with the only limit point at $+\infty$ and that the corresponding eigenfunctions can be chosen to form an orthonormal basis in $L^2(\Omega)$. A similar result holds also for Neumann (1.1.13) boundary conditions, under some mild regularity assumptions on the boundary $\partial\Omega$. We emphasise that we have not yet formally put the eigenproblem (1.1.9) subject to either (1.1.10) or (1.1.13) in the framework of operator theory and for now consider eigenvalues and eigenfunctions as those of a boundary value problem — we will call the corresponding spectral theorems the *strong spectral theorems* and postpone their formulation until §2.2.7.

The analysis behind the strong spectral theorems is somewhat delicate, and we perform it in the following steps. First of all, we switch to the so-called *weak spectral problems* (i.e., understood in the distributional sense), introduced first for the Dirichlet boundary value problem in §2.1.2, together with required preliminaries from the theory of *Sobolev spaces*. The Dirichlet case is easier as no conditions on the boundary are required; this allows us to formulate and prove the weak Dirichlet spectral theorem, Theorem 2.1.20, in §2.1.4. Along the way we give a brief reminder of basic spectral theory of unbounded selfadjoint operators in §2.1.5 and use it to put the Dirichlet spectral problem in the operator-theoretic framework via the construction of the *Friedrichs extension* in §2.1.6.

The formulation of the weak spectral theorem for the Neumann problem is a bit more subtle and is dealt with in §2.1.7; see Theorem 2.1.36.

In §2.1.8 we establish the weak spectral theorem for the Laplacian acting on a Riemannian manifold. This would allow us to treat the strong spectral theorem in this case later on within the general framework.

The strong spectral theorems do not follow immediately from the weak ones. The essential missing ingredient is the so-called *elliptic regularity*, which we review in §2.2. In essence, this fundamental property of elliptic PDEs allows us to establish that the weak eigenfunctions of either the Dirichlet Laplacian, the Neumann Laplacian, or the Laplace–Beltrami operator on a closed Riemannian manifold, for which we have already deduced some minimal regularity in weak spectral theorems, are in fact infinitely smooth in the interior. Together with the results on regularity near the boundary (which may require some additional conditions on $\partial\Omega$) this allows us to show that the weak eigenfunctions are in fact the strong ones, finally leading to the strong spectral theorem (Theorem 2.2.21).

2.1.2. Weak derivatives and Sobolev spaces.

Definition 2.1.1. Let $\Omega \subset \mathbb{R}^d$ be a domain. Let $u, v \in L^1_{\text{loc}}(\Omega)$. Suppose that for any $\varphi \in C^1_0(\Omega)$,

$$\int_\Omega u \, \partial_j \varphi \, \mathrm{d}x = -\int_\Omega v \, \varphi \, \mathrm{d}x,$$

where $\partial_j := \frac{\partial}{\partial x_j}$. Then we say that $\partial_j u$ exists in Ω in the *weak sense* and is equal to v.

Remark 2.1.2. If $u \in C^1(\Omega)$, then the weak and the classical derivatives coincide. In the theory of distributions, weak derivatives are also referred to as *distributional derivatives*. To keep the presentation more accessible, in what follows we do not use the language of distributions.

Definition 2.1.3. Set $H^0(\Omega) := L^2(\Omega)$. Let $m \in \mathbb{N}$. The *Sobolev spaces* $H^m(\Omega)$ are defined recursively as

$$H^m(\Omega) := \{u \in L^2(\Omega) : \partial_j u \text{ exists in the weak sense}$$
$$\text{and } \partial_j u \in H^{m-1}(\Omega) \text{ for all } j = 1, \ldots, d\}.$$

Equipped with the inner products

$$(u, v)_{H^1(\Omega)} := \int_\Omega uv \, \mathrm{d}x + \int_\Omega \langle \nabla u, \nabla v \rangle \, \mathrm{d}x,$$

$$(u, v)_{H^m(\Omega)} := (u, v)_{L^2(\Omega)} + \sum_{j=1}^d (\partial_j u, \partial_j v)_{H^{m-1}(\Omega)}, \qquad m \geq 2,$$

and the induced norms

(2.1.1)
$$\|u\|_{H^1(\Omega)}^2 := \|u\|_{L^2(\Omega)}^2 + \|\nabla u\|_{L^2(\Omega)}^2,$$

$$\|u\|_{H^m(\Omega)}^2 := \|u\|_{L^2(\Omega)}^2 + \sum_{j=1}^d \|\partial_j u\|_{H^{m-1}(\Omega)}^2, \qquad m \geq 2,$$

the Sobolev spaces $H^m(\Omega)$ become Hilbert spaces.

Remark 2.1.4. As we mostly deal with real-valued functions, we omit the complex conjugation in the definition of the Sobolev inner product and elsewhere.

Remark 2.1.5. The Sobolev norm may be alternatively defined as

(2.1.2)
$$\|u\|_{H^m(\Omega)}^2 = \sum_{|\alpha| \leq m} \|\partial^\alpha u\|_{L^2(\Omega)}^2,$$

where we use the multi-index notation (B.2.1). It can be easily checked that the norms (2.1.1) and (2.1.2) are equivalent and in fact coincide for $m = 1$.

Remark 2.1.6. It turns out that one may also define the Sobolev space $H^m(\Omega)$ as the completion of

$$\{u \in C^\infty(\Omega) : \|u\|_{H^m(\Omega)} < \infty\}.$$

This result is due to Meyers and Serrin [**MeySer64**]; see also [**AdaFou03**, Theorem 3.17].

We denote by $H_0^m(\Omega)$ the closure of $C_0^m(\Omega)$ with respect to the norm (2.1.1). The following important compactness result holds.

Theorem 2.1.7. Let $\Omega \subset \mathbb{R}^d$ be a bounded domain.

   (i) The space $H_0^1(\Omega)$ is compactly embedded in $L^2(\Omega)$.
  (ii) If, in addition, $\partial\Omega$ is Lipschitz, then $H^1(\Omega)$ is compactly embedded in $L^2(\Omega)$.

Theorem 2.1.7 is the Rellich–Kondrachov compactness theorem; see [**AdaFou03**, Theorem 6.3]. To get some intuition, one can compare it with a version of the Arzela–Ascoli theorem which states that for a bounded domain Ω, the Banach space $C^1(\overline{\Omega})$ is compactly embedded in the Banach space $C(\overline{\Omega})$. In fact, one can prove the Rellich–Kondrachov theorem by mollifying and reducing it to the Arzela–Ascoli theorem; see [**Bre11**, Theorem 9.16] or [**Eva10**, §II.5.7].

Remark 2.1.8. Statement (ii) of Theorem 2.1.7 is still valid under some weaker conditions on the regularity of the boundary $\partial\Omega$, namely that Ω satisfies the so-called *extension property*. For a comprehensive discussion of the extension property see, e.g., [**EdmEva18**, §V.4].

In many cases, the notion of a weak derivative is much more convenient to work with than the notion of a classical derivative. The remarkable Sobolev embedding theorem below connects these two notions. In particular, it shows that classical derivatives of all orders exist in a domain Ω if and only if weak derivatives of all orders belong to $L_{\text{loc}}^2(\Omega)$.

Theorem 2.1.9 (The Sobolev embedding theorem [**AdaFou03**, Theorem 4.12]). Let $\Omega \subset \mathbb{R}^d$ be a bounded domain. Then for $m > \frac{d}{2} + k$ we have a continuous embedding $H_0^m(\Omega) \subset C^k(\overline{\Omega})$. If, in addition, $\partial\Omega$ is Lipschitz, then $H^m(\Omega)$ is continuously embedded in $C^k(\overline{\Omega})$.

The following characterisation of Sobolev spaces in terms of the Fourier transform can be used to give a proof of Theorem 2.1.9. Here the Fourier transform of a function $u \in L^1(\mathbb{R}^d)$ is defined as

$$(2.1.3) \qquad (\mathcal{F}u)(\xi) := (2\pi)^{-\frac{d}{2}} \int_{\mathbb{R}^d} e^{-i\langle x,\xi\rangle} u(x)\,dx, \qquad \xi \in \mathbb{R}^d.$$

It can be shown (see, e.g., [**Eva10**, §4.3.1]) that the formula (2.1.3) defines an isometry $\mathcal{F}: L^1(\mathbb{R}^d) \cap L^2(\mathbb{R}^d) \to L^2(\mathbb{R}^d)$ and that $\mathcal{F}$ extends to an isometric isomorphism $\mathcal{F}: L^2(\mathbb{R}^d) \to L^2(\mathbb{R}^d)$. Its inverse on $L^1(\mathbb{R}^d) \cap L^2(\mathbb{R}^d)$ is given by

$$\left(\mathcal{F}^{-1}v\right)(x) = (2\pi)^{-\frac{d}{2}} \int_{\mathbb{R}^d} e^{i\langle x,\xi\rangle} v(\xi)\,d\xi\,, \qquad x \in \mathbb{R}^d.$$

Proposition 2.1.10 ([**Shu20**, Proposition 8.3]). *Let $u \in L^2(\mathbb{R}^d)$. Then $u \in H^m(\mathbb{R}^d)$ if and only if its Fourier transform $\mathcal{F}u$ satisfies*

$$(2.1.4) \qquad \left(1+|\xi|^2\right)^{\frac{m}{2}} (\mathcal{F}u)(\xi) \in L^2(\mathbb{R}^d)\,.$$

Now, if $m > d/2$ and $(1+|\xi|^2)^{\frac{m}{2}}(\mathcal{F}u) \in L^2(\Omega)$, then $\mathcal{F}u \in L^1(\Omega)$, which easily follows from the Cauchy–Schwarz inequality and the fact that $(1+|\xi|^2)^{-m} \in L^1(\Omega)$. Then u is the inverse Fourier transform of an $L^1(\Omega)$-function $\mathcal{F}u$, and in particular it is continuous. This gives an idea of the proof of Theorem 2.1.9.

It is sometimes desirable to define Sobolev spaces H^m for fractional (nonnegative) values of the parameter m. The characterisation (2.1.4) leads to a natural definition of $H^m(\mathbb{R}^d)$ for fractional m; see [**Fol95**, Chapter 6] or [**McL00**, Chapter 3].

For a domain $\Omega \subset \mathbb{R}^d$ and $m \in \mathbb{N}$ we define $H^{-m}(\Omega)$ as the dual Hilbert space of $H_0^m(\Omega)$.

In what follows, we will also say that $u \in H_{\mathrm{loc}}^m(\Omega)$ if $u|_U \in H^m(U)$ for any open set $U \Subset \Omega$.

We also need to define Sobolev spaces $H^m(\partial\Omega)$ on the boundary $\partial\Omega$ of a Lipschitz domain $\Omega \subset \mathbb{R}^d$. This is a delicate and technically involved construction, and we refer to [**Gri11**] and in particular to [**ChWGLS12**, §A.3] for full details. Let us briefly explain the main ideas.

First, if $\Omega = \mathbb{R}^{d-1}\times\mathbb{R}_+$ is a half-space, then $\partial\Omega = \mathbb{R}^{d-1}$ and no additional work is required. Second, let

$$\Omega = \left\{(x', x_d) \in \mathbb{R}^{d-1} \times \mathbb{R} : x_d > f(x')\right\}$$

be a "curved" half-space whose boundary $\partial\Omega = \left\{(x', f(x')) : x' \in \mathbb{R}^{d-1}\right\}$ is represented as the graph of a Lipschitz function $f : \mathbb{R}^{d-1} \to \mathbb{R}$. Given $u \in L^2(\partial\Omega)$ we define $u_f \in L^2(\mathbb{R}^{d-1})$ by $u_f(x') = u(x', f(x'))$, $x' \in \mathbb{R}^{d-1}$. Then we set

$$(2.1.5) \qquad H^m(\partial\Omega) = \left\{u \in L^2(\partial\Omega) : u_f \in H^m(\mathbb{R}^{d-1})\right\}.$$

One can check that for a Lipschitz hypersurface $\partial\Omega$ this definition makes sense only if $0 \leq m \leq 1$, whereas for smooth hypersurfaces one can take an arbitrary $m \geq 0$.

Finally, in order to define Sobolev spaces on $\partial\Omega$ for a bounded Lipschitz domain $\Omega \subset \mathbb{R}^d$, we represent the boundary locally using graphs of Lipschitz functions as in §B.3, and we use (2.1.5) together with a partition of unity argument; cf. §2.1.8 below.

The following trace theorem gives a natural example where the boundary Sobolev spaces appear.

Theorem 2.1.11 (The trace theorem [**Eva10**, §5.5], [**Gri11**, §1.5]). Let $\Omega \subset \mathbb{R}^d$ be a bounded domain with a Lipschitz boundary. There exists a bounded linear operator $T : H^1(\Omega) \to H^{\frac{1}{2}}(\partial\Omega)$ (called the *trace operator*) such that $Tu = u|_{\partial\Omega}$ if $u \in H^1(\Omega) \cap C(\overline{\Omega})$.

Note that in view of Theorem 2.1.9, functions from $H^m(\Omega)$ have pointwise boundary values for $m > d/2$.

2.1.3. Weak solutions. We will use the following standard integration by parts formula.

Lemma 2.1.12 (Integration by parts). Let $\Omega \subset \mathbb{R}^d$ be a smooth bounded domain. Let $u, v \in C^1(\overline{\Omega})$. Then

$$(2.1.6) \qquad \int_\Omega u(\partial_k v)\, dx = -\int_\Omega (\partial_k u)v\, dx + \int_{\partial\Omega} uvn_k\, d\sigma,$$

where n_k is the kth coordinate of the outward unit normal vector on $\partial\Omega$.

Lemma 2.1.12 remains valid also for Lipschitz domains [**Nec12**, §3.1, Theorem 1.1]. It implies

Lemma 2.1.13 (Green's formula [**EvaGar15**, §4.3], [**ChWGLS12**, formula (A.26)]). For a bounded domain $\Omega \subset \mathbb{R}^d$ with a Lipschitz boundary, and for any real-valued $u \in H^2(\Omega), v \in H^1(\Omega)$,

$$(2.1.7) \qquad -\int_\Omega \Delta u\, v\, dx = \int_\Omega \langle \nabla u, \nabla v \rangle\, dx - \int_{\partial\Omega} (\partial_n u)\, v\, ds.$$

■ **Exercise 2.1.14.** Prove (2.1.6) and (2.1.7) for a smooth domain Ω.

Of course, formula (2.1.7) rewritten as

$$(-\Delta u, v)_{L^2(\Omega)} = (\nabla u, \nabla v)_{L^2(\Omega)} - (\partial_n u, v)_{L^2(\partial\Omega)}$$

remains valid for complex-valued $u \in H^2(\Omega), v \in H^1(\Omega)$ as well.

Remark 2.1.15. If $v \in H_0^1(\Omega)$, a simple argument shows that for any bounded domain Ω, with no regularity assumptions on its boundary, and for any $v \in H_0^1(\Omega)$, Green's formula is still valid in the form

$$(2.1.8) \qquad -\int_\Omega \Delta u \, v \, \mathrm{d}x = \int_\Omega \langle \nabla u, \nabla v \rangle \, \mathrm{d}x.$$

We leave the proof of (2.1.8) as an exercise for the reader.

The concept of a weak solution of a boundary value problem is standard and can be found in numerous textbooks; see for example [**Shu20**]. Let $\Omega \subset \mathbb{R}^d$ be a bounded domain, let $f \in C(\Omega)$, and suppose that $u \in C^2(\Omega) \cap C(\overline{\Omega})$ is a solution of the boundary value problem

$$(2.1.9) \qquad \begin{cases} -\Delta u + u = f, \\ u|_{\partial\Omega} = 0. \end{cases}$$

Then for any test function $v \in C_0^1(\Omega)$ we get

$$(2.1.10) \qquad -\int_\Omega \Delta u \, v \, \mathrm{d}x + \int_\Omega u \, v \, \mathrm{d}x = \int_\Omega f \, v \, \mathrm{d}x.$$

Applying now Green's formula (2.1.8) with $v \in C_0^1(\Omega)$ to (2.1.10), we obtain

$$(2.1.11) \qquad \int_\Omega uv \, \mathrm{d}x + \int_\Omega \langle \nabla u, \nabla v \rangle \, \mathrm{d}x = \int_\Omega f \, v \, \mathrm{d}x.$$

Note that both sides of (2.1.11) are well-defined if $u \in H^1(\Omega)$, $v \in H_0^1(\Omega)$, and $f \in L^2(\Omega)$. A function $u \in H^1(\Omega)$ which satisfies (2.1.11) for any test function $v \in H_0^1(\Omega)$ is called a weak solution of the equation $-\Delta u + u = f$. To make it a weak solution of the Dirichlet boundary value problem we also require $u \in H_0^1(\Omega)$.

Definition 2.1.16 (Weak Dirichlet solution and weak Dirichlet spectral problem). We say that $u \in H_0^1(\Omega)$ is a *weak solution* of the boundary value problem (2.1.9) if (2.1.11) holds for all $v \in H_0^1(\Omega)$ (or, equivalently, for all $v \in C_0^1(\Omega)$). The weak *Dirichlet spectral problem* is to find $\lambda \in \mathbb{R}$ and $u \in H_0^1(\Omega) \setminus \{0\}$ such that

$$(2.1.12) \qquad \int_\Omega \langle \nabla u, \nabla v \rangle \, \mathrm{d}x = \lambda \int_\Omega uv \, \mathrm{d}x \qquad \text{for all } v \in H_0^1(\Omega).$$

■ **Exercise 2.1.17.** Prove that a weak solution of (2.1.9) always exists and is unique. *Hint:* Apply the Riesz representation theorem to the linear functional $F(v) = \int_\Omega fv \, dx$ defined on $H_0^1(\Omega)$.

2.1.4. The weak spectral theorem for the Dirichlet Laplacian. Existence and uniqueness of a weak solution of (2.1.9) allow us to define the solution operator $\widetilde{K} : L^2(\Omega) \to H_0^1(\Omega)$, $\widetilde{K}f := u$. Now we let K be the composition of $\widetilde{K}$ with the inclusion $H_0^1(\Omega) \subset L^2(\Omega)$. Note that (informally) $K = (-\Delta + 1)^{-1}$, and hence it is a resolvent of the Dirichlet Laplacian (see §2.1.6 for a formal definition). It is easy to check that $\widetilde{K}$ is bounded and, due to Theorem 2.1.7, the operator K is compact. Moreover, the following proposition holds; see §2.1.5 for a brief overview of notions in functional analysis.

Proposition 2.1.18. The operator $K : L^2(\Omega) \to L^2(\Omega)$ is compact, symmetric, and positive.

■ **Exercise 2.1.19.** Prove that K is positive and symmetric and that $\|K\| \leq 1$.

Compactness of the resolvent operator K is a crucial ingredient of the proof of the spectral theorem. By the Hilbert–Schmidt theorem, $L^2(\Omega)$ admits an orthonormal basis consisting of eigenfunctions of the compact symmetric operator K. The corresponding eigenvalues form a sequence of positive real numbers converging to zero. Note that if w is an eigenfunction of K with an eigenvalue μ, we get from (2.1.9) that $w \in H_0^1(\Omega)$ is a weak solution of the equation

$$(2.1.13) \qquad\qquad -\mu\Delta w + \mu w = w.$$

Dividing now (2.1.13) by μ and rearranging, we deduce that w is a weak solution of

$$-\Delta w = \frac{1 - \mu}{\mu} w.$$

We therefore arrive at

Theorem 2.1.20 (The weak spectral theorem for the Dirichlet Laplacian). Let $\Omega \subset \mathbb{R}^d$ be a bounded domain. There exists an orthonormal basis of $L^2(\Omega)$ composed of weak eigenfunctions of the Dirichlet spectral problem. The corresponding eigenvalues are nonnegative and form a nondecreasing sequence which tends to $+\infty$.

In fact, we additionally have

Proposition 2.1.21. The first eigenvalue of the weak Dirichlet spectral problem (2.1.12) is strictly positive.

To prove Proposition 2.1.21 we rely on the following important bound.

Proposition 2.1.22 (Poincaré's inequality; see, e.g., [**Shu20**, Proposition 8.8]). If $\Omega \subset \mathbb{R}^d$ is a bounded domain, then there exists a constant $C_\Omega > 0$ such that

$$(2.1.14) \qquad \int_\Omega |u|^2 \, \mathrm{d}x \leq C_\Omega \int_\Omega |\nabla u|^2 \, \mathrm{d}x$$

for all $u \in H_0^1(\Omega)$.

The integral in the right-hand side of (2.1.14) is called the *Dirichlet energy* of u.

■ **Exercise 2.1.23.** Prove Poincaré's inequality, first for functions in $C_0^1(\Omega)$. Show that in fact a stronger version of (2.1.14) holds: for any $j = 1, \ldots, d$,

$$\int_\Omega |u|^2 \, \mathrm{d}x \leq C_\Omega \int_\Omega |\partial_j u|^2 \, \mathrm{d}x$$

for all $u \in H_0^1(\Omega)$.

Substituting $\lambda := \lambda_1$ and $u = v := u_1$ into the weak Dirichlet spectral problem (2.1.12) (where λ_1 and u_1 are its first eigenvalue and eigenfunction), we immediately deduce from Poincaré's inequality that

$$\lambda_1 = \frac{\|\nabla u_1\|_{L^2(\Omega)}^2}{\|u_1\|_{L^2(\Omega)}^2} \geq \frac{1}{C_\Omega} > 0,$$

thus proving Proposition 2.1.21.

2.1.5. Selfadjoint unbounded linear operators. We very briefly review a few basic notions from functional analysis. For more details see, e.g., [**Lax02**].

Let $\mathcal{H}$ be a complex Hilbert space with an inner product $(\cdot, \cdot)_{\mathcal{H}}$. By a (possibly *unbounded*) *linear operator* A in $\mathcal{H}$ we understand a linear map $A : \mathrm{Dom}(A) \to \mathcal{H}$ defined on a dense (but not necessarily closed) subspace $\mathrm{Dom}(A) \subset \mathcal{H}$ called the *domain* of A. If two linear operators A, B in $\mathcal{H}$ satisfy $\mathrm{Dom}(A) \subset \mathrm{Dom}(B)$ and $Bu = Au$ whenever $u \in \mathrm{Dom}(A)$, we say that B is an *extension* of A and write $A \subset B$.

The adjoint operator A^* of A is defined to have the domain

(2.1.15)
$$\mathrm{Dom}(A^*) := \{u \in \mathcal{H} : \text{ there exists } f \in \mathcal{H} \text{ such that}$$
$$(u, Av)_{\mathcal{H}} = (f, v)_{\mathcal{H}} \text{ for all } v \in \mathrm{Dom}(A)\}$$

and then by setting $A^* u := f$, where u, f are as above (f is uniquely defined since $\mathrm{Dom}(A)$ is dense in $\mathcal{H}$). Therefore, we have

$$(A^* u, v)_{\mathcal{H}} = (u, Av)_{\mathcal{H}} \qquad \text{for all } u \in \mathrm{Dom}(A^*), v \in \mathrm{Dom}(A).$$

An operator A is called *symmetric* if

$$(Au, v)_{\mathcal{H}} = (u, Av)_{\mathcal{H}} \qquad \text{for all } u, v \in \mathrm{Dom}(A).$$

Observe that if A is symmetric, then $A \subset A^*$. A symmetric operator A is called *selfadjoint* if $\mathrm{Dom}(A) = \mathrm{Dom}(A^*)$ (so $A = A^*$). Note that not all symmetric unbounded operators are selfadjoint, as seen in the following.

Example 2.1.24. Consider the operator $A_0 := -\frac{\mathrm{d}^2}{\mathrm{d}x^2}$ with the domain $C_0^2(\mathbb{R})$ acting in the Hilbert space $L^2(\mathbb{R})$. It is easily checked that A_0 is symmetric; however it is not selfadjoint as the function e^{-x^2} belongs to the domain of A_0^* but not to the domain of A_0.

The *resolvent set* of a linear operator A in $\mathcal{H}$ is the set of complex numbers $\lambda \in \mathbb{C}$ such that the operator $A - \lambda I$ maps its domain bijectively to $\mathcal{H}$ and such that $R(\lambda) = (A - \lambda)^{-1} : \mathcal{H} \to \mathcal{H}$ is bounded. The operator $R(\lambda)$ is called the *resolvent operator*. The *spectrum* of an operator A, denoted by $\mathrm{Spec}(A)$, is defined as the complement of the resolvent set. A number $\lambda \in \mathrm{Spec}(A)$ is called an *eigenvalue* of A if $\dim \mathrm{Ker}(A - \lambda I) > 0$ — in other words, if there exists a nontrivial solution $u \in \mathcal{H} \setminus \{0\}$ of the equation

$$Au = \lambda u.$$

This dimension is then called the *multiplicity* of the eigenvalue λ, and nontrivial elements of $\mathrm{Ker}(A - \lambda I)$ are called the *eigenvectors* (or the *eigenfunctions* if $\mathcal{H}$ consists of functions) of A corresponding to the eigenvalue λ. The *discrete spectrum* of A is the set of all isolated eigenvalues of A of finite multiplicity. The complement of the discrete spectrum inside the full spectrum is called the *essential spectrum* of A. We say that the operator A has discrete spectrum if its essential spectrum is empty. Importantly, the spectrum of a selfadjoint operator is always real. Additionally, the spectrum is discrete if there is at least one point of the resolvent set λ_0 at which the resolvent $(A - \lambda_0 I)^{-1}$ is compact.

Suppose that an operator A_0 is symmetric and *semi-bounded from below*; that is, there exists a constant c (not necessarily positive) such that

(2.1.16) $$\qquad (A_0 u, u)_{\mathcal{H}} \geq c (u, u)_{\mathcal{H}} \qquad \text{for all } u \in \mathrm{Dom}(A_0).$$

If $c > 0$, we say that the operator A_0 is *positive*. We want to specify a particular selfadjoint extension A of A_0. Without loss of generality we will assume that $c = 1$ in (2.1.16); if this is not the case, we may consider instead the operator $A_0 + (1 - c)I$ and subtract $(1 - c)I$ at the end. We introduce a new inner product on $\mathrm{Dom}(A_0)$ by using the bilinear form of A_0,

$$(u, v)_{A_0} := (A_0 u, v)_{\mathcal{H}} = (u, A_0 v)_{\mathcal{H}} \qquad \text{for all } u, v \in \mathrm{Dom}(A_0).$$

Let $\mathcal{H}_0$ be the completion of $\mathrm{Dom}(A_0)$ with respect to $(\cdot, \cdot)_{A_0}$. Then there is a natural embedding $\mathcal{H}_0 \subset \mathcal{H}$ with the norm of the embedding operator not greater than one.

We now define the *Friedrichs extension* A of A_0 by setting

(2.1.17)
$$\mathrm{Dom}(A) := \{u \in \mathcal{H}_0 : \text{ there exists } f \in \mathcal{H} \text{ such that}$$
$$(u, v)_{A_0} = (f, v)_{\mathcal{H}} \text{ for all } v \in \mathcal{H}_0\}$$

and $Au := f$ for $u \in \mathrm{Dom}(A)$ and f as above.

Remark 2.1.25. Let us compare the definition of the Friedrichs extension (2.1.17) with the definition of the adjoint operator (2.1.15). They look similar, but we note that in (2.1.17) we take u from $\mathcal{H}_0$ instead of a larger space $\mathcal{H}$, and we take v also from $\mathcal{H}_0$ rather than from a smaller space $\mathrm{Dom}(A_0)$. We therefore have

$$A_0 \subset A \subset A_0^*.$$

The following result holds.

Theorem 2.1.26 ([**Lax02**, §33.3]). The Friedrichs extension of a symmetric semi-bounded from below operator is selfadjoint.

Remark 2.1.27. The construction of the Friedrichs extension shows that every symmetric semi-bounded from below operator has at least one selfadjoint extension. There exist, however, symmetric operators which are not semi-bounded and which have no selfadjoint extensions at all; see [**Lax02**, §33.2].

2.1.6. The Dirichlet Laplacian via the Friedrichs extension. We start by describing explicitly the construction of the Dirichlet Laplacian via the Friedrichs extension following the general scheme given in §2.1.5. Let Ω be an open bounded set in $\mathbb{R}^d$, and let A_0 be the operator $-\Delta$ defined on $\mathrm{Dom}(A_0) := C_0^2(\Omega) \subset L^2(\Omega)$. Green's formula (2.1.8) immediately implies that A_0 is symmetric; it is not however selfadjoint; cf. Example 2.1.24.

Proposition 2.1.22 together with Green's formula (2.1.8) also implies that A_0 is semi-bounded from below, since then

$$\|u\|_{A_0}^2 = (A_0 u, u)_{L^2(\Omega)} = (-\Delta u, u)_{L^2(\Omega)} = \|\nabla u\|_{L^2(\Omega)}^2$$

$$\geq \frac{1}{C_\Omega} \|u\|_{L^2(\Omega)}^2 \qquad \text{for all } u \in C_0^2(\Omega).$$

Therefore,

$$(1 + C_\Omega)\|u\|_{A_0}^2 \geq \|u\|_{H^1(\Omega)}^2 = \|\nabla u\|_{L^2(\Omega)}^2 + \|u\|_{L^2(\Omega)}^2 \geq \|u\|_{A_0}^2,$$

and so the norms $\|\cdot\|_{A_0}$ and $\|\cdot\|_{H^1(\Omega)}$ are equivalent on $\mathrm{Dom}(A_0)$. Hence, the completion $\mathcal{H}_0$ of $C_0^2(\Omega)$ with respect to the norm $\|\cdot\|_{A_0}$, appearing in the construction of the Friedrichs extension, is the Sobolev space $H_0^1(\Omega)$.

Using now (2.1.17), we deduce that the Friedrichs extension of A_0 is the operator A, which we from now on will denote as $-\Delta^{\mathrm{D}} := -\Delta_\Omega^{\mathrm{D}}$ and will call the *Dirichlet Laplacian* on Ω, with the domain (see also Definition 2.1.16)

(2.1.18)
$$\mathrm{Dom}(-\Delta^{\mathrm{D}}) = \Big\{ u \in H_0^1(\Omega) : \text{ there exists } f \in L^2(\Omega) \text{ such that}$$

$$(\nabla u, \nabla v)_{L^2(\Omega)} = (f, v)_{L^2(\Omega)} \text{ for all } v \in H_0^1(\Omega) \Big\}$$

$$= \Big\{ u \in H_0^1(\Omega) : \text{ there exists } f \in L^2(\Omega) \text{ such that}$$

$$-\Delta u = f \text{ in the weak sense} \Big\}$$

$$= \Big\{ u \in H_0^1(\Omega) : \Delta u \in L^2(\Omega) \Big\}.$$

Repeating now word for word the construction of the compact operator K from §2.1.4, we conclude that we indeed have $K = (-\Delta^{\mathrm{D}} - 1)^{-1}$. Therefore we arrive at the following equivalent formulation of Theorem 2.1.20.

Theorem 2.1.28. Let $\Omega \subset \mathbb{R}^d$ be a bounded domain. The Dirichlet Laplacian $-\Delta^{\mathrm{D}}$ defined as the Friedrichs extension with domain (2.1.18) is a selfadjoint operator in $L^2(\Omega)$ with a discrete spectrum of eigenvalues accumulating to $+\infty$ and the first eigenvalue being positive. The eigenfunctions can be chosen to form an orthonormal basis in $L^2(\Omega)$.

Remark 2.1.29. Theorem 2.1.28 remains valid if Ω is just an open subset of $\mathbb{R}^d$ of a finite volume, not necessarily bounded. Moreover, the spectrum of $-\Delta^{\mathrm{D}}$ is still discrete if an even less restrictive condition

$$\limsup_{\substack{|x| \to \infty \\ x \in \Omega}} \big|B_{x,1} \cap \Omega\big|_d = 0$$

is satisfied; see [**EdmEva18**, Remark V.5.18(4)].

The following simple results will be needed often later on.

Lemma 2.1.30. Given a bounded open set $\Omega \subset \mathbb{R}^d$, denote by Ω_ρ a homothety of Ω with the coefficient $\rho > 0$. Then $\lambda \in \mathrm{Spec}\left(-\Delta_\Omega^{\mathrm{D}}\right)$ if and only if $\rho^{-2}\lambda \in \mathrm{Spec}\left(-\Delta_{\Omega_\rho}^{\mathrm{D}}\right)$.

Lemma 2.1.31. Let $\Omega \in \mathbb{R}^d$ be a disjoint union of two bounded domains Ω_1 and Ω_2. Then $\mathrm{Spec}\left(-\Delta_\Omega^{\mathrm{D}}\right) = \mathrm{Spec}\left(-\Delta_{\Omega_1}^{\mathrm{D}}\right) \cup \mathrm{Spec}\left(-\Delta_{\Omega_2}^{\mathrm{D}}\right)$ with account of multiplicities.

■ **Exercise 2.1.32.** Prove Lemmas 2.1.30 and 2.1.31.

2.1.7. The weak spectral theorem: Neumann case. In this section we discuss the analog of the weak Dirichlet spectral theorem (Theorem 2.1.20) in the case of the Neumann boundary condition. Unlike the Dirichlet case, in the Neumann case some regularity conditions need to be imposed on the boundary from the start, and we will assume throughout that the boundary is Lipschitz; see the discussion below.

Definition 2.1.33 (Weak Neumann solution and weak Neumann spectral problem). Let $\Omega \subset \mathbb{R}^d$ be a bounded domain with a Lipschitz boundary. Let $f \in L^2(\Omega)$. We say that $u : \Omega \to \mathbb{R}$ is a *weak solution* of the boundary value problem

$$(2.1.19) \qquad \begin{cases} -\Delta u + u = f, \\ \partial_n u = 0 \end{cases}$$

if $u \in H^1(\Omega)$ and

$$\int_\Omega \langle \nabla u, \nabla v \rangle \, \mathrm{d}x + \int_\Omega uv \, \mathrm{d}x = \int_\Omega fv \, \mathrm{d}x \qquad \text{for all } v \in H^1(\Omega).$$

The weak *Neumann spectral problem* is to find $\lambda \in \mathbb{R}$ and $u \in H^1(\Omega) \setminus \{0\}$ such that

$$(2.1.20) \qquad \int_\Omega \langle \nabla u, \nabla v \rangle \, \mathrm{d}x = \lambda \int_\Omega uv \, \mathrm{d}x \qquad \text{for all } v \in H^1(\Omega).$$

Remark 2.1.34. We note that the boundary condition $\partial_n u = 0$ "disappears" in the weak statement. However, note that (2.1.19) is required to hold for all $v \in H^1(\Omega)$, not only for $v \in H_0^1(\Omega)$ (cf. the Dirichlet case in Definition 2.1.16). We also note that the Neumann spectrum always starts with $\lambda_1 = 0$, with the corresponding eigenfunction u_1 being a constant.

■ **Exercise 2.1.35.** Prove that if $u \in C^2(\Omega) \cap C^1(\overline{\Omega})$ is a weak solution of the problem (2.1.19), then it is also a classical solution of it.

The argument given in the Dirichlet case for the existence of weak solutions works in the Neumann case as well. The Riesz representation theorem guarantees the existence of a unique solution $u \in H^1(\Omega)$ for any given

$f \in L^2(\Omega)$. The composition of the solution operator $\widetilde{K} : L^2(\Omega) \to H^1(\Omega)$ with the inclusion $H^1(\Omega) \subset L^2(\Omega)$ is compact; see Remark 2.1.37 below. As a result, we prove

Theorem 2.1.36 (The weak spectral theorem for the Neumann Laplacian). Let $\Omega \subset \mathbb{R}^d$ be a bounded domain with a Lipschitz boundary. There exists an orthonormal basis of $L^2(\Omega)$ composed of weak eigenfunctions of the Neumann spectral problem. The corresponding eigenvalues are nonnegative and form a nondecreasing sequence which tends to $+\infty$.

As in the Dirichlet case, we can equivalently reformulate the spectral theorem in the operator theory sense by constructing the Neumann Laplacian using the Friedrichs extension; see [**Hel13**, §4.4.4] or [**AreCSVV18**, §7.4]. Given $u \in H^1(\Omega)$ such that $-\Delta u \in L^2(\Omega)$, we say that $\partial_n u \sim 0$ on $\partial\Omega$ if

$$- \int_\Omega \Delta u \, v \, dx = \int_\Omega \langle \nabla u, \nabla v \rangle \, dx \qquad \text{for all } v \in H^1(\Omega).$$

Note that while $\partial_n u$ may not be defined in $L^2(\partial\Omega)$ even weakly, both the right- and the left-hand sides of this formula are well-defined, and hence the relation $\partial_n u \sim 0$ still makes sense. This allows us to define a selfadjoint operator $-\Delta^N := -\Delta_\Omega^N$, which we call the *Neumann Laplacian* on Ω, as the weak Laplacian with the domain (cf. (2.1.18))

$$\mathrm{Dom}(-\Delta^N) = \Big\{ u \in H^1(\Omega) : \text{ there exists } f \in L^2(\Omega) \text{ such that}$$
$$(\nabla u, \nabla v)_{L^2(\Omega)} = (f, v)_{L^2(\Omega)} \text{ for all } v \in H^1(\Omega) \Big\}$$
$$= \big\{ u \in H^1(\Omega) : -\Delta u \in L^2(\Omega) \text{ and } \partial_n u \sim 0 \text{ on } \partial\Omega \big\}.$$

For a slightly different approach to Neumann boundary value problems see [**Fol95**] or [**Dav95**].

Remark 2.1.37. We emphasise that our assumption that the boundary $\partial\Omega$ is Lipschitz is crucial for the validity of Theorem 2.1.36. It guarantees that Theorem 2.1.7(ii) holds, and therefore $-\Delta^N$ has a compact resolvent, thus ensuring the discreteness of the spectrum. Although this condition can be slightly relaxed, see [**Dav95**, Chapter 7] for details, it cannot be dismissed altogether: without any regularity assumptions on $\partial\Omega$ one can construct examples of bounded domains for which the spectrum of the Neumann Laplacian is no longer discrete; see, e.g., [**HemSecSim91**].

■ **Exercise 2.1.38.** Prove the analogues of Lemmas 2.1.30 and 2.1.31 for the Neumann Laplacian.

2.1.8. The weak spectral theorem: Riemannian manifolds. Let (M, g) be a smooth compact Riemannian manifold, possibly with boundary. If the boundary is nonempty, we assume that either Dirichlet or Neumann boundary conditions are imposed on ∂M, and we recall Remark 1.2.8.

By Green's identity, the Laplacian acting on functions from $C^2(M)$ is a symmetric differential operator in the space $L^2(M)$:

$$\int\limits_M (-\Delta u)v \, \mathrm{d}V = \int\limits_M \langle \nabla u, \nabla v \rangle_g \, \mathrm{d}V = \int\limits_M u(-\Delta v) \, \mathrm{d}V.$$

Note that the boundary term vanishes due to the boundary conditions. Setting $u = v$ we also observe that the Laplacian is a nonnegative symmetric operator.

Let us introduce the Sobolev space

$$H^1(M) = \{u \in L^2(M), \nabla u \in L^2(M)\},$$

where the gradient is understood in the weak sense. The norm in $H^1(M)$ is defined by

$$\|u\|^2_{H^1(M)} := \int\limits_M u^2 \, \mathrm{d}V + \int\limits_M |\nabla u|^2 \, \mathrm{d}V.$$

Moreover, for any $m \in \mathbb{N}$, one can extend the definition of the Sobolev space $H^m(M)$ to manifolds using coordinate charts and a partition of unity. We refer to [**Tay11**, §4.3] and [**Shu01**, §I.7] for details.

Let us define also the space $H^1_0(M)$ as the closure of the space $C^1_0(M)$ in the norm of $H^1(M)$. Clearly, $H^1_0(M) \subset H^1(M) \subset L^2(M)$.

We can define the weak spectral problem for the Laplace operator on a closed manifold, or the weak Neumann spectral problem on a manifold with boundary, by analogy with (2.1.20), and the weak Dirichlet spectral problem on a manifold with boundary by analogy with (2.1.12). Acting similarly to Theorems 2.1.36 and 2.1.20, we obtain

Theorem 2.1.39 (The weak spectral theorem for a Riemannian manifold). Let (M, g) be a smooth compact Riemannian manifold, possibly with boundary. If the boundary is nonempty, we assume that either Dirichlet or Neumann boundary conditions are imposed on ∂M. In each of these cases, there exists an orthonormal basis of $L^2(M)$ composed of weak eigenfunctions of the corresponding Laplace spectral problem. The corresponding eigenvalues are nonnegative and form a nondecreasing sequence tending to $+\infty$.

■ **Exercise 2.1.40.** Show that on a compact connected Riemannian manifold the only harmonic function is a constant. In particular, this implies that the Laplace eigenvalue zero on a compact connected manifold always has multiplicity one.

Notation 2.1.41. Let M be a closed Riemannian manifold. We will be enumerating the eigenvalues of the Laplace–Beltrami operator on M starting with $\lambda_0 = 0$ as

$$0 = \lambda_0 \leq \lambda_1 \leq \cdots .$$

In particular, for a connected manifold, $\lambda_1 > 0$ by Exercise 2.1.40. This enumeration is traditional and is motivated by the importance of the first *nonzero* eigenvalue λ_1.

■ **Exercise 2.1.42.** Let (M, g) be a compact Riemannian manifold of dimension d. Show that for any $\rho > 0$, $j \in \mathbb{N}$,

(2.1.21) $$\lambda_j(M, \rho g) = \frac{\lambda_j(M, g)}{\rho}, \qquad j \in \mathbb{N},$$

and, consequently, the quantity $\lambda_j(M, g) \operatorname{Vol}(M, g)^{2/d}$ is invariant under rescaling. This is a Riemannian analogue of Lemma 2.1.30; see also Exercise 2.1.38.

2.2. Elliptic regularity and strong spectral theorems

2.2.1. Elliptic regularity for the Dirichlet Laplacian. We want to show that the weak eigenfunctions of the Dirichlet problem (2.1.12) found in Theorem 2.1.20 are in fact smooth. This is due to an important phenomenon known as *elliptic regularity*. We present an overview of this theory below.

   We have

Theorem 2.2.1 (Smoothness of Dirichlet eigenfunctions). Let $\Omega \subset \mathbb{R}^d$ be a bounded open set, and let $u_1, u_2, \ldots$ be the eigenfunctions of the weak Dirichlet spectral problem (2.1.12). Then:

   (i) Each eigenfunction u_j, $j \in \mathbb{N}$, belongs to $C^\infty(\Omega)$.

   (ii) Moreover, each eigenfunction is real analytic in Ω.

   (iii) The eigenfunctions are smooth up to the boundary near the smooth parts of the boundary; if $\partial_\infty \Omega$ is the C^∞ part of $\partial\Omega$, then u_j and all its derivatives can be continuously extended to $\Omega \cup \partial_\infty \Omega$.

   (iv) If $\partial\Omega$ is Lipschitz, $u_j \in C(\overline{\Omega})$, and $u_j|_{\partial\Omega} = 0$ pointwise.

Parts (i) and (ii) of Theorem 2.2.1 follow from what is usually referred to as *local* or *interior elliptic regularity*. Clearly, (ii) implies (i); however we present an independent proof of the latter statement, as it can be generalised to the setting of smooth Riemannian manifolds. Parts (iii) and (iv) follow from *global elliptic regularity*, or *regularity up to the boundary*.

2.2.2. Proof of the local regularity. The goal of this subsection is to prove parts (i) and (ii) of Theorem 2.2.1. We start with the proof of the latter as it can be easily deduced from the real analyticity of harmonic functions.

Proof of Theorem 2.2.1, part (ii). We use the so-called *lifting trick* (cf. Exercise 4.3.17) and consider the harmonic function $h(x,t) := u(x)e^{\sqrt{\lambda}t}$ in $\Omega \times \mathbb{R} \subset \mathbb{R}^{d+1}$. Since harmonic functions are real analytic [**AxlBouWad01**, Theorem 1.28], it follows that $u(x) = h(x,0)$ is real analytic. $\qquad\square$

We note that this argument can be adjusted to work for solutions of $-\Delta u - \lambda u = 0$ with negative λ and in any case does not require any boundary conditions. Let us also remark that the analogue of this statement holds for uniformly elliptic operators with real analytic coefficients, see [**Fri69**, Theorem III.1.2], [**Joh81**, Ch. 7], [**MorNir57**], and hence applies to the Laplace–Beltrami operators on Riemannian manifolds with real analytic metrics.

In order to prove part (i) we use a fundamental regularity result from the theory of elliptic partial differential equations. First, we need to define weak solutions for a wider class of problems.

Consider an open set $\Omega \subset \mathbb{R}^d$ and a uniformly elliptic equation in divergence form,

$$(2.2.1) \qquad\qquad -\operatorname{div} A\nabla u = f \qquad \text{in } \Omega,$$

where $f \in L^2(\Omega)$ and $A = \left(a^{ij}\right)_{i,j=1}^d$ is a positive definite symmetric matrix with entries $a^{ij} \in L^\infty(\Omega)$ which satisfies

$$(2.2.2) \qquad\qquad \langle A\xi, \xi \rangle \geq \alpha_0 |\xi|^2$$

for all $x \in \Omega$ and $\xi \in \mathbb{R}^d$ and some fixed $\alpha_0 > 0$.

Definition 2.2.2 (Weak solution of a uniformly elliptic equation in divergence form). We say that $u \in H^1(\Omega)$ is a *weak solution* of the equation (2.2.1) (or alternatively that u satisfies the equation (2.2.1) *in the weak sense*) if

$$\int_{\Omega} \langle A \nabla u, \nabla v \rangle \, \mathrm{d}x = \int_{\Omega} fv \, \mathrm{d}x$$

for all $v \in H_0^1(\Omega)$.

Remark 2.2.3. If we take A to be the identity matrix, then equation (2.2.1) becomes the standard Laplace equation $-\Delta u = f$.

The fundamental result mentioned above is

Theorem 2.2.4 (Local elliptic regularity for the Laplacian [**GilTru01**, Theorem 8.10], [**Fol95**, Lemma 6.32], [**Eva10**, §6.3.1, Theorem 2]). Let $\Omega \subset \mathbb{R}^d$ be an open set. Suppose that for some $m \geq 0$ and $f \in H_{\mathrm{loc}}^m(\Omega)$, a function $u \in H^1(\Omega)$ satisfies the equation $-\Delta u = f$ in Ω in the weak sense. Then $u \in H_{\mathrm{loc}}^{m+2}(\Omega)$.

Assuming this result for the moment, let us show how it implies what we need.

Proof of Theorem 2.2.1, part (i). Let $u \in H^1(\Omega)$ be a weak solution of the equation $-\Delta u = \lambda u$. Then applying iteratively Theorem 2.2.4 to u we conclude that $u \in H_{\mathrm{loc}}^k(\Omega)$ for all $k \geq 1$ (this procedure is called *elliptic bootstrapping*). Therefore, by Theorem 2.1.9 it follows that $u \in C^\infty(U)$ for any open set $U \Subset \Omega$, and hence $u \in C^\infty(\Omega)$. $\square$

Remark 2.2.5 (Local regularity of eigenfunctions for other eigenvalue problems). Note that the proof does not use boundary conditions, and hence local regularity holds also for Neumann eigenfunctions. Moreover, arguments of elliptic regularity are robust in a sense that Theorem 2.2.4 can be extended to second-order elliptic operators with smooth coefficients such as the Laplace–Beltrami operator; see Theorem 2.2.17 below.

2.2.3. A priori estimates and the method of difference quotients. The proof of Theorem 2.2.4 uses two key ideas: an *a priori estimate* and *Nirenberg's method of difference quotients*.

Let us start with the latter. Following [**Nir59**], let the *difference quotient* be defined as

$$D_h u(x) := \frac{u(x+h) - u(x)}{|h|},$$

where $h \in \mathbb{R}^d \setminus \{0\}$. Since D_h commutes with differentiations, we get

$$-\Delta(D_h u) = D_h f.$$

Louis **Nirenberg**
(1925–2020)

Given $f \in H^1(\mathbb{R}^d)$, it is not difficult to verify that if $t > 0$ and e_k is the kth unit coordinate vector in $\mathbb{R}^d$,

$$(2.2.3) \qquad \|D_{te_k}\|_{L^2(\mathbb{R}^d)} \leq \|\partial_k f\|_{L^2(\mathbb{R}^d)},$$

and hence

$$(2.2.4) \qquad \|D_h f\|_{L^2(\mathbb{R}^d)} \leq \|\nabla f\|_{L^2(\mathbb{R}^d)}.$$

■ **Exercise 2.2.6.** Prove estimate (2.2.3). *Hint:* Prove it first for C_0^1-functions using the fundamental theorem of calculus and Fubini's theorem, and then use the fact that $C_0^1(\mathbb{R}^d)$ is dense in $H^1(\mathbb{R}^d)$ (see [**GilTru01**, Lemma 7.23] or [**Bre11**, Proposition 9.3]).

The following important theorem gives a condition for showing that an $L^2(\mathbb{R}^d)$ function belongs to the Sobolev space $H^1(\mathbb{R}^d)$. It is proved using weak compactness of closed bounded sets in $L^2(\mathbb{R}^d)$ (cf. proof of Lemma 2.2.14 for a similar argument).

Theorem 2.2.7 (The method of difference quotients [**GilTru01**, Lemma 7.24], [**Bre11**, Proposition 9.3]). Let $u \in L^2(\mathbb{R}^d)$. If there exists $C > 0$ such that for all $h \in \mathbb{R}^d$ with $0 < |h| \leq 1$ we have

$$\|D_h u\|_{L^2(\mathbb{R}^d)} \leq C,$$

then $u \in H^1(\mathbb{R}^d)$. In particular, if $\|D_{te_k} u\|_{L^2(\mathbb{R}^d)} \leq C$ for all $|t| < 1$, then $\partial_k u \in L^2(\mathbb{R}^d)$.

◾ **Exercise 2.2.8** (Leibniz rule and integration by parts for difference quotients). Show that for $u, v \in H^1(\mathbb{R}^d)$ and $h \in \mathbb{R}^d$,

(i)
$$D_h(uv) = (D_h u)v + u(D_h v) + |h|(D_h u)(D_h v);$$

(ii)
$$\int_{\mathbb{R}^d} (D_h u)v \, dx = - \int_{\mathbb{R}^d} u(D_{-h}v) \, dx.$$

Let us move to the second part of the argument. In order to formulate an a priori estimate we recall that we have defined the negative order Sobolev space $H^{-1}(\Omega)$ as the dual space of $H_0^1(\Omega)$, with the usual norm of the dual Hilbert space.

Example 2.2.9. Let $f \in L^2(\Omega)$. The formula

$$F_f(v) := \int_\Omega fv \, dx, \qquad v \in H_0^1(\Omega),$$

defines an element of $H^{-1}(\Omega)$. Moreover, $\|F_f\|_{H^{-1}(\Omega)} \leq \|f\|_{L^2(\Omega)}$. Slightly abusing notation, we write $\|f\|_{H^{-1}(\Omega)} \leq \|f\|_{L^2(\Omega)}$.

We can now state the following.

Lemma 2.2.10 (An a priori estimate in $H^1(\mathbb{R}^d)$ [**Fol95**, Theorem 6.28]). Let $f \in L^2(\mathbb{R}^d)$, and let $u \in H^1(\mathbb{R}^d)$ be a weak solution of the equation $-\Delta u = f$ in $\mathbb{R}^d$. Then

(2.2.5) $$\|u\|_{H^1(\mathbb{R}^d)}^2 \leq 2 \left(\|u\|_{L^2(\mathbb{R}^d)}^2 + \|f\|_{H^{-1}(\mathbb{R}^d)}^2 \right).$$

Proof. We have

$$\int_{\mathbb{R}^d} |\nabla u|^2 \, dx = \int_{\mathbb{R}^d} fu \, dx \leq \|f\|_{H^{-1}(\mathbb{R}^d)} \|u\|_{H^1(\mathbb{R}^d)}$$

$$\leq \frac{1}{2}\|f\|_{H^{-1}(\mathbb{R}^d)}^2 + \frac{1}{2}\|u\|_{L^2(\mathbb{R}^d)}^2 + \frac{1}{2}\|\nabla u\|_{L^2(\mathbb{R}^d)}^2.$$

Rearranging the terms yields the result. □

Remark 2.2.11. One way to think about a priori estimate (2.2.5) is as follows: an L^2 bound on a function and H^{-1} bound on its Laplacian imply L^2 bounds on *all* its first derivatives. Another illustration of a similar phenomenon is a more elementary estimate

$$(2.2.6) \qquad \int_{\mathbb{R}^d} \left| \frac{\partial^2 u}{\partial x_k \partial x_l} \right|^2 \, \mathrm{d}x \leq \int_{\mathbb{R}^d} |\Delta u|^2 \, \mathrm{d}x,$$

which holds for any $u \in C_0^2(\mathbb{R}^d)$ and $k, l, = 1, \ldots, d$.

The reason why (2.2.6) holds is the ellipticity of the Laplace operator. Consider also an a priori estimate for the first-order elliptic Cauchy–Riemann operator; for a complex-valued $u \in C_0^1(\mathbb{R}^2)$,

$$(2.2.7) \qquad \int_{\mathbb{R}^2} \left| \frac{\partial u}{\partial x_j} \right|^2 \, \mathrm{d}x \leq \int_{\mathbb{R}^2} \left| \frac{\partial u}{\partial x_1} + \mathrm{i} \frac{\partial u}{\partial x_2} \right|^2 \, \mathrm{d}x, \qquad j = 1, 2.$$

On the other hand, no a priory estimate is possible for the operator $\mathcal{A} := u \mapsto \frac{\partial^2 u}{\partial x_1^2} - \frac{\partial^2 u}{\partial x_2^2}$, which is not elliptic:

for any $C > 0$ there exists $u \in C_0^2(\mathbb{R}^2)$ such that

$$(2.2.8) \qquad \int_{\mathbb{R}^d} \left| \frac{\partial^2 u}{\partial x_1^2} \right|^2 \, \mathrm{d}x \geq C \int_{\mathbb{R}^d} |\mathcal{A}u|^2 \, \mathrm{d}x.$$

We leave the proofs of (2.2.6)–(2.2.8) as an exercise for the reader

In the proof of Theorem 2.2.4 we will require the following.

Lemma 2.2.12. Let $f \in L^2(\mathbb{R}^d)$ and $h \in \mathbb{R}^d$. Then

$$\|D_h f\|_{H^{-1}(\mathbb{R}^d)} \leq \|f\|_{L^2(\mathbb{R}^d)}.$$

Proof. For any $v \in H^1(\mathbb{R}^d)$ we have

$$\left| \int_{\mathbb{R}^d} D_h f \, v \, \mathrm{d}x \right| = \left| \int_{\mathbb{R}^d} f D_{-h} v \, \mathrm{d}x \right| \leq \|f\|_{L^2(\mathbb{R}^d)} \|D_{-h} v\|_{L^2(\mathbb{R}^d)}$$

$$\leq \|f\|_{L^2(\mathbb{R}^d)} \|v\|_{H^1(\mathbb{R}^d)},$$

which implies the desired estimate. Here the first equality follows from Exercise 2.2.8 and the last inequality follows from (2.2.4). $\square$

We now have all the required tools to prove Theorem 2.2.4.

Proof of Theorem 2.2.4. Assume first that $u \in H^1(\mathbb{R}^d)$ is a weak solution of the equation $-\Delta u = f$ in $\mathbb{R}^d$ with $f \in L^2(\mathbb{R}^d)$. Taking difference quotients we obtain the equation

$$-\Delta D_h u = D_h f,$$

and after applying Lemma 2.2.10 on this new equation we obtain

$$\|D_h\partial_k u\|^2_{L^2(\mathbb{R}^d)} \leq \|D_h u\|^2_{H^1(\mathbb{R}^d)} \leq 2\left(\|D_h u\|^2_{L^2(\mathbb{R}^d)} + \|D_h f\|^2_{H^{-1}(\mathbb{R}^d)}\right)$$
$$\leq 2\|u\|^2_{H^1(\mathbb{R}^d)} + 2\|f\|^2_{L^2(\mathbb{R}^d)},$$

for any $k = 1,\ldots,d$. Here we have used (2.2.4) and Lemma 2.2.12 to estimate the right-hand side. Applying Theorem 2.2.7, we deduce that $\partial_k u \in H^1(\mathbb{R}^d)$ for any $k = 1,\ldots,d$, and hence $u \in H^2(\mathbb{R}^d)$. This proves the assertion of the theorem for $\Omega = \mathbb{R}^d$ and $m = 0$. Since $-\Delta\partial_k u = \partial_k f$ we deduce the result by induction for any $m \geq 1$.

Now, in order to prove the theorem for an arbitrary Ω, we use the standard localisation argument. Suppose that $u \in H^1_{\text{loc}}(\Omega)$ satisfies $-\Delta u = f$ in Ω in the weak sense with $f \in L^2_{\text{loc}}(\Omega)$. Take a cut-off function $\varphi \in C_0^\infty(\Omega)$. It is immediate that the function φu, extended by zero onto $\mathbb{R}^d$, belongs to $H^1(\mathbb{R}^d)$. Then, $-\Delta(\varphi u) = g$ in the weak sense, where $g = \varphi f - 2\langle\nabla\varphi, \nabla u\rangle - (\Delta\varphi)u$. Note that $g \in L^2(\mathbb{R}^d)$, and we deduce that $\varphi u \in H^2(\mathbb{R}^d)$, and hence $u \in H^2_{\text{loc}}(\Omega)$. Iterating the argument as above completes the proof of the theorem. □

2.2.4. Global regularity of Dirichlet eigenfunctions. So far, we have shown that if u is a weak solution of the equation $-\Delta u = \lambda u$ in Ω, then $u \in H^k_{\text{loc}}(\Omega)$ for all $k \in \mathbb{N}$, and hence $u \in C^\infty(\Omega)$. Note that the boundary conditions as well as boundary regularity are irrelevant for this property. Our goal is to prove part (iii) of Theorem 2.2.1 which states u is smooth up to the boundary near smooth parts of the boundary, provided the Dirichlet condition is imposed.

After a partition of unity argument, we can assume that u is localised in a small neighbourhood of the boundary. Using an appropriate change of variables we can "straighten" the smooth part of the boundary, i.e., transform it into a part of a hyperplane. At the same time, the Euclidean Laplacian is transformed into a certain Laplace–Beltrami operator. Indeed, if $-\Delta u(x) = f(x)$, $x = \varphi(y)$ denotes a change of variables, and $\mathfrak{u} = u \circ \varphi$, $\mathfrak{f} = f \circ \varphi$, then $\mathfrak{u}$ satisfies the equation $-\mathfrak{L}\mathfrak{u} = \mathfrak{f}$, where

$$(2.2.9) \qquad \mathfrak{L}\mathfrak{u} := \frac{1}{\sqrt{\det g}}\sum_{i=1}^d\sum_{j=1}^d \partial_i\left(g^{ij}\sqrt{\det g}\,\partial_j\mathfrak{u}\right),$$

the matrix $g := \{g_{ij}\} = \{\langle\partial_i\varphi, \partial_j\varphi\rangle\}^d_{i,j=1}$, $\{g^{ij}\} = g^{-1}$, and $\sqrt{\det g} = |\det(\text{Jac}\,\varphi)|$, where $\text{Jac}\,\varphi$ is the Jacobian matrix of φ.

As before, we would like to show that $u \in H^k(\Omega)$ for all k, and hence, by Theorem 2.1.9, $u \in C^\infty(\overline{\Omega})$. Similarly to the local regularity, the main tools are an a priori estimate in a half-space $\mathbb{R}^d_+ := \{(x_1,\ldots,x_d) \in \mathbb{R}_d : x_d > 0\}$ and an appropriately adjusted Nirenberg method of difference quotients.

The equation

$$\left(\sqrt{\det g} \, \mathfrak{L} \right) \mathfrak{u} = \sqrt{g} \, \mathfrak{f}$$

is of divergence type as in (2.2.1).

Proposition 2.2.13 (An a priori estimate in half-space). Let $m \geq 0$. Consider the equation (2.2.1), where we additionally assume that the entries of the matrix A satisfy $a^{ij} \in C^m \left(R_+^d \right)$ and have compact supports.

Let u be a weak solution of this equation, and suppose $u \in H^{m+1} \left(\mathbb{R}_,^d + \right)$ $\cap H_0^1 \left(\mathbb{R}_+^d \right)$ and $f \in H^m \left(\mathbb{R}_+^d \right)$. Then

$$\|u\|_{H^{m+1}\left(\mathbb{R}_+^d\right)} \leq C \left(\|u\|_{H^m\left(\mathbb{R}_+^d\right)} + \|f\|_{H^{m-1}\left(\mathbb{R}_+^d\right)} \right),$$

with some constant $C > 0$ which depends only on the constant α_0 from (2.2.2) and bounds on $\|a^{ij}\|_{C^m\left(\mathbb{R}_+^d\right)}$.

Proof. Consider first the case $m = 0$. Then

$$\alpha_0 \int_{\mathbb{R}_+^d} |\nabla u|^2 \, dx \leq \int_{\mathbb{R}_+^d} \langle A\nabla u, \nabla u \rangle \, dx = \int_{\mathbb{R}_+^d} fu \, dx < \|f\|_{H^{-1}\left(\mathbb{R}_+^d\right)} \|u\|_{H^1\left(\mathbb{R}_+^d\right)}$$

$$\leq \frac{1}{2\alpha_0} \|f\|_{H^{-1}\left(\mathbb{R}_+^d\right)}^2 + \frac{\alpha_0}{2} \left(\|u\|_{L^2\left(\mathbb{R}_+^d\right)}^2 + \|\nabla u\|_{L^2\left(\mathbb{R}_+^d\right)}^2 \right).$$

After rearranging and collecting terms, one gets

$$\|\nabla u\|_{L^2\left(\mathbb{R}_+^d\right)}^2 \leq \frac{1}{\alpha_0^2} \|f\|_{H^{-1}\left(\mathbb{R}_+^d\right)}^2 + \|u\|_{L^2\left(\mathbb{R}_+^d\right)}^2,$$

or, equivalently,

$$\|u\|_{H^1\left(\mathbb{R}_+^d\right)}^2 \leq \frac{1}{\alpha_0^2} \|f\|_{H^{-1}\left(\mathbb{R}_+^d\right)}^2 + 2\|u\|_{L^2\left(\mathbb{R}_+^d\right)}^2.$$

For $m > 0$, we use an inductive argument. By differentiating equation (2.2.1), it is easy to check that the equation

$$- \operatorname{div} A\nabla \partial_k u = \partial_k f + \operatorname{div} \left((\partial_k A)\nabla u \right)$$

is satisfied in $\mathbb{R}_+^d$ in the weak sense. Let $1 \leq k \leq d - 1$; i.e., consider tangential directions with respect to the hyperplane $\mathbb{R}^{d-1} \times \{0\}$ bounding $\mathbb{R}_d^+$. Using Lemma 2.2.14 below we get that $\partial_k u \in H^m \left(\mathbb{R}_+^d \right) \cap H_0^1 \left(\mathbb{R}_+^d \right)$, and by induction

$$\|\partial_k u\|_{H^m(\mathbb{R}^d_+)} \leq C \left(\|\partial_k u\|_{H^{m-1}(\mathbb{R}^d_+)} + \|\partial_k f\|_{H^{m-2}(\mathbb{R}^d_+)} \right.$$

(2.2.10)
$$+ \left. \|\operatorname{div}((\partial_k A)\nabla u)\|_{H^{m-2}(\mathbb{R}^d_+)} \right)$$

$$\leq C \left(\|u\|_{H^m(\mathbb{R}^d_+)} + \|f\|_{H^{m-1}(\mathbb{R}^d_+)} \right).$$

Equivalently, we have an estimate on $\|\partial_i \partial_j u\|_{H^{m-1}(\mathbb{R}^d_+)}$ for all $1 \leq i \leq d$ and $1 \leq j \leq d-1$.

It remains to estimate $\|\partial_d^2 u\|_{H^{m-1}(\mathbb{R}^d_+)}$, which can be done by using the partial differential equation (2.2.1) once more. We isolate this derivative,

$$(2.2.11) \qquad -a^{dd}\partial_d^2 u = f + \sum_{i=1}^{d}\sum_{j=1}^{d-1} \partial_i(a^{ij}\partial_j u) + (\partial_d a^{dd})(\partial_d u).$$

Hence, the desired estimate follows from the fact that $a^{dd} \geq \alpha_0$ and the existence of the bounds on $\|\partial_i \partial_j u\|_{H^{m-1}(\mathbb{R}^d_+)}$ for $1 \leq i \leq d$ and $1 \leq j \leq d-1$ given by (2.2.10). $\qquad\square$

It remains to state and prove the auxiliary lemma used in the proof of Proposition 2.2.13.

Lemma 2.2.14 ([**Bre11**, Lemma 9.7]). Let $u \in H^2(\mathbb{R}^d_+) \cap H_0^1(\mathbb{R}^d_+)$. Then $\partial_k u \in H_0^1(\mathbb{R}^d_+)$ for $1 \leq k \leq d-1$.

Proof. Given $1 \leq k \leq d-1$, we set $h = te_k$, where e_k is the kth unit coordinate vector. Then, for $v \in H_0^1(\mathbb{R}^d_+)$, we have $D_h v \in H_0^1(\mathbb{R}^d_+)$, since e_k is parallel to the hyperplane bounding $\mathbb{R}^d_+$. Due to (2.2.3) and the weak compactness of the unit ball in $H_0^1(\mathbb{R}^d_+)$, we can find $w \in H_0^1(\mathbb{R}^d_+)$ and a sequence $(h_n)_{n=1}^{\infty} = (t_n e_k)_{n=1}^{\infty}$ such that $D_{h_n} v \to w$ weakly in $H_0^1(\mathbb{R}^d_+)$ as $n \to \infty$. On the other hand, for all $\varphi \in C_0^{\infty}(\mathbb{R}^d_+)$ we have

$$\int_{\mathbb{R}^d_+} (D_{h_n}v)\varphi \,dx = -\int_{\mathbb{R}^d_+} v D_{-h_n}\varphi \,dx \xrightarrow{t_n \to 0} -\int_{\mathbb{R}^d_+} v\partial_k\varphi \,dx.$$

It follows that

$$\int_{\mathbb{R}^d_+} w\varphi \,dx = -\int_{\mathbb{R}^d_+} v\partial_k\varphi \,dx.$$

Hence, $\partial_k v = w$, and in particular $\partial_k v \in H_0^1(\mathbb{R}^d_+)$. $\qquad\square$

Combining the a priori estimate in Proposition 2.2.13 with Nirenberg's argument we obtain the global regularity statement.

Theorem 2.2.15 (Global regularity in half-space). Let u be a weak solution of equation (2.2.1), where we assume the conditions on the entries a^{ij} imposed in Proposition 2.2.13. Suppose that $u \in H^{m+1}\left(\mathbb{R}^d_+\right) \cap H^1_0\left(\mathbb{R}^d_+\right)$ and $f \in H^m\left(\mathbb{R}^d_+\right)$ for some $m \geq 0$. Then

$$u \in H^{m+2}\left(\mathbb{R}^d_+\right).$$

Proof. Let $h = te_k$ as above. For $1 \leq k \leq d-1$, we have $D_h u \in H^{m+1}\left(\mathbb{R}^d_+\right) \cap H^1_0\left(\mathbb{R}^d_+\right)$, and therefore we can apply the a priori estimate of Proposition 2.2.13 to the equation

$$-\operatorname{div} A\nabla D_h u = D_h f + \operatorname{div}\left((D_h A)\nabla u\right) + |h|\operatorname{div}\left((D_h A)\nabla D_h u\right).$$

Here we have used the analogue of the Leibniz rule; see Exercise 2.2.8. Hence, for small enough $|h|$ we have

$$\|D_h u\|_{H^{m+1}\left(\mathbb{R}^d_+\right)} \leq C\left(\|D_h u\|_{H^m\left(\mathbb{R}^d_+\right)} + \|D_h f\|_{H^{m-1}\left(\mathbb{R}^d_+\right)} \right.$$
$$\left. + \|\operatorname{div}\left((D_h A)\nabla u\right)\|_{H^{m-1}\left(\mathbb{R}^d_+\right)}\right)$$
$$+ \frac{1}{2}\|D_h u\|_{H^{m+1}\left(\mathbb{R}^d_+\right)}.$$

Rearranging terms and recalling that the norms $\|D_h f\|_{H^{m-1}}$ are bounded by $\|f\|_{H^m}$ in view of (2.2.4), we obtain that $\|D_h u\|_{H^{m+1}\left(\mathbb{R}^d_+\right)}$ is bounded independently of h. It follows from Theorem 2.2.7 that $\partial_k u \in H^{m+1}\left(\mathbb{R}^d_+\right)$ for all $1 \leq k \leq d-1$, or, equivalently, $\partial_i \partial_j u \in H^m\left(\mathbb{R}^d_+\right)$ for all $1 \leq i \leq d$ and $1 \leq j \leq d-1$. Finally, we can express $\partial_d^2 u$ as in (2.2.11) and deduce that $\partial_d^2 u \in H^m\left(\mathbb{R}^d_+\right)$. Summarising, we have shown that $u \in H^{m+2}\left(\mathbb{R}^d_+\right)$. $\qquad\square$

Corollary 2.2.16 (Global regularity for the Dirichlet problem). Let $m \geq 0$, and let $\Omega \subset \mathbb{R}^d$ be a bounded domain with C^{m+2} boundary. Let $u \in H^1_0(\Omega)$ satisfy

$$-\Delta u = f \text{ in } \Omega$$

in the weak sense, where $f \in H^m(\Omega)$. Then, $u \in H^{m+2}(\Omega)$.

Proof. A partition of unity argument and a change of coordinates leading to (2.2.9) reduces the problem to Theorem 2.2.15 . We obtain an equation

$$-\operatorname{div} A\nabla v = wg \text{ in } \mathbb{R}^d_+,$$

with a positive definite $C^{m+1}\left(\overline{\mathbb{R}^d_+}\right)$ matrix A (see (2.2.1)), a positive $C^{m+1}\left(\overline{\mathbb{R}^d_+}\right)$ weight function w, and $g \in H^m\left(\mathbb{R}^d_+\right)$. Hence $wg \in H^m\left(\mathbb{R}^d_+\right)$, and we can apply Theorem 2.2.15. It follows that $v \in H^{m+2}\left(\mathbb{R}^d_+\right)$ and finally that $u \in H^{m+2}(\Omega)$. $\square$

We can now finish the proof of Theorem 2.2.1(iii).

Proof of Theorem 2.2.1, part (iii). Applying Corollary 2.2.16 with $f = \lambda u$ and elliptic bootstrapping shows that if Ω has C^{m+2} boundary, then $u \in H^{m+2}(\Omega)$. If the boundary of Ω is C^∞, the Sobolev embedding theorem (Theorem 2.1.9) shows that $u \in C^\infty(\overline{\Omega})$. $\square$

Recall that the Laplace–Beltrami operator on a Riemannian manifold is a multiple of a second-order elliptic operator in divergence form. Hence, the proof of Corollary 2.2.16 can be extended to this case verbatim, and we obtain

> **Theorem 2.2.17** (Smoothness of eigenfunctions of the Laplace–Beltrami operator). Let (M, g) be a closed Riemannian manifold. Then the eigenfunctions of the weak spectral problem for the Laplace–Beltrami operator are C^∞ on M.

Clearly, global regularity also holds for Dirichlet eigenfunctions on compact Riemannian manifolds with boundary with the same proof as in the Euclidean case.

2.2.5. Continuity up to the boundary of Dirichlet eigenfunctions on Lipschitz domains. Boundary regularity may fail near corners of piecewise-smooth domains. A standard example is a domain with a re-entrant corner.

■ **Exercise 2.2.18.** Let $\Omega = \left\{(r, \varphi) : 0 < r < 1, 0 < \varphi < \frac{3\pi}{2}\right\}$ be a three-quarter disk. Find its Dirichlet eigenfunctions by separation of variables, and show that they do *not* lie in $H^2(\Omega)$.

Still, Dirichlet eigenfunctions on Lipschitz domains are continuous up to the boundary. The proof of this result uses a different set of ideas from the usual boundary regularity. We present them below.[1]

Sketch of the proof of Theorem 2.2.1, part (iv). First, one can show that a Dirichlet eigenfunction $u \in L^\infty(\Omega)$. One way to do it is due to Moser [**Mos60**] (the so-called *Moser iteration method*); see [**GilTru01**, Theorem 8.15]. Another approach uses the fact that the *heat kernel* (to be defined in Chapter 6) in Ω at any fixed positive time is bounded; see [**Dav89**, Example 2.1.8].

---

[1]We are grateful to Dorin Bucur for outlining this argument.

Let B be a ball containing Ω. Let us extend u to B by zero, and denote this extension $\widetilde{u}$. Observe that due to the boundedness of u the extension $\widetilde{u} \in L^p(B)$ for any p. Let us solve the Dirichlet problem $-\Delta\theta = \lambda\widetilde{u}$ in B with $\theta|_{\partial B} = 0$. By an L^p analogue of the local elliptic regularity theorem (Theorem 2.2.4), it follows that $\theta \in W^{2,p}(B)$ (see [**GilTru01**, Theorem 9.15]; here $W^{2,p}$ is an L^p analogue of the Sobolev space $H^2 = W^{2,2}$). In particular, by the Sobolev embedding theorem ([**Eva10**, §5.6.3]), $\theta \in C^1(B)$.

Consider now the unique harmonic function h in Ω such that $h - \theta \in H_0^1(\Omega)$. In other words h is a weak solution of the Dirichlet problem $-\Delta h = 0$ in Ω and $h = \theta$ on $\partial\Omega$. Since $\partial\Omega$ is Lipschitz, all its boundary points are regular in the sense that $h \in C(\overline{\Omega})$ and the boundary values of h are given by θ (see [**HeiKilMar93**, Theorems 6.31 and 6.27], [**ArmGar01**, Theorem 6.6.15]).

Finally, set $v = \theta - h$. Note that $v \in H_0^1(\Omega)$, while $-\Delta v = \lambda u$ in Ω. Since $-\Delta u = \lambda u$ in Ω, we conclude that $-\Delta(v - u) = 0$ in Ω for $v - u \in H_0^1(\Omega)$, and hence $u = v = \theta - h$ in Ω. Since $\theta, h \in C(\overline{\Omega})$ and $\theta = h$ on $\partial\Omega$, we obtain that $u \in C(\overline{\Omega})$, and it vanishes on $\partial\Omega$. $\qquad\square$

2.2.6. Regularity of Neumann eigenfunctions.
Consider now the Neumann Laplacian. We have

Theorem 2.2.19 (Elliptic regularity in the Neumann case). Let $\Omega \subset \mathbb{R}^d$ be a bounded domain with a Lipschitz boundary. Then:

(i) The eigenfunctions of the weak Neumann spectral problem (2.1.20) are C^∞ in Ω. Moreover, they are real analytic in Ω.

(ii) The eigenfunctions are smooth up to the boundary near the smooth parts of the boundary.

The local regularity has already been established in §2.2.2. The proof of the global regularity for the Neumann problem is essentially the same as that for the Dirichlet problem. One observes that an a priori estimate in $H^1(\mathbb{R}_+^d)$ still holds due to the ellipticity assumption and that $H^k(\mathbb{R}_+^d)$ is invariant under translations or differentiation along the boundary (see (2.2.10)) and proceeds in the same way. Moreover, if Ω is smooth, any eigenfunction $u \in H^k(\Omega)$ for all k. It follows that the eigenfunctions are smooth near the smooth parts of the boundary.

Remark 2.2.20. We can additionally deduce that for a bounded domain Ω with a Lipschitz boundary, every weak Neumann eigenfunction $u \in H^1(\Omega)$ corresponding to an eigenvalue λ in fact belongs to the Sobolev space

$H^{3/2}(\Omega)$. To do so, consider an auxiliary problem

$$\begin{cases} -\Delta v = \lambda u & \text{in } \Omega, \\ v = 0 & \text{on } \partial\Omega. \end{cases}$$

This is an inhomogeneous Dirichlet problem for v, and by [**JerKen95**, Theorem B, part 2] we have $v \in H^{3/2}(\Omega)$. Then by [**ChWGLS12**, Lemma A.10], $\partial_n v \in L^2(\partial\Omega)$. Set $w = v - u$; then w solves

$$\begin{cases} -\Delta w = 0 & \text{in } \Omega, \\ \partial_n w = \partial_n v \in L^2(\partial\Omega) & \text{on } \partial\Omega. \end{cases}$$

Hence by [**JerKen81**], $w \in H^{3/2}(\Omega)$, which implies $u \in H^{3/2}(\Omega)$. We also note that the weak Dirichlet eigenfunctions belong to $H^{3/2}(\Omega)$ as follows directly from [**JerKen95**].

2.2.7. Strong spectral theorems. Elliptic regularity theorems (Theorems 2.2.1, 2.2.19, and 2.2.17), together with the weak spectral theorems (Theorems 2.1.20, 2.1.36, and 2.1.39), immediately imply that subject to some assumptions on the regularity of the boundary, where applicable, the eigenvalues and eigenfunctions of the corresponding weak spectral problems in fact satisfy the relevant equations and boundary conditions in the strong sense. More precisely, we have

Theorem 2.2.21 (Strong spectral theorem). Consider one of the following eigenvalue problems:

- The Dirichlet eigenvalue problem

(2.2.12)
$$\begin{cases} -\Delta u = \lambda u & \text{in } \Omega, \\ u = 0 & \text{on } \partial\Omega \end{cases}$$

for a bounded Lipschitz domain $\Omega \subset \mathbb{R}^d$.

- The Neumann eigenvalue problem

$$\begin{cases} -\Delta u = \lambda u & \text{in } \Omega, \\ \partial_n u = 0 & \text{on } \partial\Omega \end{cases}$$

for a smooth bounded domain $\Omega \subset \mathbb{R}^d$.

- The Laplace–Beltrami eigenvalue problem

$$-\Delta_g u = \lambda u$$

for a closed Riemannian manifold (M, g).

Then the eigenvalues and the eigenfunctions of each eigenvalue problem understood in the strong sense (i.e., the eigenvalue equations and boundary conditions are satisfied pointwise) coincide with those of the corresponding weak eigenvalue problem.

One can also show that the same result holds for the Dirichlet and Neumann eigenvalue problems on a compact Riemannian manifold with boundary; see [**Tay11**, §§5.1 and 5.7].

Remark 2.2.22. A Neumann eigenfunction u of a Lipschitz domain Ω can be thought to satisfy the Neumann condition pointwise almost everywhere in the following sense. Given a boundary point $x \in \partial\Omega$ at which the normal derivative exists, consider a sequence of points $y_i \in \Omega$ which approach x nontangentially. Then $\partial_n u(x) = \lim_{y_i \to x} \langle n, \nabla u(y_i) \rangle = 0$. We refer to [**JerKen81, ChWGLS12**] for details including the formal definition of nontangential convergence.

The immediate corollary of Theorem 2.2.21 is that in each case the "strong" spectrum is discrete, consists of eigenvalues of finite multiplicity accumulating only to $+\infty$, and the eigenfunctions can be chosen to form a basis in $L^2(\Omega)$ or $L^2(M)$, as appropriate.

It is important to emphasise that a restriction on the smoothness of the boundary in the Euclidean case is essential. We assume the boundary to be Lipschitz, which is not optimal and can be slightly relaxed at a cost of extra technicalities — but this condition cannot be omitted altogether. We have already remarked that omitting it in the Neumann problem may lead to undesirable consequences even in the weak form: the spectrum may no longer be discrete. Although the weak Dirichlet spectral theorem works without any restrictions on the smoothness of the boundary, this may not be the case for the strong one, as the following example indicates.

Example 2.2.23. Consider the eigenvalue problem (2.2.12) in a punctured disk $\Omega = \mathbb{D} \setminus \{0\}$. The weak eigenvalues and eigenfunctions of this problem are the same as for the whole disk; see Example 1.1.17. However, the eigenfunctions $J_0(j_{0,k}r)$ *do not* satisfy the boundary condition at the origin in the strong (pointwise) sense.

Chapter 3

Variational principles and applications

John William
Strutt,
3rd Baron
Rayleigh
(1842–1919)

In this chapter, we introduce the variational principles for eigenvalues and discuss their applications. These include domain monotonicity, Dirichlet–Neumann bracketing, and Weyl's law for general domains. Along the way, we also introduce the Robin and Zaremba eigenvalue problems and consider some applications of variational principles on symmetric domains. We also prove the Friedlander–Filonov inequalities between Dirichlet and Neumann eigenvalues, and the Berezin–Li–Yau inequalities, and we discuss Pólya's conjecture.

Hermann
Klaus Hugo
Weyl
(1885–1955)

3.1. Variational principles for Laplace eigenvalues

3.1.1. The Rayleigh quotient. Let $\mathcal{H}$ be a real (or complex) Hilbert space with an inner product $(\cdot, \cdot)_{\mathcal{H}}$. Consider a symmetric bilinear (respectively, sesquilinear) form $\mathcal{Q}[u, v]$, $\mathcal{Q} : U \times U \to \mathbb{R}$, defined on a dense linear subspace $U =: \operatorname{Dom}(\mathcal{Q})$ of $\mathcal{H}$, which we from now on refer to as the *domain* of the form $\mathcal{Q}$. Of particular importance to us is the corresponding *quadratic form* $\mathcal{Q}[u, u]$, $u \in \operatorname{Dom}(\mathcal{Q})$.

Definition 3.1.1 (Semi-bounded quadratic form). We say that the quadratic form $\mathcal{Q}[u, u]$ is *semi-bounded from below* if there exists a constant $c \in \mathbb{R}$ such that

$$\mathcal{Q}[u, u] \geq c\, (u, u)_{\mathcal{H}} \qquad \text{for all } u \in \mathrm{Dom}(\mathcal{Q}).$$

In what follows, we assume that $U = \mathrm{Dom}(\mathcal{Q})$ is complete in the norm induced by the inner product

$$(3.1.1) \qquad\qquad (u, v)_U := \mathcal{Q}[u, v] + (1 - c)\, (u, v)_{\mathcal{H}}.$$

Consider an abstract eigenvalue problem for a symmetric semi-bounded from below bilinear form $\mathcal{Q}$; we are looking for $\lambda \in \mathbb{R}$ and $u \in \mathrm{Dom}(\mathcal{Q}) \setminus \{0\}$ such that

$$(3.1.2) \qquad\qquad \mathcal{Q}[u, v] = \lambda\, (u, v)_{\mathcal{H}} \qquad \text{for all } v \in \mathrm{Dom}(\mathcal{Q}).$$

Assume in addition that the embedding $U \hookrightarrow \mathcal{H}$ is compact (here U is endowed with the norm induced by (3.1.1)). Then all the eigenvalues of (3.1.2) are of finite multiplicity, their sequence may have an accumulation point only at $+\infty$, and the corresponding eigenfunctions may be chosen to form an orthogonal basis in $\mathcal{H}$ (see [**Ban80**, §III.1.2]). We enumerate the eigenvalues of (3.1.2) in nondecreasing order with account of multiplicities as

$$\lambda_1(\mathcal{Q}) := \lambda_1 \leq \lambda_2 \leq \cdots.$$

The basic examples are the forms Q^{D} and Q^{N} for the weak Dirichlet spectral problem (2.1.12) and the weak Neumann spectral problem (2.1.20), respectively, in a bounded Euclidean domain Ω (which we assume to be Lipschitz in the Neumann case). These forms are defined by the same differential expression

$$(3.1.3) \qquad Q_{\Omega}^{\mathrm{D}}[u, v] = Q_{\Omega}^{\mathrm{N}}[u, v] := (\nabla u, \nabla v)_{L^2(\Omega)} = \int_{\Omega} \langle \nabla u, \nabla v \rangle \, \mathrm{d}x$$

and act in the same Hilbert space $\mathcal{H} = L^2(\Omega)$ but have different domains: $\mathrm{Dom}(Q^{\mathrm{D}}) = H_0^1(\Omega)$, and $\mathrm{Dom}(Q^{\mathrm{N}}) = H^1(\Omega)$.

The following simple result is often useful; see Remark 3.1.21 below for particular applications.

Proposition 3.1.2. Let $\{u_j\}$ be a basis of eigenfunctions of the eigenvalue problem (3.1.2), chosen to be orthogonal in $\mathcal{H}$. Then distinct eigenfunctions are also orthogonal in U equipped with the inner product (3.1.1).

Proof. Take $u = u_j$ and $v = u_k$ in (3.1.1) and (3.1.2), with $k \neq j$. Then

$$(u_j, u_k)_U = \mathcal{Q}[u_j, u_k] + (1 - c)\,(u_j, u_k)_{\mathcal{H}} = (\lambda_j + 1 - c)\,(u_j, u_k)_{\mathcal{H}} = 0. \qquad \square$$

For each $u \in \operatorname{Dom}(\mathcal{Q}) \setminus \{0\}$, we define its *Rayleigh quotient*

$$(3.1.4) \qquad\qquad R[u] := \frac{\mathcal{Q}[u, u]}{\|u\|_{\mathcal{H}}^2}.$$

Then the following *variational* (or *min-max*) *principle* (variously associated in the literature with the names of Lord Rayleigh, W. Ritz, R. Courant, and H. Poincaré, among others) for the eigenvalues of the weak spectral problem (3.1.2) holds.

Proposition 3.1.3 (The variational principle for a quadratic form [**Dav95**, §4.5], [**Ban80**, §III.1.2]). We have

$$(3.1.5) \qquad \lambda_k(\mathcal{Q}) = \min_{\substack{\mathcal{L} \subset \operatorname{Dom}(\mathcal{Q}) \\ \dim \mathcal{L} = k}}\ \max_{u \in \mathcal{L} \setminus \{0\}} R[u], \qquad k \in \mathbb{N}.$$

Remark 3.1.4. We will use the following additional properties of the variational principle.

  (i) For the principal eigenvalue λ_1, (3.1.5) becomes

$$(3.1.6) \qquad\qquad \lambda_1 = \min_{u \in \operatorname{Dom}(\mathcal{Q}) \setminus \{0\}} R[u].$$

  Any given $u_0 \in \operatorname{Dom}(\mathcal{Q}) \setminus \{0\}$ becomes a *test function* for λ_1 in the sense that

$$\lambda_1 \leq R[u_0].$$

  (ii) If $u_1, \ldots, u_{k-1}$ are eigenvectors of the weak spectral problem (3.1.2) corresponding to the eigenvalues $\lambda_1, \ldots, \lambda_{k-1}$ and if $\mathcal{L}_{k-1} := \operatorname{Span}\{u_1, \ldots, u_{k-1}\}$, then (3.1.5) with $k \geq 2$ is equivalent to

$$(3.1.7) \qquad\qquad \lambda_k = \min_{\substack{u \in \operatorname{Dom}(\mathcal{Q}) \setminus \{0\} \\ u \perp \mathcal{L}_{k-1}}} R[u].$$

  The minimum in (3.1.6) and (3.1.7) is attained by u if and only if u is an eigenvector of (3.1.2) corresponding to the eigenvalues λ_1 and λ_k, respectively. The minimum in (3.1.5) is attained by $\mathcal{L} = \operatorname{Span}\{u_1, \ldots, u_k\}$; however it may not be the only minimiser; see [**Ste70**].

Remark 3.1.5. If $\mathcal{Q}[u, v] = (Au, v)_{\mathcal{H}}$ is a bilinear form associated with a non-negative selfadjoint operator A, such as the Dirichlet or Neumann Laplacian, one has $\operatorname{Dom}(\mathcal{Q}) = \operatorname{Dom}\left(\sqrt{A}\right)$ (see [**Dav95**, §4.4]), and one can replace

Dom($\mathcal{Q}$) in the variational principles above by Dom(A), replacing at the same time min and max by inf and sup, respectively; see [**Dav95**, Theorem 4.5.3].

Let us illustrate the idea of the proof of the abstract Proposition 3.1.3. Let $u_1, u_2, \ldots$ be the eigenfunctions of $\mathcal{Q}$ chosen to form an orthonormal basis in $\mathcal{H}$. By Proposition 3.1.2, $\{u_j\}_{j=1}^\infty$ is also an orthogonal basis in $U = \text{Dom}(\mathcal{Q})$ with respect to the inner product (3.1.1). Let us take $u \in \text{Dom}(\mathcal{Q})$ and expand it in this basis,

$$u = \sum_{j=1}^\infty \alpha_j u_j, \qquad \alpha_j = (u, u_j)_{\mathcal{H}}.$$

Then it is easy to see that

$$R[u] = \frac{\mathcal{Q}[u,u]}{\|u\|_{\mathcal{H}}^2} = \frac{\sum\limits_{j=1}^\infty \lambda_j |\alpha_j|^2}{\sum\limits_{j=1}^\infty |\alpha_j|^2},$$

and Proposition 3.1.3 follows immediately.

■ **Exercise 3.1.6.** Use the method outlined above to prove Proposition 3.1.3 for $\mathcal{Q}[u,v] := \langle Au, v \rangle$, where A is a Hermitian $d \times d$ matrix acting in $\mathbb{R}^d$.

■ **Exercise 3.1.7.** Show that the eigenvectors of the weak spectral problem (3.1.2) are the critical points of the functional $u \mapsto \mathcal{Q}[u,u]$ subject to the constraint $\|u\|_{\mathcal{H}} = 1$, with the corresponding critical values being the eigenvalues of (3.1.2). We refer to [**Lau12**, Chapter 9] for a solution.

The following comparison principle immediately follows from Proposition 3.1.3.

Proposition 3.1.8 (Abstract eigenvalue comparison principle). Let $\mathcal{Q}_1$ and $\mathcal{Q}_2$ be two bilinear forms as above such that Dom($\mathcal{Q}_2$) $\subseteq$ Dom($\mathcal{Q}_1$) and

$$\mathcal{Q}_1[u,u] \leq \mathcal{Q}_2[u,u] \qquad \text{for all } u \in \text{Dom}(\mathcal{Q}_2).$$

Then the eigenvalues of the corresponding weak spectral problems satisfy

$$\lambda_k(\mathcal{Q}_1) \leq \lambda_k(\mathcal{Q}_2) \qquad \text{for all } k \in \mathbb{N}.$$

Simply speaking, Proposition 3.1.8 states that either narrowing the domain of a quadratic form or increasing the value of the form may only push all the eigenvalues up but not down.

3.1.2. Variational principles. Let $\Omega \subset \mathbb{R}^d$ be a bounded open set, and consider the weak spectral problem for the Dirichlet Laplacian $-\Delta_\Omega^D$ on Ω. As was mentioned above, the corresponding quadratic form has the domain $H_0^1(\Omega)$ and is given by

$$Q^D[u, u] = \|\nabla u\|_{L^2(\Omega)}^2 = \int_\Omega |\nabla u|^2 \, dx.$$

Hence, its Rayleigh quotient is

$$R[u] = \frac{\|\nabla u\|_{L^2(\Omega)}^2}{\|u\|_{L^2(\Omega)}^2} = \frac{\int_\Omega |\nabla u|^2 \, dx}{\int_\Omega u^2 \, dx}, \qquad u \in H_0^1(\Omega) \setminus \{0\}.$$

Proposition 3.1.3 then leads to the variational characterisation of the eigenvalues of $-\Delta_\Omega^D$.

Theorem 3.1.9 (The variational principle for the eigenvalues of the Dirichlet Laplacian). Let $\lambda_k = \lambda_k^D(\Omega)$, $k \in \mathbb{N}$, be the eigenvalues of the Dirichlet Laplacian $-\Delta_\Omega^D$ on a bounded open set $\Omega \subset \mathbb{R}^d$. Then

$$(3.1.8) \qquad \lambda_k = \min_{\substack{\mathcal{L} \subset H_0^1(\Omega) \\ \dim \mathcal{L} = k}} \max_{u \in \mathcal{L} \setminus \{0\}} \frac{\|\nabla u\|_{L^2(\Omega)}^2}{\|u\|_{L^2(\Omega)}^2}, \qquad k \in \mathbb{N}.$$

If additionally $\mathcal{L}_{k-1} := \mathrm{Span}\,\{u_1, \ldots, u_{k-1}\}$ is the linear subspace of $H_0^1(\Omega)$ spanned by the first $k-1$ eigenfunctions of $-\Delta_\Omega^D$, then we also have

$$(3.1.9) \qquad \lambda_k = \min_{\substack{u \in H_0^1(\Omega) \setminus \{0\} \\ u \perp \mathcal{L}_{k-1}}} \frac{\|\nabla u\|_{L^2(\Omega)}^2}{\|u\|_{L^2(\Omega)}^2}, \qquad k \in \mathbb{N}.$$

The minimum in (3.1.9) is attained by u if and only if u is an eigenfunction of $-\Delta_\Omega^D$ corresponding to λ_k.

■ **Exercise 3.1.10.** Finish the proof of Theorem 3.1.9 using the weak Dirichlet spectral theorem (Theorem 2.1.20).

For the Neumann Laplacian, the Rayleigh quotient is the same as in the Dirichlet case, and we have a direct analogue of Theorem 3.1.9, the only difference being that the space $H_0^1(\Omega)$ should be replaced by $H^1(\Omega)$. Note that the Neumann spectrum always starts with the eigenvalue $\mu_1 = \lambda_1^N(\Omega) = 0$, for which the corresponding eigenfunction is a constant.

Theorem 3.1.11 (The variational principle for the eigenvalues of the Neumann Laplacian). Let $\mu_k = \lambda_k^N(\Omega)$, $k \in \mathbb{N}$, be the eigenvalues of the Neumann Laplacian $-\Delta_\Omega^N$ on a bounded open set $\Omega \subset \mathbb{R}^d$ with Lipschitz boundary. Then

$$(3.1.10) \qquad \mu_k = \min_{\substack{\mathcal{L} \subset H^1(\Omega) \\ \dim \mathcal{L} = k}} \max_{u \in \mathcal{L} \setminus \{0\}} \frac{\|\nabla u\|_{L^2(\Omega)}^2}{\|u\|_{L^2(\Omega)}^2}, \qquad k \in \mathbb{N}.$$

If additionally $\mathcal{L}_{k-1} := \operatorname{Span}\{u_1, \ldots, u_{k-1}\}$ is the linear subspace of $H^1(\Omega)$ spanned by the first $k-1$ eigenfunctions of $-\Delta_\Omega^N$, then we also have

$$\mu_k = \min_{\substack{u \in H^1(\Omega) \setminus \{0\} \\ u \perp \mathcal{L}_{k-1}}} \frac{\|\nabla u\|_{L^2(\Omega)}^2}{\|u\|_{L^2(\Omega)}^2}, \qquad k \in \mathbb{N},$$

and in particular

$$(3.1.11) \qquad \mu_2 = \min_{\substack{u \in H^1(\Omega) \setminus \{0\} \\ \int_\Omega u \, dx = 0}} \frac{\|\nabla u\|_{L^2(\Omega)}^2}{\|u\|_{L^2(\Omega)}^2}.$$

The minima in (3.1.10) and (3.1.11) are attained by u if and only if u is an eigenfunction of $-\Delta_\Omega^N$ corresponding to μ_k and μ_2, respectively.

Remark 3.1.12. In practice, one can replace $\operatorname{Dom}(\mathcal{Q})$ in (3.1.5) by its dense subspace, simultaneously replacing min by inf and max by sup. In particular, $H_0^1(\Omega)$ appearing in Theorem 3.1.9 can be replaced by $C_0^\infty(\Omega)$, and $H^1(\Omega)$ appearing in Theorem 3.1.11 can be replaced by $C^\infty(\Omega)$; see also Appendix A.

Finally, the case of the Laplace–Beltrami operator on a smooth closed Riemannian manifold (M, g) is essentially identical to that of the Neumann Laplacian. We however have to remember our notational convention (Notation 2.1.41) on the enumeration of the eigenvalues of the Laplace–Beltrami operator on a closed Riemannian manifold.

Theorem 3.1.13 (The variational principle for the eigenvalues of the Laplace–Beltrami operator on a closed Riemannian manifold). Let $0 = \lambda_0(M) < \lambda_1(M) \leq \cdots$ be the eigenvalues of the Laplace–Beltrami operator $-\Delta_M$ on a smooth closed Riemannian manifold $M := (M, g)$. Then

$$\lambda_k = \min_{\substack{\mathcal{L} \subset H^1(M) \\ \dim \mathcal{L} = k+1}} \max_{u \in \mathcal{L} \setminus \{0\}} \frac{\|\nabla u\|_{L^2(M)}^2}{\|u\|_{L^2(M)}^2}, \qquad k \in \mathbb{N}_0.$$

If additionally $\mathcal{L}_k := \operatorname{Span}\{u_0 = 1, \ldots, u_{k-1}\}$ is the linear subspace of $H^1(M)$ spanned by the first k eigenfunctions of $-\Delta_M$, then we also have

$$(3.1.12) \qquad \lambda_k = \min_{\substack{u \in H^1(M)\setminus\{0\} \\ u \perp \mathcal{L}_k}} \frac{\|\nabla u\|_{L^2(M)}^2}{\|u\|_{L^2(M)}^2}, \qquad k \in \mathbb{N},$$

and in particular

$$(3.1.13) \qquad \lambda_1 = \min_{\substack{u \in H^1(M)\setminus\{0\} \\ \int_M u \, dV = 0}} \frac{\|\nabla u\|_{L^2(M)}^2}{\|u\|_{L^2(M)}^2}.$$

The minima in (3.1.12) and (3.1.13) are attained by u if and only if u is an eigenfunction of $-\Delta_M$ corresponding to λ_k and λ_1, respectively.

■ **Exercise 3.1.14.** Let Σ be a smooth surface, and let g_1 and g_2 be two Riemannian metrics on Σ which are conformally equivalent; i.e., $g_1 = \alpha(x)g_2$ for some smooth positive function α. Show that the Dirichlet energy is conformally invariant, i.e., that

$$(3.1.14) \qquad \int_\Sigma |\nabla f|_{g_1}^2 \, dV_{g_1} = \int_\Sigma |\nabla f|_{g_2}^2 \, dV_{g_2}.$$

3.1.3. The Robin and Zaremba problems. As we have briefly mentioned in Remark 1.1.20, one can consider other types of boundary conditions for the Laplacian apart from the Dirichlet and Neumann conditions. We now discuss some of the many possible generalisations.

Let $\Omega \subset \mathbb{R}^d$ be a bounded domain with a Lipschitz boundary. Fix a parameter $\gamma \in \mathbb{R}$, and consider the spectral problem

$$(3.1.15) \qquad \begin{cases} -\Delta u = \lambda u & \text{in } \Omega, \\ \partial_n u + \gamma u = 0 & \text{on } \partial\Omega. \end{cases}$$

The boundary condition in (3.1.15) is known as the *Robin* condition, and the problem (3.1.15) is known as the *Robin spectral problem* (see [**GusAbe98**] for a fascinating historical investigation into the origins of this terminology). We note that for $\gamma = 0$ the Robin condition becomes the Neumann condition.

Acting as in §2.1.7 for the Neumann Laplacian, we can construct the Robin Laplacian $-\Delta^{R,\gamma}$ as the Friedrichs extension with the domain

$$\operatorname{Dom}(-\Delta^{R,\gamma}) = \{u \in H^1\Omega) : -\Delta u \in L^2(\Omega) \text{ and } \partial_n u + \gamma u \sim 0 \text{ on } \partial\Omega\},$$

where the condition $\partial_n u + \gamma u \sim 0$ is understood in the sense of

$$\int_\Omega \Delta u \, v \, dx + \int_\Omega \langle \nabla u, \nabla v \rangle \, dx + \int_{\partial\Omega} \gamma u v \, ds = 0$$

for all $v \in H^1(\Omega)$ (see [**AreCSVV18**, §7.5]). The corresponding bilinear form is given by

$$(3.1.16) \quad Q^{\mathrm{R},\gamma}[u,v] = \left(-\Delta^{\mathrm{R},\gamma}u, v\right)_{L^2(\Omega)} = (\nabla u, \nabla v)_{L^2(\Omega)} + \gamma\,(u,v)_{L^2(\partial\Omega)}$$

and has the same form domain $H^1(\Omega)$ as the Neumann Laplacian; the corresponding quadratic form is obviously semi-bounded from below by zero for $\gamma \geq 0$. For each fixed $\gamma \geq 0$, the spectrum of the Robin Laplacian is discrete and consists of eigenvalues

$$0 \leq \lambda_1^{\mathrm{R},\gamma} \leq \lambda_2^{\mathrm{R},\gamma} \leq \cdots$$

accumulating to $+\infty$ which can be found from the variational principle analogous to (3.1.10),

$$(3.1.17) \quad \lambda_k^{\mathrm{R},\gamma}(\Omega) = \min_{\substack{\mathcal{L} \subset H^1(\Omega) \\ \dim \mathcal{L} = k}} \max_{u \in \mathcal{L}\setminus\{0\}} \frac{\|\nabla u\|_{L^2(\Omega)}^2 + \gamma\|u\|_{L^2(\partial\Omega)}^2}{\|u\|_{L^2(\Omega)}^2}, \qquad k \in \mathbb{N}.$$

Taking in (3.1.17) $k = 1$ and $\mathcal{L} = \mathrm{Span}\{1\}$ we immediately obtain the bound

$$(3.1.18) \qquad\qquad \lambda_1^{\mathrm{R},\gamma}(\Omega) \leq \gamma\frac{\mathrm{Vol}_{d-1}(\partial\Omega)}{\mathrm{Vol}_d(\Omega)}.$$

Remark 3.1.15. It can be shown using a Sobolev trace inequality [**Gri11**, Theorem 1.5.1.10] that the Robin Laplacian is semi-bounded from below also for $\gamma < 0$; see [**AreCSVV18**, Theorem 7.15] for details. It is then not hard to check that the variational formula (3.1.17) holds for $\gamma < 0$ as well; see [**BucFreKen17**, formula (4.5)]. The principal eigenvalue $\lambda_1^{\mathrm{R},\gamma}(\Omega)$ is negative for $\gamma < 0$; moreover, inequality (3.1.18) still holds.

■ **Exercise 3.1.16.** Write down transcendental equations whose roots are the eigenvalues of $-\Delta^{\mathrm{R},\gamma}$ for the interval $(0, L) \subset \mathbb{R}^1$.

▢ **Numerical Exercise 3.1.17.** By separating the variables in polar coordinates, write down transcendental equations whose roots are the eigenvalues of $-\Delta^{\mathrm{R},\gamma}$ for the unit disk $\mathbb{D}$, and hence reproduce Figure 3.1.

■ **Exercise 3.1.18.** Note that the scaling for the Robin eigenvalues is not the same as in the Dirichlet and Neumann cases; cf. Lemma 2.1.30 and Exercise 2.1.38. Namely, prove that for a scaled copy Ω_ρ, $\rho > 0$, of a Lipschitz domain $\Omega \subset \mathbb{R}^d$ and for $j \in \mathbb{N}$ we have

$$\lambda_j^{\mathrm{R},\gamma}(\Omega_\rho) = \frac{1}{\rho^2}\lambda_j^{\mathrm{R},\rho\gamma}(\Omega).$$

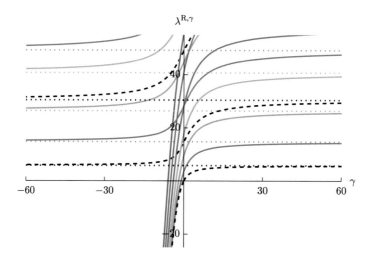

Figure 3.1. Some eigenvalues of the Robin Laplacian $-\Delta^{\mathrm{R},\gamma}$ for the unit disk as functions of γ. The dashed black curves correspond to single eigenvalues, and the solid curves correspond to double eigenvalues. The horizontal dotted lines are placed at the ordinates coinciding with the Dirichlet eigenvalues of the unit disk.

We will shortly obtain further bounds on Robin eigenvalues. For the moment, we observe only that for any fixed $k \in \mathbb{N}$,

$$(3.1.19) \qquad \lim_{\gamma \to +\infty} \lambda_k^{\mathrm{R},\gamma}(\Omega) = \lambda_k^{\mathrm{D}}(\Omega).$$

We omit a formal proof of this fact (see [**BucFreKen17**, Proposition 4.5]), but it can be easily deduced from the variational principle (3.1.17): for very large γ the minimisation procedure eliminates the dominant term $\gamma\|u\|_{L^2(\partial\Omega)}^2$ in the numerator of the Rayleigh quotient, thus forcing $u|_{\partial\Omega} = 0$.

Remark 3.1.19. An alternative approach to the Robin problem (3.1.15) is to consider λ as a given parameter and to treat γ (or, more precisely, $\sigma = -\gamma$) as a spectral parameter. This is the spectral problem for the so-called *Dirichlet-to-Neumann map* which we study extensively in Chapter 7.

We will also need to consider spectral problems with *mixed* Dirichlet–Neumann boundary conditions, often called *Zaremba problems*, which first appeared in [**Zar10**]. Let Ω be a bounded domain in $\mathbb{R}^d$ with a Lipschitz boundary $\partial\Omega$ which we decompose into the Dirichlet boundary $\Gamma := \partial_{\mathrm{D}}\Omega$ and the Neumann boundary $\partial_{\mathrm{N}}\Omega := \partial\Omega \setminus \Gamma$. To avoid unnecessary complications we assume that each of $\partial_{\mathrm{D,N}}\Omega$ consists of a finite number of connected components and that the interface between the two parts, $\overline{\partial_{\mathrm{D}}\Omega} \cap \overline{\partial_{\mathrm{N}}\Omega}$, is sufficiently regular for $d \geq 3$; see [**OttBro13**] for more precise conditions.

We consider a mixed Dirichlet–Neumann spectral problem

(3.1.20)
$$\begin{cases} -\Delta u = \lambda u & \text{in } \Omega, \\ u = 0 & \text{on } \partial_D \Omega, \\ \partial_n u = 0 & \text{on } \partial_N \Omega. \end{cases}$$

Obviously, if $\Gamma = \partial\Omega$, we get the standard Dirichlet problem, and if $\Gamma = \emptyset$ — the standard Neumann problem.

To give an operator-theoretic form of (3.1.20) and to obtain its variational formulation, we first define the space

$$C_{0,\Gamma}^\infty(\Omega) := \{u \in C^\infty(\Omega) : \text{supp } u \cap \overline{\Gamma} = \emptyset\}$$

and then the Sobolev space $H_{0,\Gamma}^1(\Omega)$ as the completion of $C_{0,\Gamma}^\infty(\Omega)$ in the $H^1(\Omega)$ norm. Then the *Zaremba* (or *mixed Dirichlet–Neumann*) Laplacian $-\Delta_{\Omega;\Gamma}^Z = -\Delta_\Gamma^Z$ can be defined via the Friedrichs extension with the domain

$$\text{Dom}\left(-\Delta_\Gamma^Z\right) = \left\{u \in H_{0,\Gamma}^1(\Omega) : \Delta u \in L^2(\Omega) \text{ and } \partial_n u \sim 0 \text{ on } \partial_N \Omega\right\},$$

where the last condition is understood in the sense that (2.1.8) holds for any $v \in H_{0,\Gamma}^1(\Omega)$. It is easy to check that the bilinear form $Q^Z[u,v]$ corresponding to the weak Zaremba problem coincides with the bilinear form for the Dirichlet and Neumann Laplacians, with the difference that its domain is given by $\text{Dom}(Q^Z) = H_{0,\Gamma}^1(\Omega)$. Hence, the eigenvalues $\lambda_k^Z(\Omega, \Gamma)$ of $-\Delta_\Gamma^Z$ can be determined from the variational principle

(3.1.21)
$$\lambda_k^Z(\Omega, \Gamma) = \min_{\substack{\mathcal{L} \subset H_{0,\Gamma}^1(\Omega) \\ \dim \mathcal{L} = k}} \max_{u \in \mathcal{L} \backslash \{0\}} \frac{\|\nabla u\|_{L^2(\Omega)}^2}{\|u\|_{L^2(\Omega)}^2}, \qquad k \in \mathbb{N},$$

which is identical to the Dirichlet variational principle (3.1.8) with $H_0^1(\Omega)$ replaced by $H_{0,\Gamma}^1(\Omega)$.

■ **Exercise 3.1.20.**

(i) Find the eigenvalues of the one-dimensional mixed Laplacian on the interval $(0, L)$ with the Dirichlet condition imposed at one end and the Neumann condition at the other.

(ii) Use (i) to find the eigenvalues of the Zaremba Laplacian $-\Delta_\Gamma^Z$ in the unit square in the following cases:
   (a) Γ is a single side of the square.
   (b) Γ is the union of two adjacent sides.
   (c) Γ is the union of two opposite sides.
   (d) Γ is the union of three sides of the square.

Remark 3.1.21. Let $\{u_j\}$ be a basis of eigenfunctions of either Dirichlet, Neumann, or Zaremba Laplacian in a bounded domain $\Omega \subset \mathbb{R}^d$, chosen to be orthogonal in $L^2(\Omega)$. It immediately follows from Proposition 3.1.2 that $(\nabla u_j, \nabla u_k)_{L^2(\Omega)} = 0$ for $j \neq k$, and therefore distinct eigenfunctions are also orthogonal in $H^1(\Omega)$. This is however not true for the eigenfunctions of the Robin Laplacian $-\Delta^{\mathrm{R},\gamma}$ with $\gamma \neq 0$.

3.2. Consequences of variational principles

3.2.1. Domain monotonicity and Dirichlet–Neumann bracketing.
We start with the following simple but immensely important application of the variational principle for the Dirichlet Laplacian.

Theorem 3.2.1 (Domain monotonicity for the Dirichlet Laplacian). Let $\Omega_1 \subset \Omega_2$ be two bounded domains. Then their Dirichlet eigenvalues satisfy $\lambda_k^{\mathrm{D}}(\Omega_2) \leq \lambda_k^{\mathrm{D}}(\Omega_1)$ for all $k \in \mathbb{N}$.

Proof. We have a natural embedding $H_0^1(\Omega_1) \subset H_0^1(\Omega_2)$; if $u \in H_0^1(\Omega_1)$, extending u by zero onto Ω_2 we obtain a function $\tilde{u} \in H_0^1(\Omega_2)$. Moreover $R_{\Omega_1}[u] = R_{\Omega_2}[\tilde{u}]$. The result then follows immediately from Proposition 3.1.8. $\qquad\square$

Proposition 3.2.2 (Strict domain monotonicity for the Dirichlet Laplacian). Let $\Omega \subsetneq \widetilde{\Omega} \subset \mathbb{R}^d$ be two bounded domains such that $\widetilde{\Omega}\backslash\Omega$ contains an open set. Then their Dirichlet eigenvalues satisfy $\lambda_k^{\mathrm{D}}(\widetilde{\Omega}) < \lambda_k^{\mathrm{D}}(\Omega)$ for all $k \in \mathbb{N}$.

Proof. This was first observed in [**CouHil89**, footnote on p. 409]. We mostly follow the argument in [**Wel72**]. Firstly, by the nonstrict domain monotonicity theorem (Theorem 3.2.1) we have $\lambda_k^{\mathrm{D}}(\widetilde{\Omega}) \leq \lambda_k^{\mathrm{D}}(\Omega)$ for all $k \in \mathbb{N}$. Suppose, for contradiction to the statement of proposition, that for some number k,

$$(3.2.1) \qquad \lambda := \lambda_k^{\mathrm{D}}(\widetilde{\Omega}) = \lambda_k^{\mathrm{D}}(\Omega).$$

Since the spectrum of the Dirichlet Laplacian $-\Delta_{\widetilde{\Omega}}^{\mathrm{D}}$ is unbounded above, there exists $m \in \mathbb{N}$ such that

$$(3.2.2) \qquad \lambda_m^{\mathrm{D}}(\widetilde{\Omega}) > \lambda.$$

Choose a nested sequence of m domains

$$\Omega =: \widetilde{\Omega}_1 \subsetneq \widetilde{\Omega}_2 \subsetneq \cdots \subsetneq \widetilde{\Omega}_m := \widetilde{\Omega},$$

such that $\widetilde{\Omega}_{i+1} \setminus \widetilde{\Omega}_i$ contains an open set, $i = 1, \ldots, m-1$; see Figure 3.2. By domain monotonicity and (3.2.1),

$$\lambda = \lambda_k^{\mathrm{D}}(\widetilde{\Omega}) \le \lambda_k^{\mathrm{D}}(\widetilde{\Omega}_i) \le \lambda_k^{\mathrm{D}}(\Omega) = \lambda,$$

and therefore $\lambda_k^{\mathrm{D}}(\widetilde{\Omega}_i) = \lambda$ for all $i = 1, \ldots, m$.

Let $u_i \in H_0^1(\widetilde{\Omega}_i)$ be an eigenfunction of $-\Delta_{\widetilde{\Omega}_i}^{\mathrm{D}}$ corresponding to the eigenvalue λ, and let $\widetilde{u}_i \in H_0^1(\widetilde{\Omega})$ be its extension by zero onto $\widetilde{\Omega}$. We claim that the set $\{\widetilde{u}_i\}_{i=1}^m$ is linearly independent. Indeed, suppose that

(3.2.3)
$$f := \sum_{i=1}^{m} \alpha_i \widetilde{u}_i$$

is identically zero in $\widetilde{\Omega}$ for some coefficients $\alpha_1, \ldots, \alpha_m \in \mathbb{R}$. The restriction of f to $\widetilde{\Omega}_m \setminus \widetilde{\Omega}_{m-1}$ is equal to $\alpha_m \widetilde{u}_m = \alpha_m u_m$, and since the eigenfunction u_m cannot vanish on an open subset by real domain analyticity, we have $\alpha_m = 0$. We therefore have a shorter linear combination f; repeating the argument we at the end conclude that $\alpha_m = \alpha_{m-1} = \cdots = \alpha_1 = 0$. Thus for $\mathcal{L} = \mathrm{Span}\,\{\widetilde{u}_i\}_{i=1}^m$ we have $\dim \mathcal{L} = m$.

Let now $f \in \mathcal{L}$ be given by (3.2.3), and let us evaluate its Rayleigh quotient. We have

$$
\begin{aligned}
\|\nabla f\|_{L^2(\widetilde{\Omega})}^2 &= \sum_{i=1}^{m} \left(\alpha_i^2 \|\nabla u_i\|_{L^2(\widetilde{\Omega}_i)}^2 + 2\sum_{j=1}^{i-1} \alpha_i \alpha_j \left(\nabla u_i, \nabla u_j\right)_{L^2(\widetilde{\Omega}_j)} \right) \\
&= \sum_{i=1}^{m} \left(\alpha_i^2 \left(-\Delta u_i, u_i\right)_{L^2(\widetilde{\Omega}_i)} + 2\sum_{j=1}^{i-1} \alpha_i \alpha_j \left(-\Delta u_i, u_j\right)_{L^2(\widetilde{\Omega}_j)} \right) \\
&= \lambda \sum_{i=1}^{m} \left(\alpha_i^2 \|u_i\|_{L^2(\widetilde{\Omega}_i)}^2 + 2\sum_{j=1}^{i-1} \alpha_i \alpha_j \left(u_i, u_j\right)_{L^2(\widetilde{\Omega}_j)} \right) = \lambda \|f\|_{L^2(\widetilde{\Omega})}^2,
\end{aligned}
$$

and therefore $R_{\widetilde{\Omega}}[f] = \lambda$ for all $f \in \mathcal{L}$. Thus, by the variational principle $\lambda_m^{\mathrm{D}}(\widetilde{\Omega}) \le \lambda$, which contradicts (3.2.2), and hence our assumption (3.2.1) is incorrect. $\qquad\square$

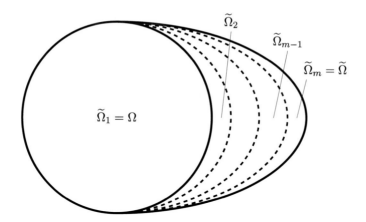

Figure 3.2. A nested sequence of domains appearing in the proof of Proposition 3.2.2.

Remark 3.2.3.

(i) For Dirichlet eigenfunctions for domains in Riemannian manifolds, the analogue of Proposition 3.2.2 holds as well, where the nonvanishing on open sets follows from the Aronszajn unique continuation property [**Aro57**]; see also Remark 4.1.14.

(ii) Strict domain monotonicity does not hold for disconnected sets. For example, if $\Omega = \Omega_1 \sqcup \Omega_2$, then

$$\lambda_1^{\mathrm{D}}(\Omega) = \min\left\{\lambda_1^{\mathrm{D}}(\Omega_1), \lambda_1^{\mathrm{D}}(\Omega_2)\right\}.$$

■ **Exercise 3.2.4.** Use domain monotonicity and Exercise 1.2.21 to find explicit two-sided estimates, in terms of $d = 2, 3, \ldots$, for the first positive zero $j_{\frac{d}{2}-1,1}$ of the Bessel function $J_{\frac{d}{2}-1}(x)$.

■ **Exercise 3.2.5.** Show that domain monotonicity does not generally hold for Neumann eigenvalues. *Hint:* Compare Neumann eigenvalues of a square and of a thin rectangle inscribed along a diagonal of the square; see [**Lau12**].

Example 3.2.6. Despite the result of Exercise 3.2.5, there are particular situations when Neumann domain monotonicity holds and can be once more deduced from Proposition 3.1.8. Consider a family of planar domains

$$\Omega_f := \{(x, y) : 0 < x < 1, -f(x) < y < f(x)\},$$

where f is a positive Lipschitz continuous function on $(0, 1)$ such that $\partial\Omega_f$ is Lipschitz as well. Fix any such function f and a number $\rho > 1$; obviously $\Omega_f \subset \Omega_{\rho f}$; see Figure 3.3.

We claim that

(3.2.4) $\lambda_k^N(\Omega_{\rho f}) \le \lambda_k^N(\Omega_f),$ for all $k \in \mathbb{N}$.

Indeed, we first of all can establish a bijection between the spaces $H^1(\Omega_f)$ and $H^1(\Omega_{\rho f})$ by identifying $u \in H^1(\Omega_f)$ with $\widetilde{u} := u(x, \rho y) \in H^1(\Omega_{\rho f})$. Moreover, a simple change of variables shows the monotonicity of the Rayleigh quotients:

$$R_{\Omega_{\rho f}}[\widetilde{u}] = \frac{\iint\limits_{\Omega_f} \left(\left(\frac{\partial u}{\partial x} \right)^2 + \frac{1}{\rho^2} \left(\frac{\partial u}{\partial y} \right)^2 \right) \mathrm{d}x\,\mathrm{d}y}{\iint\limits_{\Omega_f} u^2 \,\mathrm{d}x\,\mathrm{d}y} \le R_{\Omega_f}[u].$$

Then (3.2.4) follows from a simple rewording of Proposition 3.1.8.

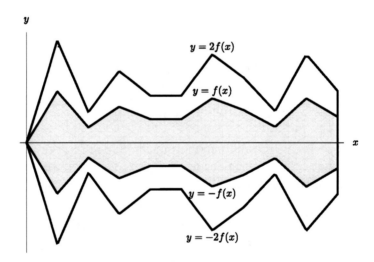

Figure 3.3. An example of the domains Ω_f (shaded) and Ω_{2f}, with $\lambda_k^N(\Omega_{2f}) \le \lambda_k^N(\Omega_f)$, $k \in \mathbb{N}$.

■ **Exercise 3.2.7.** As a particular application of Example 3.2.6, consider the ellipse $E_\rho := \left\{ (x, y) : x^2 + \frac{y^2}{\rho^2} < 1 \right\}$, $\rho > 1$. By using the above construction and recalling Lemma 2.1.30 as well as Exercise 2.1.38, prove that

$$\frac{1}{\rho^2} \lambda_k^N(\mathbb{D}) \le \lambda_k^N(E_\rho) \le \lambda_k^N(\mathbb{D}) \qquad \text{for all } k \in \mathbb{N}.$$

Note that similar inequalities hold for the eigenvalues of the Dirichlet Laplacian in E_ρ directly by domain monotonicity $\mathbb{D} \subset E_\rho \subset B_{0,\rho}^2$ and Lemma 2.1.30.

▣ **Numerical Exercise 3.2.8.** Verify the inequalities in Exercise 3.2.7 for the first few k and $\rho = 2$ numerically.

Another important corollary of Proposition 3.1.8 is the following result establishing the inequalities between the eigenvalues of the Dirichlet and Robin Laplacians in the same region.

Theorem 3.2.9. Let $\Omega \subset \mathbb{R}^d$ be a bounded open set with a Lipschitz boundary, and let $\gamma_2 \geq \gamma_1$. Then

$$\lambda_k^{\mathrm{R},\gamma_1} \leq \lambda_k^{\mathrm{R},\gamma_2} \leq \lambda_k^{\mathrm{D}} \qquad \text{for } k \in \mathbb{N}.$$

Proof. The inequality between the eigenvalues of the Robin Laplacians follows directly from Proposition 3.1.8; they have the same domains, and the quadratic form (3.1.16) is monotone increasing in γ. To establish the inequality between the Robin and the Dirichlet eigenvalues, we rewrite the Dirichlet quadratic form as

$$\left(-\Delta^{\mathrm{D}} u, u\right)_{L^2(\Omega)} = \left(-\Delta^{\mathrm{R},\gamma} u, u\right)_{L^2(\Omega)}$$

for any $u \in H_0^1(\Omega)$ and any $\gamma \in \mathbb{R}$ since in this case $u|_{\partial\Omega} = 0$, and we use the fact that $H_0^1(\Omega) \subset H^1(\Omega)$. □

Taking $\gamma_2 = 0$ in Theorem 3.2.9 immediately implies the following.

Corollary 3.2.10. Let $\Omega \subset \mathbb{R}^d$ be a bounded open set with a Lipschitz boundary. Then $\lambda_k^{\mathrm{N}}(\Omega) \leq \lambda_k^{\mathrm{D}}(\Omega)$.

In fact, as we will show in §3.2.4, a much stronger inequality holds between the Dirichlet and Neumann eigenvalues.

Let us now discuss the *Dirichlet–Neumann bracketing*. Informally, its idea is as follows: given a Laplacian on a domain, adding some extra Dirichlet conditions yields higher eigenvalues, and adding some extra Neumann conditions yields lower eigenvalues. Let us illustrate this by two specific examples.

The first result illustrates the effect of changing the boundary conditions from Dirichlet to Neumann (or vice versa) on a part of the boundary.

Proposition 3.2.11 (Dirichlet–Neumann bracketing, version 1). Let $\Omega \subset \mathbb{R}^d$ be a bounded domain with a Lipschitz boundary, and let $\Gamma_1 \subset \Gamma_2 \subset \partial\Omega$. Then

$$\lambda_k^Z(\Omega, \Gamma_1) \leq \lambda_k^Z(\Omega, \Gamma_2) \qquad \text{for all } k \in \mathbb{N}.$$

This result follows immediately from the variational principle for a mixed eigenvalue problem (3.1.21) and Proposition 3.1.8 due to the inclusion $H_{0,\Gamma_2}^1(\Omega) \subset H_{0,\Gamma_1}^1(\Omega)$.

The second version illustrates the effect of adding Dirichlet or Neumann conditions on a hypersurface inside the domain. Namely, let $\Omega \subset \mathbb{R}^d$ be a bounded domain, and consider the Dirichlet Laplacian $-\Delta_\Omega^D$ in Ω. Let $\Gamma \subset \Omega$ be a Lipschitz hypersurface. Let $\widetilde{\Omega} = \Omega \setminus \Gamma$, so that $\partial\widetilde{\Omega} = \partial\Omega \cup \Gamma$; see Figure 3.4 for some possible configurations of Γ within Ω. (In particular, Γ may separate Ω into two subdomains. This case will be particularly important, for example in §3.2.2.) We consider the Dirichlet Laplacian $-\Delta_{\widetilde{\Omega}}^D$, obtained from $-\Delta_\Omega^D$ by imposing the additional Dirichlet conditions on Γ, and the mixed Laplacian $-\Delta_{\widetilde{\Omega},\partial\Omega}^Z$ on $\widetilde{\Omega}$, obtained from $-\Delta_\Omega^D$ by imposing the additional Neumann conditions on Γ and preserving the Dirichlet conditions on $\partial\Omega \subset \partial\widetilde{\Omega}$.

Proposition 3.2.12 (Dirichlet–Neumann bracketing, version 2). In the geometry described above, we have

$$\lambda_k^Z(\widetilde{\Omega}, \partial\Omega) \leq \lambda_k^D(\Omega) \leq \lambda_k^D(\widetilde{\Omega}) \qquad \text{for all } k \in \mathbb{N}.$$

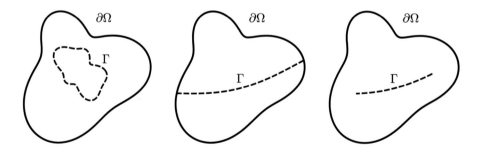

Figure 3.4. Three possible configurations of a hypersurface Γ inside Ω. On the left, Γ is a closed hypersurface; in the middle, $\partial\Gamma \subset \partial\Omega$; and on the right, $\partial\Gamma \subset \Omega$.

Remark 3.2.13. We note that for the middle and the right domains in Figure 3.4, the boundary part Γ of $\widetilde{\Omega}$ is not Lipschitz with respect to $\widetilde{\Omega}$ at the points of $\partial\Gamma$. Nevertheless, the extension property, see Remark 2.1.8, still holds, and therefore all the operators are well-defined and have discrete spectra.

■ **Exercise 3.2.14.**

  (i) Prove Proposition 3.2.12.

 (ii) Prove a version of Proposition 3.2.12 in which some arbitrary combination of Dirichlet, Neumann, and Robin conditions is originally imposed on parts of $\partial\Omega$.

(iii) Suppose that a Lipschitz domain Ω is partitioned into N disjoint Lipschitz domains Ω_n, $n = 1, \ldots, N$, in the sense that Ω is the interior of the closure of the union of Ω_n; see Figure 3.5. Prove that

$$\lambda_k^{\mathrm{D}}(\Omega) \leq \lambda_k^{\mathrm{D}} \left(\bigcup_{n=1}^{N} \Omega_n \right), \qquad k \in \mathbb{N},$$

and

$$\lambda_k^{\mathrm{N}}(\Omega) \geq \lambda_k^{\mathrm{N}} \left(\bigcup_{n=1}^{N} \Omega_n \right), \qquad k \in \mathbb{N}.$$

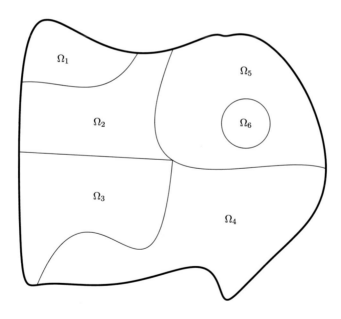

Figure 3.5. An example of partitioning a domain into subdomains. Note that in the spectral problems on $\bigcup_{n=1}^{N} \Omega_n$, the boundary conditions are imposed both on the exterior and the interior boundaries.

Remark 3.2.15. Imposing boundary conditions on sets of codimension two or higher does not affect the eigenvalues. Indeed, such sets have zero *capacity* (see Definition 4.1.8) and hence do not influence the spectrum (see [**RauTay75**]).

■ **Exercise 3.2.16.** Use domain monotonicity and Dirichlet–Neumann bracketing to derive two-sided estimates on the first few Dirichlet and Neumann eigenvalues of the L-shaped domain and the Π-shaped domain shown in Figure 3.6.

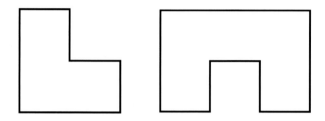

Figure 3.6. The L-shaped domain constructed from three unit squares and the Π-shaped domain constructed from five unit squares.

▢ **Numerical Exercise 3.2.17.** Compute the first ten Dirichlet and Neumann eigenvalues for the L-shaped and the Π-shaped domains and compare them with bounds you have derived in Exercise 3.2.16.

3.2.2. Symmetry tricks. Let Ω be a Euclidean domain which is symmetric with respect to a hyperplane S. Consider a Laplacian in Ω subject to some combination of Dirichlet, Neumann, and Robin boundary conditions which are also imposed symmetrically with respect to S. It turns out that one can choose a basis of eigenfunctions of the Laplacian on Ω in such a way that each eigenfunction is either symmetric with respect to S (and therefore satisfies the Neumann condition on $S \cap \Omega$) or antisymmetric with respect to S (and therefore satisfies the Dirichlet condition on $S \cap \Omega$). In this way, the spectral problem for the Laplacian on Ω decomposes into two mixed problems on a half Ω' of Ω lying to one side of S, with the Neumann and Dirichlet conditions, respectively, imposed on $S \cap \Omega$; see Figure 3.7.

The spectral decomposition described above is a consequence of the following abstract result.

Theorem 3.2.18. Let A be a selfadjoint operator with a discrete spectrum acting in a Hilbert space $\mathcal{H}$, and let J be a selfadjoint involution in $\mathcal{H}$ which commutes with A on Dom A; that is, $J^2 = \text{Id}$, and $JA - AJ = 0$.

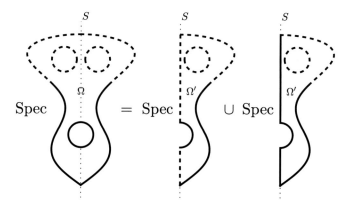

Figure 3.7. An example of a spectrum decomposition for a symmetric domain. The solid lines denote the Dirichlet conditions, and the dashed ones the Neumann conditions. The union of spectra is understood in the sense of multisets, with account of multiplicities.

Then one can choose an orthogonal basis of eigenfunctions of A in such a way that every eigenfunction u of A is either symmetric with respect to J, i.e., $Ju = u$, or antisymmetric with respect to J, i.e., $Ju = -u$.

Proof. Fix an eigenvalue λ of A, and denote by $\mathcal{U}$ the corresponding eigenspace. We start with the case of a simple eigenvalue λ, so that $\dim \mathcal{U} = 1$. If u is a corresponding eigenfunction, $Au = \lambda u$, and since J commutes with A, we also have $AJu = JAu = \lambda Ju$. Therefore, u and Ju should be linearly dependent, $Ju = cu$, $c = \text{const}$. As $J^2 u = u$, we have $c = \pm 1$, and either $Ju - u$ or $Ju + u$ vanishes identically. Therefore, an eigenfunction corresponding to a simple eigenvalue is automatically either symmetric or antisymmetric.

If $\dim \mathcal{U} > 1$, we first remark that any $u \in \mathcal{U}$ can be decomposed into a sum of symmetric and antisymmetric elements with respect to J:

$$u = \frac{u + Ju}{2} + \frac{u - Ju}{2}.$$

Let $\mathcal{U}_\pm := \{v \in \mathcal{U} : Jv = \pm v\}$, and note that the subspaces $\mathcal{U}_\pm$ are orthogonal: for any $u_\pm \in \mathcal{U}_\pm$ we have

$$(u_+, u_-)_{\mathcal{H}} = (Ju_+, u_-)_{\mathcal{H}} = (u_+, Ju_-)_{\mathcal{H}} = -(u_+, u_-)_{\mathcal{H}},$$

which implies $(u_+, u_-)_{\mathcal{H}} = 0$. Since A commutes with J, the finite-dimensional operator $A|_{\mathcal{U}}$ decomposes into the direct sum

$$A|_{\mathcal{U}} = A|_{\mathcal{U}_+} \oplus A|_{\mathcal{U}_-}$$

of two selfadjoint operators, and the result follows immediately. $\square$

■ **Exercise 3.2.19.** Let $\Omega \subset \mathbb{R}^d$ be an open set which is symmetric with respect to either a hyperplane or a point in $\mathbb{R}^d$. If $\tau : \Omega \to \Omega$ is a corresponding symmetry reflection, prove that the operator $J : L^2(\Omega) \to L^2(\Omega)$ defined by $Ju = u \circ \tau$ is a selfadjoint involution which commutes with the Laplacian on $H^1(\Omega)$.

Let us now return to the example considered in the beginning of this subsection and illustrated in Figure 3.7, assuming for definiteness that the Dirichlet conditions are imposed on $\partial\Omega$. Let $\tau_S : \Omega \to \Omega$ be the mirror symmetry with respect to S. We choose the involution J on $H_0^1(\Omega)$ to be $Ju = u \circ \tau_S$. Applying now Theorem 3.2.18, we immediately obtain

(3.2.5) $\operatorname{Spec}\left(-\Delta_\Omega^{\mathrm{D}}\right) \subseteq \operatorname{Spec}\left(-\Delta_{\Omega'}^{\mathrm{D}}\right) \cup \operatorname{Spec}\left(-\Delta_{\Omega';\partial_1\Omega'}^{\mathrm{Z}}\right),$

where we set $\partial_1\Omega' = \partial\Omega' \setminus S$ to be the part of the boundary of Ω' excluding the extra "cut" along S. We recall that $-\Delta_{\Omega';\partial_1\Omega'}^{\mathrm{Z}}$ denotes the mixed, or Zaremba, Laplacian, with the Dirichlet condition imposed on $\partial_1\Omega'$, and the Neumann condition on the rest of the boundary; see §3.1.3.

To show the opposite inclusion, we need to demonstrate that every eigenfunction of the Laplacian on Ω' subject to the Dirichlet or Neumann condition on $S \cap \Omega$ can be reflected antisymmetrically or symmetrically, respectively, across S to produce an eigenfunction on the whole domain Ω.

Proposition 3.2.20 (Reflection principle). Let $\Omega \subset \mathbb{R}^d$ be a domain symmetric with respect to a hyperplane S which divides it into two disjoint parts Ω' and $\tau_S\Omega'$. Decompose the boundary of Ω' into $\partial_1\Omega' = \partial\Omega' \setminus S$ and $\partial_2\Omega' = \partial\Omega' \cap S$. Then:

(i) If $u \in H_0^1(\Omega')$ is an eigenfunction of the Dirichlet Laplacian $-\Delta_{\Omega'}^{\mathrm{D}}$ corresponding to an eigenvalue λ, then

$$v(x) = \begin{cases} u(x) & \text{if } x \in \Omega', \\ -u(\tau_S x) & \text{if } x \in \tau_S(\Omega'), \\ 0 & \text{if } x \in \partial_2\Omega' \end{cases}$$

is an eigenfunction of the Dirichlet Laplacian on Ω corresponding to the same eigenvalue.

(ii) If $u \in H_{0,\partial_1\Omega'}^1(\Omega')$ is an eigenfunction of the mixed Laplacian $-\Delta_{\Omega',\partial_1\Omega'}^{\mathrm{D}}$ with the Dirichlet condition imposed on $\partial_1\Omega'$ and the Neumann condition on $\partial_2\Omega'$, then

$$v(x) = \begin{cases} u(x) & \text{if } x \in \Omega', \\ u(\tau_S x) & \text{if } x \in \tau_S(\Omega'), \end{cases}$$

extended by continuity to $\partial_2\Omega'$, is an eigenfunction of the Dirichlet Laplacian on Ω corresponding to the same eigenvalue.

■ **Exercise 3.2.21.** Prove this proposition by showing first that in both cases $v(x)$ is a weak eigenfunction of the Dirichlet problem in Ω, and then apply elliptic regularity.

Remark 3.2.22. Note that the elliptic regularity of eigenfunctions is essential in the above argument, and a reflection of an arbitrary smooth function does not necessarily yield a smooth function. For example, consider in $(0, +\infty)$ the function $u(x) = x^2 + x$, which satisfies the Dirichlet condition at the origin. Reflecting this function in an odd fashion with respect to the origin yields

$$f(x) = \begin{cases} x^2 + x & \text{if } x \geq 0, \\ -x^2 + x & \text{if } x < 0, \end{cases}$$

which is a $C^1(\mathbb{R})$ function but does not belong to C^2 near the origin.

Proposition 3.2.20 immediately implies

$$(3.2.6) \qquad \mathrm{Spec}\left(-\Delta_\Omega^{\mathrm{D}}\right) \supseteq \mathrm{Spec}\left(-\Delta_{\Omega'}^{\mathrm{D}}\right) \cup \mathrm{Spec}\left(-\Delta_{\Omega';\partial_1\Omega'}^{\mathrm{Z}}\right).$$

Combining (3.2.5) and (3.2.6) gives the *symmetry decomposition* (or *symmetry reduction*) formula for symmetric domains:

$$(3.2.7) \qquad \mathrm{Spec}\left(-\Delta_\Omega^{\mathrm{D}}\right) = \mathrm{Spec}\left(-\Delta_{\Omega'}^{\mathrm{D}}\right) \cup \mathrm{Spec}\left(-\Delta_{\Omega',\partial_1\Omega'}^{\mathrm{Z}}\right).$$

Remark 3.2.23. The same symmetry reduction method is applicable on a Riemannian manifold; for example, the spectrum of the Laplace–Beltrami operator on the sphere $\mathbb{S}^d$ decomposes into the union of the spectra of the Dirichlet and Neumann problems on the hemisphere. It also works for other boundary conditions on $\partial\Omega$ (for example, in a Robin or in a Zaremba problem) as long as they are imposed symmetrically with respect to S.

Remark 3.2.24. As an immediate application of the reflection principle, consider the Dirichlet problem for the right isosceles triangle with legs of length π. By Proposition 3.2.20(i), any eigenfunction on the triangle, reflected antisymmetrically with respect to the hypothenuse, extends to an eigenfunction of the Dirichlet Laplacian on the square of side π. Therefore, the Dirichlet eigenvalues of this triangle coincide with those of the square corresponding to an eigenfunction antisymmetric with respect to the diagonal. It is easy to verify that these eigenvalues are given by

$$\lambda_{k,m} = k^2 + m^2, \qquad k, m \in \mathbb{N}, \quad k > m.$$

A similar approach works in the Neumann case, as well as for the equilateral triangles; see [**Lam33**], [**Mak70**], [**Pin80**], and [**Pin85**]. We refer to [**McC11**] for a historical overview of the reflection method in application to polygons.

There are two main applications of the symmetry decomposition. One is pretty straightforward and is often used in numerical analysis for reducing the underlying mesh sizes (since one can consider a smaller domain).

▣ **Numerical Exercise 3.2.25.** Compute the eigenvalues of the Dirichlet Laplacian on an ellipse by two methods: first, directly, and second, by decomposing the problem into four problems on a quarter-ellipse, with Dirichlet and Neumann conditions imposed on the semi-axes.

The second application of the symmetry decomposition is often used in conjunction with the Dirichlet–Neumann bracketing.

> **Proposition 3.2.26.** Let $\Omega \subset \mathbb{R}^d$ be a domain symmetric with respect to a hyperplane S, and consider a Laplacian in Ω with some boundary conditions imposed symmetrically with respect to S. Then its first eigenfunction is symmetric with respect to S.

Proof. By (3.2.7) and Remark 3.2.23, the first eigenfunction will satisfy either the Dirichlet or the Neumann condition on S. However, imposing the Dirichlet condition on S increases the eigenvalues compared to imposing the Neumann condition; therefore the eigenfunction corresponding to the minimal eigenvalue is symmetric. We note that in the case of the Neumann problem in Ω, the result is trivially true since the first eigenfunction is a constant. □

■ **Exercise 3.2.27.** Let Ω be a planar domain symmetric with respect to a line S passing through the origin O and such that the set $\Omega \cap S$ is centrally symmetric with respect to O. Impose some boundary conditions on $\partial\Omega$ symmetrically with respect to S, and denote the first eigenvalue of the corresponding problem by $\lambda_1(\Omega)$. Now take a half Ω' of Ω lying to one side of S, and let $\widetilde{\Omega}$ be the union of Ω' and its centrally symmetric reflection $\tau(\Omega')$ around O; reflect the boundary conditions in the same way; see Figure 3.8. Show that $\lambda_1(\widetilde{\Omega}) \geq \lambda_1(\Omega)$. A solution can be found in [**JakLNP06**].

■ **Exercise 3.2.28.** Modify the argument in Example 3.2.27 to show that $\lambda_1(\widetilde{\Omega}) \geq \lambda_1(\Omega)$, where $\lambda_1(\Omega)$ and $\lambda_1(\widetilde{\Omega})$ refer to two boundary value problems on the quarter-sphere shown in Figure 3.9.

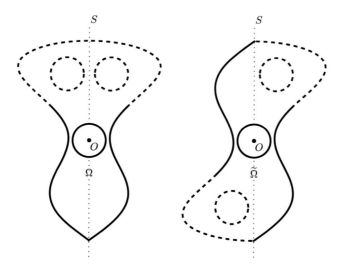

Figure 3.8. An example of a symmetric domain Ω and a centrally symmetric domain $\widetilde{\Omega}$, obtained by adding to the right half of Ω its copy reflected with respect to the point O. The solid lines denote the Dirichlet conditions, and the dashed lines denote the Neumann conditions.

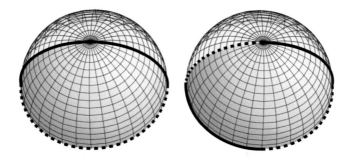

Figure 3.9. Two boundary value problems on a quarter-sphere, Ω on the left and $\widetilde{\Omega}$ on the right. The solid lines denote the Dirichlet conditions, and the dashed lines denote the Neumann conditions.

3.2.3. Counting functions. We have already encountered the counting function of eigenvalues of a flat torus in §1.2.2. Studying counting functions as opposed to individual eigenvalues provides an alternative, and often more convenient, approach to certain problems in spectral geometry.

Definition 3.2.29 (Eigenvalue counting function). Let A be a selfadjoint semi-bounded from below operator with a discrete spectrum consisting of eigenvalues $\lambda_1 \le \lambda_2 \le \cdots$. The *eigenvalue counting function* of A is the function $\mathcal{N} : \mathbb{R} \to \mathbb{N}_0$ defined as

$$\mathcal{N}(\lambda) = \mathcal{N}^A(\lambda) := \#\{j : \lambda_j(A) \le \lambda\}.$$

It is clear that $\mathcal{N}(\lambda)$ is right-continuous and monotone nondecreasing. Importantly, knowing $\mathcal{N}^A(\lambda)$ for all $\lambda \in \mathbb{R}$ we can recover the eigenvalues of A: if $\mathcal{N}^A(\lambda + 0) - \mathcal{N}^A(\lambda - 0) = 0$, then $\lambda \notin \mathrm{Spec}(A)$, and if $\mathcal{N}^A(\lambda + 0) - \mathcal{N}^A(\lambda - 0) = m > 0$, then λ is an eigenvalue of A of multiplicity m.

Sometimes, we will deal instead with the left-continuous eigenvalue counting function

(3.2.8) $$\widetilde{\mathcal{N}}(\lambda) = \widetilde{\mathcal{N}}^A(\lambda) := \#\{j : \lambda_j(A) < \lambda\},$$

whose values differ from those of $\mathcal{N}(\lambda)$ only at eigenvalues of A: if λ is an eigenvalue of A of multiplicity m_λ, then $\mathcal{N}^A(\lambda) = \widetilde{\mathcal{N}}^A(\lambda) + m_\lambda$.

Remark 3.2.30. Given any two selfadjoint semi-bounded from below operators A and B with discrete spectra, the inequalities $\lambda_k(A) \le \lambda_k(B)$, $k \in \mathbb{N}$, could be equivalently rewritten as $\mathcal{N}^A(\lambda) \ge \mathcal{N}^B(\lambda)$ for all $\lambda \in \mathbb{R}$; indeed, the smaller the eigenvalues are, the larger the counting function is. This simple observation will be very useful in the sequel.

Similarly, one can define an eigenvalue counting function $\mathcal{N}^{\mathcal{Q}}(\lambda)$ of the weak spectral problem (3.1.2) associated with a bilinear form $\mathcal{Q}$. The following important result shows that the variational principle from Proposition 3.1.3 can be reformulated in terms of the eigenvalue counting function.

Lemma 3.2.31 (Glazman's lemma). Consider the weak spectral problem (3.1.2) associated with a symmetric bilinear semi-bounded from below form as defined in §3.1.1. Then the counting function of the corresponding weak eigenvalues satisfies

$$\mathcal{N}^{\mathcal{Q}}(\lambda) = \max_{\substack{\mathcal{L} \subset \mathrm{Dom}(\mathcal{Q}) \\ R[u] \le \lambda \text{ for all } u \in \mathcal{L} \setminus \{0\}}} \dim \mathcal{L},$$

where $\mathcal{L}$ is a finite-dimensional linear subspace of $\mathrm{Dom}(\mathcal{Q})$ and $R[u]$ is the Rayleigh quotient (3.1.4).

■ Exercise 3.2.32. Prove Glazman's lemma; see [**Shu20**, Proposition 9.5].

Since we will be mostly dealing with the counting functions of Dirichlet and Neumann Laplacians, we introduce a shorthand notation for them.

Notation 3.2.33 (Eigenvalue counting functions for the Laplacians). If Ω is a bounded domain (with sufficiently regular boundary in the Neumann case) in a Euclidean space or in a Riemannian manifold, we will write for brevity

$$\mathcal{N}_\Omega^D(\lambda) := \mathcal{N}^{-\Delta_\Omega^D}(\lambda), \qquad \mathcal{N}_\Omega^N(\lambda) := \mathcal{N}^{-\Delta_\Omega^N}(\lambda),$$

and so on. Similarly, for a closed Riemannian manifold (M, g) we will write

$$\mathcal{N}_{(M,g)}(\lambda) = \mathcal{N}_M(\lambda) = \mathcal{N}_g(\lambda) := \mathcal{N}^{-\Delta(M,g)}(\lambda),$$

depending on the context.

▣ **Numerical Exercise 3.2.34.** Plot $\mathcal{N}^D(\lambda)$ and $\mathcal{N}^N(\lambda)$ for the planar unit disk, unit square, or any other domain of your choice, with eigenvalues computed either analytically or numerically.

3.2.4. Inequalities between the Dirichlet and Neumann eigenvalues for Euclidean domains. The goal of this subsection is to prove

Theorem 3.2.35 (The Friedlander–Filonov inequality). Let $\Omega \subset \mathbb{R}^d$, $d \geq 2$, be a bounded open set with Lipschitz boundary, and let $\lambda_k := \lambda_k^D(\Omega)$, $\mu_k := \lambda_k^N(\Omega)$. Then

$$(3.2.9) \qquad \mu_{k+1}(\Omega) < \lambda_k(\Omega), \qquad k \in \mathbb{N}.$$

This inequality was first proposed by L. Payne in 1955 [**Pay55**]. Its nonstrict version was proved by L. Friedlander in 1991 [**Fri91**] for C^1 domains. Friedlander's original proof is very instructive, and we will revisit it in §7.4.3. In 2004, N. Filonov [**Fil04**] found a strikingly simple and elegant argument that proved Theorem 3.2.35 as stated above.

Before proceeding to Filonov's proof, we start with the following simple lemma.

Lemma 3.2.36. Let u be an eigenfunction of the Neumann Laplacian on $\Omega \subset \mathbb{R}^d$. Then $u \notin H_0^1(\Omega)$.

Proof. Suppose, for contradiction, that u is an eigenfunction of the Neumann Laplacian in Ω corresponding to an eigenvalue μ and $u \in H_0^1(\Omega)$. Let w be an extension of u by zero to the whole $\mathbb{R}^d$. Then $w \in H^1(\mathbb{R}^d)$, and, given $v \in C_0^\infty(\mathbb{R}^d)$, we have

$$(3.2.10) \qquad \begin{aligned} (\nabla w, \nabla v)_{L^2(\mathbb{R}^d)} &= (\nabla u, \nabla v)_{L^2(\Omega)} = (-\Delta u, v)_{L^2(\Omega)} \\ &= \mu\,(u, v)_{L^2(\Omega)} = \mu\,(w, v)_{L^2(\mathbb{R}^d)}\,. \end{aligned}$$

Note that the boundary term vanishes because u is a Neumann eigenfunction. Comparing the left- and the right-hand sides of (3.2.10) we deduce that w is a weak solution of the equation $-\Delta w = \mu w$ in $\mathbb{R}^d$. By elliptic regularity it is therefore real analytic, and since $w|_{\mathbb{R}^d \setminus \Omega} = 0$, w is identically zero. Hence u is identically zero and therefore not an eigenfunction. $\qquad\square$

■ **Exercise 3.2.37.** Modify the proof of Lemma 3.2.36 to show that a Neumann eigenfunction on Ω cannot belong to the space $H^1_{0,\Gamma}(\Omega)$, where Γ is an open subset of $\partial\Omega$.

Let us also state the following exercise which we will use later.

■ **Exercise 3.2.38.** Let Ξ be any finite nonempty subset of $\mathbb{R}^d$. Prove that the set of exponential functions $\{e^{i\langle\omega,x\rangle} : \omega \in \Xi\}$ is linearly independent over $\mathbb{C}$.

Proof of Theorem 3.2.35. In this proof we, exceptionally, work with complex-valued functions, and therefore all scalar products are understood over $\mathbb{C}$.

By Glazman's lemma (Lemma 3.2.31) applied to the Dirichlet Laplacian on Ω, we have

$$(3.2.11) \qquad \mathcal{N}^{\mathrm{D}}(\lambda) = \max_{\substack{\mathcal{L} \subset H^1_0(\Omega) \\ R[u] \leq \lambda \text{ for all } u \in \mathcal{L} \setminus \{0\}}} \dim \mathcal{L}.$$

Fix $\lambda \geq \lambda_1$, and let $V_\lambda \subset H^1_0(\Omega)$ be a maximising $\mathcal{L}$ in (3.2.11), that is, a linear subspace of $H^1_0(\Omega)$ such that $\dim V_\lambda = \mathcal{N}^{\mathrm{D}}(\lambda)$, and $R[u] \leq \lambda$ for all $u \in V_\lambda \setminus \{0\}$. Let also $F_\lambda = \mathrm{Ker}(-\Delta^{\mathrm{N}} - \lambda) \subset H^1(\Omega)$; that is, $F_\lambda = \{0\}$ if $\lambda \notin \mathrm{Spec}(-\Delta^{\mathrm{N}})$; otherwise F_λ is the eigenspace of dimension m_λ corresponding to the Neumann eigenvalue λ of multiplicity $m_\lambda \geq 1$. According to Lemma 3.2.36, $F_\lambda \cap V_\lambda = \{0\}$; also $V_\lambda + F_\lambda = V_\lambda \oplus F_\lambda$ is finite dimensional: $\dim(V_\lambda + F_\lambda) = \mathcal{N}^{\mathrm{D}}(\lambda) + m_\lambda$.

Consider now the set of functions $\{e^{i\langle\omega,x\rangle} : \omega \in \mathbb{R}^d, |\omega|^2 = \lambda\}$. By the result of Exercise 3.2.38, this set is infinite dimensional if $d \geq 2$, and we therefore can choose a particular vector ω with $|\omega|^2 = \lambda$ in such a way that $g := e^{i\langle\omega,x\rangle}$ does not belong to $V_\lambda \oplus F_\lambda$. Set

$$W_\lambda := V_\lambda + F_\lambda + \{cg : c \in \mathbb{C}\},$$

and consider an arbitrary $w \in W_\lambda \setminus \{0\}$, $w = v + f + cg$, where $v \in V_\lambda$ and $f \in F_\lambda$.

Let us estimate the Rayleigh quotient $R[w]$, taking into account, firstly, that by the definition of V_λ we have $\|\nabla v\|^2 \leq \lambda \|v\|^2$ for any $v \in V_\lambda$, secondly that $\|\nabla f\|^2 = \lambda \|f\|^2$ for any $f \in F_\lambda$, and lastly that $\nabla g = i g \omega$ and $-\Delta g = |\omega|^2 g = \lambda g$ (all norms and inner products here and for the rest of the proof are in $L^2(\Omega)$).

In the numerator of $R[w]$ we have

$$\|\nabla(v + f + cg)\|^2 = \underbrace{\|\nabla v\|^2 + \|\nabla f\|^2 + \|c\nabla g\|^2}_{=:\mathcal{I}_1}$$
$$+ \underbrace{2\operatorname{Re}\left((\nabla f, \nabla(v + cg)) + (\nabla(cg), \nabla v)\right)}_{=:\mathcal{I}_2}.$$

We further simplify

(3.2.12)
$$\mathcal{I}_1 = \|\nabla v\|^2 + \lambda\|f\|^2 + |c|^2|\omega|^2\operatorname{Vol}_d(\Omega)$$
$$= \|\nabla v\|^2 + \lambda\|f\|^2 + |c|^2\lambda\operatorname{Vol}_d(\Omega),$$

and, using Green's formula,

(3.2.13)
$$\mathcal{I}_2 = 2\operatorname{Re}\left((-\Delta f, v + cg) + c\,(-\Delta g, v)\right)$$
$$= 2\lambda\operatorname{Re}\left((f, v + cg) + c\,(g, v)\right)$$

(the boundary terms vanish since f is a Neumann eigenfunction or zero and $v \in H_0^1(\Omega)$).

In the denominator of $R[w]$ we have

(3.2.14) $\quad \|v + f + cg\|^2 = \underbrace{\|v\|^2 + \|f\|^2 + \|cg\|^2}_{=:\mathcal{I}_1'} + \underbrace{2\operatorname{Re}\left((f, v + cg) + (cg, v)\right)}_{=:\mathcal{I}_2'},$

where after a simplification

(3.2.15)
$$\mathcal{I}_1' = \|v\|^2 + \|f\|^2 + |c|^2\operatorname{Vol}_d(\Omega).$$

Note that due to $\|\nabla v\|^2 \le \lambda\|v\|^2$, the comparison of (3.2.12) and (3.2.15) yields

$$\mathcal{I}_1 \le \lambda\mathcal{I}_1',$$

and the comparison of (3.2.13) and (3.2.14) yields

$$\mathcal{I}_2 = \lambda\mathcal{I}_2'.$$

Thus, we deduce the bound on the Rayleigh quotient,

(3.2.16)
$$R[w] = \frac{\mathcal{I}_1 + \mathcal{I}_2}{\mathcal{I}_1' + \mathcal{I}_2'} \le \lambda,$$

valid for all $w \in W_\lambda \setminus \{0\}$.

We now restate Glazman's lemma for the Neumann Laplacian in Ω:

(3.2.17)
$$\mathcal{N}^{\mathrm{N}}(\lambda) = \max_{\substack{\mathcal{L} \subset H^1(\Omega) \\ R[u] \le \lambda \text{ for all } u \in \mathcal{L}\setminus\{0\}}} \dim\mathcal{L}.$$

By (3.2.16), we can take $\mathcal{L} = W_\lambda$ in (3.2.17), giving

$$\mathcal{N}^{\mathrm{N}}(\lambda) \ge \dim W = N^{\mathrm{D}}(\lambda) + m_\lambda + 1.$$

Substituting into this inequality $\lambda = \lambda_k$, for which we have $N^{\mathrm{D}}(\lambda_k) \geq k$, we obtain

$$\mathcal{N}^{\mathrm{N}}(\lambda_k) = \widetilde{\mathcal{N}}^{\mathrm{N}}(\lambda_k) + m_{\lambda_k} \geq k + 1 + m_{\lambda_k},$$

or $\widetilde{\mathcal{N}}^{\mathrm{N}}(\lambda_k) \geq k + 1$. In other words, on the semi-open interval $[0, \lambda_k)$ there are at least $k + 1$ Neumann eigenvalues, which means that $\mu_{k+1} < \lambda_k$.   □

Remark 3.2.39. Note that the proof hinges upon the existence of a function g such that $-\Delta g = \lambda g$ and $\|\nabla g\| \leq \sqrt{\lambda}\|g\|$. For Euclidean domains, one can take an exponential function as we do. As was shown in [**Maz91**], such a function does not always exist on Riemannian manifolds, and the Friedlander–Filonov inequality may fail there. For instance, it fails for spherical caps that are larger than a hemisphere.

Remark 3.2.40. In dimension $d = 1$ the inequality (3.2.9) turns into an equality for each $k \geq 1$.

■ **Exercise 3.2.41.** Inspect the proof of Theorem 3.2.35 and explain why the strict inequality (3.2.9) fails in dimension one.

Let us conclude this section with the following open problem, which gives a stronger version of (3.2.9).

Conjecture 3.2.42. For any bounded domain $\Omega \subset \mathbb{R}^d$, we have $\mu_{k+d} \leq \lambda_k$, $k \geq 1$.

This result was proved by H. A. Levine and H. F. Weinberger [**LevWei86**] for convex domains, but for arbitrary domains it remains a challenging open question.

3.3. Weyl's law and Pólya's conjecture

3.3.1. Weyl's law. Weyl's law for the asymptotic distribution of eigenvalues is one of the most important results in spectral geometry. In its original form it was proved by Hermann Weyl in 1911, confirming a conjecture proposed in 1905 by Lord Rayleigh (with a constant corrected by J. H. Jeans; see [**SafVas97**] for a discussion).

Weyl's law is quite universal, in a sense that its versions apply to a wide variety of situations: Riemannian manifolds, Euclidean domains, various selfadjoint boundary conditions, and different elliptic operators. Below we prove Weyl's law for the Dirichlet Laplacian on Euclidean domains and leave its generalisations for later.

Theorem 3.3.1. Let $-\Delta_\Omega^D$ be the Dirichlet Laplacian on a bounded domain $\Omega \subset \mathbb{R}^d$. Then its eigenvalue counting function $\mathcal{N}_\Omega^D(\lambda)$ satisfies the asymptotic formula

$$(3.3.1) \qquad \mathcal{N}_\Omega^D(\lambda) = C_d \operatorname{Vol}_d(\Omega) \lambda^{\frac{d}{2}} + R(\lambda),$$

where $R(\lambda) = o\left(\lambda^{\frac{d}{2}}\right)$ as $\lambda \to +\infty$. Here

$$(3.3.2) \qquad C_d := \frac{\omega_d}{(2\pi)^d} = \frac{1}{(4\pi)^{\frac{d}{2}} \Gamma\left(\frac{d}{2}+1\right)}$$

is the *Weyl constant*, and ω_d denotes the volume of the unit ball in $\mathbb{R}^d$; see (B.1.1).

Proof. Let us split the proof into three steps. First, arguing in a similar way as in the proof of the asymptotic formula (1.2.14) for the flat torus, we prove (3.3.1) for cubes with either the Dirichlet or the Neumann boundary condition. The only difference compared to the torus case is that now one needs to take into account points with *positive* integer coordinates in the Dirichlet case, and *nonnegative* ones in the Neumann case. We leave the details as an exercise.

The next step is to consider domains that could be represented as an *almost disjoint union* of cubes (this means that if $\mathcal{K}$ is a finite collection of disjoint open cubes, then Ω is the interior of the closure of $\mathcal{K}$, and therefore $\partial\Omega \subset \partial\mathcal{K}$). Let Ω be such a domain; see Figure 3.10. Consider its partition into cubes (that is, the region $\widetilde{\Omega} := \Omega \setminus \partial\mathcal{K}$) and impose the Dirichlet (respectively, the Neumann) boundary conditions on $\partial\widetilde{\Omega}$.

By Dirichlet–Neumann bracketing and Remark 3.2.30 we then have, for all λ,

$$\mathcal{N}_{\widetilde{\Omega}}^D(\lambda) \le \mathcal{N}_\Omega^D(\lambda) \le \mathcal{N}_{\widetilde{\Omega}}^N(\lambda).$$

The result then follows by noticing that the counting functions $\mathcal{N}_{\widetilde{\Omega}}^D(\lambda)$ and $\mathcal{N}_{\widetilde{\Omega}}^N(\lambda)$ are sums of the corresponding counting functions for the cubes and applying the first step of the argument.

Finally, let Ω be an arbitrary bounded domain. Let $\Omega_{E,a}$ and $\Omega_{I,a}$ be two domains that can be represented as almost disjoint unions of cubes of side $a > 0$, such that $\Omega_{I,a} \subset \Omega \subset \Omega_{E,a}$; see Figure 3.11.

By the domain monotonicity for the Dirichlet eigenvalues,

$$\mathcal{N}_{\Omega_{I,a}}^D(\lambda) \le \mathcal{N}_\Omega^D(\lambda) \le \mathcal{N}_{\Omega_{E,a}}^D(\lambda).$$

Therefore, applying the result obtained in step two, we get

$$\limsup_{\lambda \to \infty} \frac{\mathcal{N}_\Omega^D(\lambda)}{\lambda^{\frac{d}{2}}} \le C_d \operatorname{Vol}_d\left(\Omega_{E,a}\right)$$

and

$$\liminf_{\lambda \to \infty} \frac{\mathcal{N}_\Omega^{\mathrm{D}}(\lambda)}{\lambda^{\frac{d}{2}}} \geq C_d \, \mathrm{Vol}_d \left(\Omega_{I,a} \right).$$

The result then follows by taking the limit $a \to 0$ and observing that one can choose $\Omega_{E,a}$ and $\Omega_{I,a}$ in such a way that

$$\lim_{a \to 0} \mathrm{Vol}_d \left(\Omega_{E,a} \right) = \lim_{a \to 0} \mathrm{Vol}_d \left(\Omega_{I,a} \right) = \mathrm{Vol}_d(\Omega).$$

This completes the proof of the theorem in the Dirichlet case. $\qquad\qquad$ $\square$

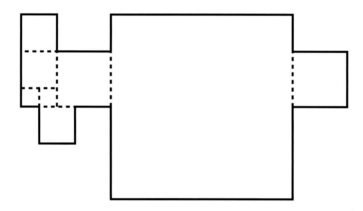

Figure 3.10. An almost disjoint union of open squares.

Remark 3.3.2. This proof could be found, for instance, in [**ReeSim75**, Chapter XIII], [**CouHil89**, Chapter VI.4], [**Bér86**, Chapter 3]. As shown in [**Roz72**] (see also [**Fri21**]), Theorem 3.3.1 holds in fact for arbitrary Euclidean domains of finite volume.

Remark 3.3.3 (Weyl's law for the Neumann Laplacian). An analogue of Theorem 3.3.1 holds for the Neumann eigenvalue problem in bounded Euclidean domains with Lipschitz boundary; see [**NetSaf05**] for a detailed discussion. In fact, for piecewise C^2 planar domains one can prove Weyl's law for the Neumann Laplacian using a modification of the argument presented above; see [**CouHil89**, §VI.4.4]. Note that a direct generalisation of the proof to the Neumann case does not work, as the last step involves domain monotonicity for the Dirichlet eigenvalues. Instead, one can approximate Ω by a union of cubes (in the interior) and right triangles (near the boundary) and show that small perturbations of triangles do not change the asymptotics of the eigenvalue counting function assuming that the boundary is sufficiently regular.

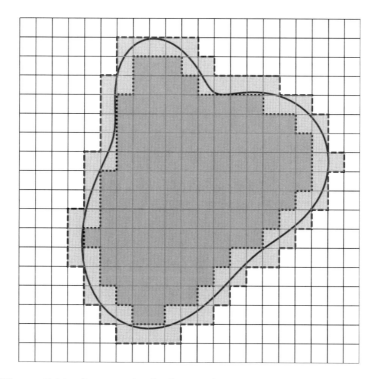

Figure 3.11. An example of a domain Ω, with its boundary shown as a solid curve, and corresponding domains $\Omega_{I,a}$ (darker shading, boundary is shown as a dotted line) and $\Omega_{E,a}$ (lighter shading, boundary is shown as a dashed line.)

Theorem 3.3.1 admits various extensions and improvements. In particular, for Euclidean domains with piecewise smooth boundaries, the remainder estimate can be improved to

$$R(\lambda) = O\left(\lambda^{\frac{d-1}{2}}\right),$$

for both Dirichlet and Neumann boundary conditions; see [**Vas86**]. Further improvements of the remainder estimates will be discussed in the next subsection. There exist also remainder estimates for domains with very rough boundaries, including some fractal ones; see, e.g., [**Mét77**], [**Lap91**], [**LevVas96**].

Weyl's law holds also in the Riemannian setting.

Theorem 3.3.4. Let M be a d-dimensional smooth compact Riemannian manifold. If $\partial M \neq \emptyset$, assume that either the Dirichlet or the Neumann boundary conditions are imposed on the boundary. Then the eigenvalue counting function for M has the asymptotics

(3.3.3) $$\mathcal{N}_M(\lambda) = C_d \operatorname{Vol}(M)\lambda^{\frac{d}{2}} + O\left(\lambda^{\frac{d-1}{2}}\right),$$

where C_d is again defined by (3.3.2).

Remark 3.3.5. The error estimate in (3.3.3) is sharp, as follows from the eigenvalue asymptotics on the round sphere; see (1.2.26). The proof of the sharp Weyl law uses the theory of pseudodifferential operators and is beyond the scope of this book. We refer to [**Shu01**], [**Tré82**], [**SafVas97**] for further details. We will revisit Weyl's law on manifolds in Chapter 6 and will explain how to deduce (3.3.3) with a weaker remainder estimate from the heat trace asymptotics.

■ **Exercise 3.3.6.** Prove that Theorem 3.3.1 is equivalent to the asymptotic law

(3.3.4) $$\lambda_k^{\mathrm{D}}(\Omega) = (C_d \operatorname{Vol}_d(\Omega))^{-\frac{2}{d}} k^{\frac{2}{d}} + o\left(k^{\frac{2}{d}}\right) \qquad \text{as } k \to \infty.$$

The same asymptotics also holds for the Neumann eigenvalue $\lambda_k^{\mathrm{N}}(\Omega)$, and the remainder estimates can be improved.

3.3.2. The two-term asymptotic formula and Weyl's conjecture. Let us recall Weyl's law on a square: can one get a better remainder estimate in this case? Now we will be more careful and take boundary conditions into account.

Consider a square K_π of side π. In the Dirichlet case, the eigenvalues correspond to integer points inside the cirle of radius $\sqrt{\lambda}$ lying in the first quadrant *excluding* the coordinate axes; in the Neumann case, the points on the axes (i.e., points having zero as one of the coordinates) should be counted as well.

How many integer points lie on the coordinate axes inside the circle of radius $\sqrt{\lambda}$? Approximately, $\sqrt{\lambda}$ on each semi-axis. There are four semi-axes, and therefore the quarter of integer points inside a circle is equal to the number of integer points in the interior of a quadrant plus the number of integer points on a single semi-axis. Therefore, in the Dirichlet case we need to take a quarter of integer points inside a circle and subtract the contribution of one semi-axis, while in the Neumann case we need to add the contribution of one semi-axis. Therefore, for a square K_π we get

$$\mathcal{N}^{\mathrm{D}}(\lambda) = \frac{\pi}{4}\lambda - \sqrt{\lambda} + R^{\mathrm{D}}(\lambda), \qquad \mathcal{N}^{\mathrm{N}}(\lambda) = \frac{\pi}{4}\lambda + \sqrt{\lambda} + R^{\mathrm{N}}(\lambda).$$

Note that these two-term asymptotic formulas would be meaningful only if the remainders $R^{\mathrm{D,N}}(\lambda)$ are of order $o\left(\sqrt{\lambda}\right)$. This is indeed true and could be deduced from the number-theoretic results on Gauss's circle problem; see discussion after Conjecture 1.2.13.

In 1911, H. Weyl conjectured that a similar two-term asymptotic formula holds for an arbitrary Euclidean domain and that the second term arises from the boundary.

Conjecture 3.3.7 (Weyl's conjecture). Let $\Omega \subset \mathbb{R}^d$ be a piecewise smooth Euclidean domain. Then

$$(3.3.5) \quad \mathcal{N}(\lambda) = C_d \operatorname{Vol}_d(\Omega)\lambda^{\frac{d}{2}} \pm C_{\mathrm{b},d} \operatorname{Vol}_{d-1}(\partial\Omega)\lambda^{\frac{d-1}{2}} + o\left(\lambda^{\frac{d-1}{2}}\right),$$

where the minus sign corresponds to the Dirichlet boundary conditions, and the plus sign to the Neumann boundary conditions. Here

$$(3.3.6) \qquad\qquad C_{\mathrm{b},d} = \frac{1}{2^{d+1}\pi^{\frac{d-1}{2}}\Gamma\left(\frac{d+1}{2}\right)}.$$

The expression (3.3.6) can be deduced from the heat trace asymptotics; see Remark 6.1.11 and Exercise 6.1.12.

In dimension two, (3.3.5) takes a particularly simple form,

$$(3.3.7) \qquad \mathcal{N}(\lambda) = \frac{\operatorname{Area}(\Omega)}{4\pi}\lambda \pm \frac{\operatorname{Length}(\partial\Omega)}{4\pi}\sqrt{\lambda} + o\left(\sqrt{\lambda}\right).$$

Example 3.3.8. In practice, at least for relatively simple planar domains, both the Dirichlet and Neumann asymptotic formulas (3.3.7) work remarkably well even for low values of λ. To illustrate this, we plot in Figure 3.12 the actual Dirichlet and Neumann counting functions together with one-term Weyl asymptotics (3.3.1) and the corresponding two-term asymptotics (3.3.7) for the rectangle $R_{\pi,2\pi}$ and for the unit disk $\mathbb{D}$.

In full generality Weyl's conjecture remains open. There has been significant progress on it in the past decades, in particular, due to V. Ivrii [**Ivr80**], R. Melrose [**Mel84**], and Yu. Safarov and D. Vassiliev [**SafVas97**]. The key observation here is that the growth of the error term is closely linked to the dynamical properties of the billiard flow (or, in a more general setting of a Riemannian manifold, of the geodesic flow). From the physical standpoint, this can be explained via Bohr's correspondence principle in quantum mechanics. Mathematically, the connection could be made via the wave trace. A rigorous treatment of this subject is way beyond the scope of this book, and we refer the reader to [**SafVas97**] for details. We shall simply state the main result of this theory, which is essentially due to V. Ivrii [**Ivr80**] with some improvements and generalisations due to D. Vassiliev [**Vas86**].

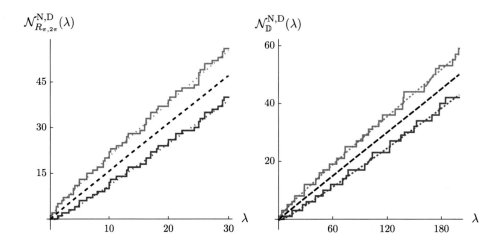

Figure 3.12. The actual counting functions and the one- and two-term Weyl's asymptotics for the rectangle $R_{\pi,2\pi}$ (left) and for the unit disk $\mathbb{D}$ (right). In both figures, blue curves correspond to the Dirichlet Laplacian and the magenta curves to the Neumann Laplacian. The graphs of the actual $\mathcal{N}(\lambda)$ are shown as solid, and the graphs of the two-term asymptotics as dotted lines. The dashed black line corresponds to the one-term Weyl's asymptotics.

A billiard trajectory satisfying the usual law of reflection in a bounded Euclidean domain $\Omega \subset \mathbb{R}^d$ is uniquely determined by the initial point $x \in \Omega$ and the initial direction $\xi \in \mathbb{S}^{d-1}$. Consider the *Liouville measure* on the unit (co)tangent bundle of Ω, which in this case can be simply viewed as the measure $dxd\xi$ on $\Omega \times \mathbb{S}^{d-1}$. We say that Ω satisfies the *nonperiodicity condition* if the set of pairs (x, ξ) corresponding to periodic billiard trajectories has Liouville measure zero.

Theorem 3.3.9. Let $\Omega \subset \mathbb{R}^d$ be a bounded domain with piecewise smooth boundary satisfying the nonperiodicity condition. Then the two-term asymptotics (3.3.5) holds.

Remark 3.3.10. It was conjectured by V. Ivrii (see also [**SafVas97**, Conjecture 1.3.35]) that the nonperiodicity condition holds for *any* Euclidean domain. This is an outstanding open problem in billiard dynamics. The affirmative answer is known for just a few specific classes of domains, such as convex analytic domains, piecewise-concave domains, and polygons.

■ **Exercise 3.3.11.** Show that a rectangle satisfies the nonperiodicity condition.

Remark 3.3.12. Under conditions of Theorem 3.3.9, the Neumann two-term asymptotic formula (3.3.5) remains valid for the eigenvalue counting function $\mathcal{N}^{\mathrm{R},\gamma}(\lambda)$ of the Robin Laplacian for any fixed γ. This is due to the fact that the second Weyl asymptotic term (for an elliptic boundary value problem in general) depends only upon the leading order differentiations in the boundary conditions and ignores the lower-order differentiations; see [**SafVas97**].

Theorem 3.3.9 admits a generalisation to Riemannian manifolds with boundary. However, in this case the nonperiodicity condition is not always satisfied and is essential for the two-term asymptotics (3.3.5) to hold.

■ **Exercise 3.3.13.**

(i) Show that all the trajectories of the geodesic flow on a hemisphere are periodic.

(ii) Using Theorem 1.2.16 and formula (1.2.26) show that the two-term asymptotics does not hold for a hemisphere with either the Dirichlet or the Neumann boundary conditions. *Hint:* Show that the eigenfunctions on a hemisphere with the Dirichlet (respectively, the Neumann) conditions are precisely the eigenfunctions of the full sphere which are antisymmetric (respectively, symmetric) with respect to the equatorial plane bounding the hemisphere. Full details can be found in [**BérBes80**].

Finally, there is a version of Theorem 3.3.9 for closed Riemannian manifolds; see [**DuiGui75**]. In this case the second term is equal to zero, and we simply obtain a refinement of the error term (big O is replaced by little o) in (3.3.3) under the nonperiodicity assumption. Once again, this assumption is essential, as we have already seen in the example of the round sphere.

3.3.3. Pólya's conjecture. Assuming that (3.3.5) holds (say, under the conditions of Theorem 3.3.9), it follows immediately that for $\Omega \subset \mathbb{R}^d$ with a sufficiently regular boundary and for a sufficiently large λ, we have

$$(3.3.8) \qquad \mathcal{N}^{\mathrm{D}}(\lambda) \le C_d \operatorname{Vol}_d(\Omega)\lambda^{\frac{d}{2}} \le \mathcal{N}^{\mathrm{N}}(\lambda).$$

In 1954, George Pólya [**Pól54**] conjectured that the inequalities (3.3.8) hold for *all* $\lambda \ge 0$. In fact Pólya's original conjecture was stated only for planar domains and in a slightly different form.

There exist other versions of these inequalities, rewritten, for example, using strict inequalities in (3.3.8). We will state Pólya's conjecture as the inequalities for the kth Dirichlet eigenvalue $\lambda_k = \lambda_k^{\mathrm{D}}(\Omega)$ and the kth nonzero

Neumann eigenvalue $\mu_{k+1} = \lambda_{k+1}^{\mathrm{N}}(\Omega)$:

(3.3.9) $$\mu_{k+1} \leq \left(\frac{1}{C_d \operatorname{Vol}_d(\Omega)} \right)^{\frac{2}{d}} k^{\frac{2}{d}} \leq \lambda_k;$$

cf. (3.3.4).

George **Pólya**

(1887–1985)

Conjecture 3.3.14 (Pólya's conjecture). The inequalities (3.3.9) hold for any $k \geq 1$.

In fact, it is expected that (3.3.9) holds with strict inequalities; see [**FreLagPay21**].

We will start by showing that the two forms of Pólya's conjecture, the bounds on the eigenvalue counting functions and the bounds on eigenvalues, are in fact equivalent.

Lemma 3.3.15. The inequalities (3.3.8) hold for all $\lambda \geq 0$ if and only if the inequalities (3.3.9) hold for all $k \geq 1$.

Proof. Obviously, the Dirichlet and Neumann cases can be treated independently. We will give the proof in the Dirichlet case only and will leave the Neumann case as an exercise. First, assume that (3.3.8) holds. Substitute, for any $k \geq 1$, $\lambda = \lambda_k$ into (3.3.8), and note that $\mathcal{N}^{\mathrm{D}}(\lambda_k) \geq k$. Then we have

$$k \leq \mathcal{N}^{\mathrm{D}}(\lambda_k) \leq C_d \operatorname{Vol}_d(\Omega) \lambda_k^{\frac{d}{2}},$$

and the second inequality (3.3.9) follows by raising both sides to the power $\frac{2}{d}$. Thus, (3.3.8) implies (3.3.9).

Assume now that (3.3.9) holds for all $k \geq 1$. We will prove (3.3.8) by induction in the intervals of the nonnegative λ-axis between consecutive distinct Dirichlet eigenvalues. To start with, as $\lambda_1 > 0$, we automatically get

$$0 = \mathcal{N}^{\mathrm{D}}(\lambda) \le C_d \operatorname{Vol}_d(\Omega)\lambda^{\frac{d}{2}} \qquad \text{for } \lambda \in [0, \lambda_1).$$

Assume now additionally that (3.3.8) holds for $\lambda \in [0, \lambda_k)$ with some $k \ge 1$. Let $\lambda_k = \cdots = \lambda_{k+m} < \lambda_{k+m+1}$ be a Dirichlet eigenvalue of multiplicity $m + 1$, where $m \ge 0$. Then by (3.3.9),

$$(3.3.10) \qquad \lambda_k = \cdots = \lambda_{k+m} \ge \left(\frac{1}{C_d \operatorname{Vol}_d(\Omega)}\right)^{\frac{2}{d}} (k + m)^{\frac{2}{d}}.$$

Moreover,

$$\mathcal{N}^{\mathrm{D}}(\lambda) = k + m \qquad \text{for } \lambda \in [\lambda_k, \lambda_{k+m+1}),$$

giving, on account of (3.3.10),

$$\mathcal{N}^{\mathrm{D}}(\lambda) \le C_d \operatorname{Vol}_d(\Omega)\lambda_k^{\frac{d}{2}} \le C_d \operatorname{Vol}_d(\Omega)\lambda^{\frac{d}{2}} \qquad \text{for } \lambda \in [\lambda_k, \lambda_{k+m+1}).$$

This completes the induction step; therefore (3.3.9) implies (3.3.8). $\qquad\square$

■ **Exercise 3.3.16.**

  (i) Prove that the original inequalities (3.3.8) are equivalent to their analogues for the left-continuous counting functions $\widetilde{\mathcal{N}}^{\mathrm{D}}(\lambda)$ and $\widetilde{\mathcal{N}}^{\mathrm{N}}(\lambda)$; see (3.2.8).

  (ii) Prove Lemma 3.3.15 in the Neumann case. You may find it easier to work with $\widetilde{\mathcal{N}}^{\mathrm{N}}(\lambda)$ instead of $\mathcal{N}^{\mathrm{N}}(\lambda)$ and use the result of part (i) at the end.

In a paper [**Pól61**] written a few years after stating his conjecture, G. Pólya proved Conjecture 3.3.14 for any *tiling domain* $\Omega \subset \mathbb{R}^d$ — that is, a domain such that the whole space $\mathbb{R}^d$ is an almost disjoint union of an infinite number of nonintersecting copies (shifted and possibly rotated) of Ω, with some additional restrictions in the Neumann case (these restrictions were later removed in [**Kel66**]). We emphasise that Pólya's conjecture (Conjecture 3.3.14) still remains open in full generality.

Theorem 3.3.17 (Pólya conjecture holds for tiling domains). Let $\Omega \subset \mathbb{R}^d$ be a tiling domain. Then the inequalities (3.3.9) hold for any $k \ge 1$.

Proof. We present Pólya's proof in the Dirichlet case and refer to [**Kel66**] for the Neumann case.

Suppose that $\Omega \subset \mathbb{R}^d$ is a tiling domain; somewhat abusing notation, we will denote its shifted (and possibly rotated) nonintersecting copies by the same symbol. Let also Ω_h denote a copy of Ω scaled with a factor $h > 0$. Obviously, if Ω tiles the space, so does Ω_h (for a fixed h); also,

$$\mathrm{Vol}_d(\Omega_h) = h^d \, \mathrm{Vol}_d(\Omega).$$

Fix for the moment $h > 0$ and some tiling of $\mathbb{R}^d$ by Ω_h. Let K be a unit cube, let

$$\boldsymbol{\Omega}_h := \bigsqcup_{\Omega_h \subset K} \Omega_h$$

be a disjoint union of copies of Ω_h fully inside K, and let

$$M_h := \#\{\Omega_h \subset K\}$$

be the number of such copies.

By the Dirichlet domain monotonicity and Dirichlet–Neumann bracketing, we have

$$\lambda_\ell(K) \leq \lambda_\ell(\boldsymbol{\Omega}_h)$$

for any $\ell \in \mathbb{N}$. Fix now $k \in \mathbb{N}$, and choose $\ell = kM_h$. As $\boldsymbol{\Omega}_h$ is a disjoint union of M_h copies of Ω_h, we have

$$\lambda_\ell(\boldsymbol{\Omega}_h) = \lambda_{kM_h}(\boldsymbol{\Omega}_h) = \lambda_k(\Omega_h) = \frac{1}{h^2}\lambda_k(\Omega),$$

and so

(3.3.11) $$h^2 \lambda_{kM_h}(K) \leq \lambda_k(\Omega).$$

We now take the limit as $h \to 0^+$, noting two limiting identities. Firstly, we have

$$\lim_{h \to 0^+} M_h h^d \, \mathrm{Vol}_d(\Omega) = \lim_{h \to 0^+} M_h \, \mathrm{Vol}_d(\Omega_h) = \mathrm{Vol}_d(K) = 1.$$

Secondly, by the one-term Weyl law for the eigenvalues of the cube,

$$\lim_{h \to 0^+} \frac{\lambda_{kM_h}(K)}{(kM_h)^{\frac{2}{d}}} = \left(\frac{1}{C_d \, \mathrm{Vol}_d(K)} \right)^{\frac{2}{d}} = \frac{1}{C_d^{\frac{2}{d}}}.$$

Passing now to the limit $h \to 0^+$ in the left-hand side of (3.3.11) and using the two limiting identities above, we obtain

$$\lambda_k(\Omega) \geq \lim_{h \to 0^+} h^2 \lambda_{kM_h}(K) = \frac{1}{C_d^{\frac{2}{d}}} \lim_{h \to 0^+} h^2 (kM_h)^{\frac{2}{d}}$$

$$= \frac{k^{\frac{2}{d}}}{(C_d \, \mathrm{Vol}_d(\Omega))^{\frac{2}{d}}} \lim_{h \to 0^+} h^2 (h^{-d})^{\frac{2}{d}} = \frac{k^{\frac{2}{d}}}{(C_d \, \mathrm{Vol}_d(\Omega))^{\frac{2}{d}}},$$

proving the second inequality in (3.3.9). $\square$

▢ **Numerical Exercise 3.3.18.** Use any software capable of finding zeros of Bessel functions and their derivatives to verify that Pólya conjecture holds for the first thousand eigenvalues of the unit disk.

Remark 3.3.19 (Pólya's conjecture for disks and balls). We note that Pólya's conjecture for the planar disk and, in the Dirichlet case, for balls in $\mathbb{R}^d$, $d \geq 3$, has been recently proved in [**FilLPS23**], thus making the disk the first non-tiling planar domain for which it is known. The proofs in [**FilLPS23**] are based on relations between the Dirichlet and Neumann eigenvalue counting functions for the balls and some lattice counting problems and, in the Neumann case for the disk, is partially computer-assisted.

We cite the following result which in a sense complements Theorem 3.3.17.

Theorem 3.3.20 ([**FilLPS23**, Theorem 1.8]). *Let* $\Omega \subset \mathbb{R}^d$ *be a domain for which either the Dirichlet or the Neumann Pólya conjecture holds, and let* Ω' *be a domain which tiles* Ω. *Then the same Pólya conjecture also holds for* Ω'.

Proof. Assume that Ω can be tiled by $M \geq 2$ congruent copies of Ω', so that $\mathrm{Vol}_d(\Omega) = M \, \mathrm{Vol}_d(\Omega')$. We have, by Dirichlet–Neumann bracketing and since the eigenvalues of all the congruent copies coincide with those of Ω',

$$M\mathcal{N}^{\mathrm{D}}_{\Omega'}(\lambda) \leq \mathcal{N}^{\mathrm{D}}_{\Omega}(\lambda) \leq \mathcal{N}^{\mathrm{N}}_{\Omega}(\lambda) \leq M\mathcal{N}^{\mathrm{N}}_{\Omega'}(\lambda).$$

Assuming now (3.3.8) for all $\lambda \geq 0$, we get

$$M\mathcal{N}^{\mathrm{D}}_{\Omega'}(\lambda) \leq C_d \, \mathrm{Vol}_d(\Omega)\lambda^d = C_d M \, \mathrm{Vol}_d(\Omega')\lambda^d \leq M\mathcal{N}^{\mathrm{N}}_{\Omega'}(\lambda),$$

and the result follows by cancelling M.   □

Theorem 3.3.20 and the validity of Pólya's conjecture for the disk imply that Pólya's conjecture is also valid for planar sectors of an aperture $2\pi/n$, $n \in \mathbb{N}$ [**FilLPS23**]. A more complicated argument also shows it to be true for sectors of any aperture.

3.3.4. The Berezin–Li–Yau inequalities. Using the inequality for the sum of the first k Dirichlet eigenvalues, which originated in the studies of the Schrödinger operator, one can deduce slightly weakened (in comparison to Pólya's conjecture) bounds for the Dirichlet eigenvalues which are *always* true. Our exposition here follows [**LieLos97**, §12.11]; see also [**Nam21**, §5.1].

Theorem 3.3.21 (The Berezin–Li–Yau inequality [**Ber72**], [**LiYau83**]).
Let $\Omega \subset \mathbb{R}^d$ be a bounded domain. Then its Dirichlet eigenvalues $\lambda_m = \lambda_m^{\mathrm{D}}(\Omega)$ satisfy

$$(3.3.12) \qquad \sum_{m=1}^{k} \lambda_m \geq \frac{d}{d+2} \frac{k^{1+\frac{2}{d}}}{(C_d \operatorname{Vol}_d(\Omega))^{\frac{2}{d}}}$$

for all $k \in \mathbb{N}$.

An immediate consequence of Theorem 3.3.21, obtained from (3.3.12)
by using $\lambda_k \geq \frac{1}{k} \sum_{m=1}^{k} \lambda_m$, is

Corollary 3.3.22. For any $k \in \mathbb{N}$,

$$\lambda_k \geq \frac{d}{d+2} \left(\frac{k}{C_d \operatorname{Vol}_d(\Omega)} \right)^{\frac{2}{d}}.$$

In other words, Polya's conjecture for Dirichlet eigenvalues, that is, the
right inequality in (3.3.9), holds in a weakened form with an additional
factor $\frac{d}{d+2}$.

Before proceeding to the proof of Theorem 3.3.21, we introduce the
following notation, which we will also need further on.

Notation 3.3.23. Let $F : \mathcal{O} \to \mathbb{R}$ by a real-valued function defined on an
open set $\mathcal{O} \subset \mathbb{R}^d$. We set, for $t \in \mathbb{R}$,

$$\mathcal{L}_F(t) := \{y \in \mathcal{O} : F(y) = t\},$$

to denote its *level sets*,

$$\mathcal{U}_F(t) := \{y \in \mathcal{O} : F(y) < t\},$$

to denote its *sublevel sets*, and

$$\mathcal{V}_F(t) := \{y \in \mathcal{O} : F(y) > t\},$$

to denote its *superlevel sets*. We additionally denote the volume of a sublevel
set by

$$U_F(t) := \operatorname{Vol}_d \left(\mathcal{U}_F(t) \right),$$

and we denote the volume of a superlevel set by

$$V_F(t) := \operatorname{Vol}_d \left(\mathcal{V}_F(t) \right).$$

■ **Exercise 3.3.24.** Let $\mathcal{O} = (-2, 2) \times (-1, 1) \subset \mathbb{R}^2$, and let $F : \mathcal{O} \to \mathbb{R}$ be
defined by $F(x, y) = \sqrt{x^2 + y^2}$. Plot the graphs of $U_F(t)$ and $V_F(t)$.

We will also need the result of the following.

Proposition 3.3.25 (A variant of the Bathtub principle [**LieLos97**, Theorem 1.14]). Let $f : \mathbb{R}^d \to \mathbb{R}$ be a measurable function such that for all $t \in \mathbb{R}$, $\mathrm{Vol}_d(\mathcal{L}_f(t)) = 0$ and $U_f(t)$ is finite, and let $g : \mathbb{R}^d \to [0,1] \in L^1(\mathbb{R}^d)$. Set

$$(3.3.13) \qquad A := \int_{\mathbb{R}^d} g(\xi)\,\mathrm{d}\xi, \qquad s := \sup\{t : U_f(t) \le A\}.$$

Then

$$(3.3.14) \qquad \int_{\mathbb{R}^d} f(\xi) g(\xi)\,\mathrm{d}\xi \ge \int_{U_f(s)} f(\xi)\,\mathrm{d}\xi.$$

Proof of Proposition 3.3.25. Let $h(\xi) := \chi_{U_f(s)}(\xi)$ be the characteristic function of the set $U_f(s)$. Proving (3.3.14) is equivalent to showing that for any g satisfying the conditions of the proposition we have

$$\int_{\mathbb{R}^d} f(\xi)\,(h(\xi) - g(\xi))\,\mathrm{d}\xi \le 0.$$

Since $\mathrm{Vol}_d(\mathcal{L}_f(s)) = 0$, we can rewrite the integral above as

$$(3.3.15) \qquad \int_{\mathbb{R}^d} f(\xi)\,(h(\xi) - g(\xi))\,\mathrm{d}\xi = \left(\int_{\mathcal{V}_f(s)} + \int_{U_f(s)} \right) f(\xi)\,(h(\xi) - g(\xi))\,\mathrm{d}\xi.$$

Note that when $\xi \in \mathcal{V}_f(s)$ we have $f(\xi) \ge s$ and $h(\xi) - g(\xi) = -g(\xi) \le 0$. Similarly, for $\xi \in U_f(s)$ we have $f(\xi) \le s$ and $h(\xi) - g(\xi) = 1 - g(\xi) \ge 0$. Therefore, replacing $f(\xi)$ by s in both integrals in the right-hand side of (3.3.15) leads to an upper bound, yielding

$$\int_{\mathbb{R}^d} f(\xi)\,(h(\xi) - g(\xi))\,\mathrm{d}\xi \le s \int_{\mathbb{R}^d} (h(\xi) - g(\xi))\,\mathrm{d}\xi = s\,(U_f(s) - A) = 0$$

by (3.3.13), which completes the proof. $\qquad\qquad\square$

Proof of Theorem 3.3.21. Let $u_m = u_m^{\mathrm{D}}$ be an orthonormal sequence of Dirichlet eigenfunctions corresponding to the eigenvalues λ_m, $m \in \mathbb{N}$. Then we get

$$(3.3.16) \qquad \sum_{m=1}^{k} \lambda_m = \sum_{m=1}^{k} \|\nabla u_m\|_{L^2(\Omega)}^2 = \sum_{m=1}^{k} \|\xi(\mathcal{F}u_m)\|_{L^2(\mathbb{R}^d)}^2,$$

where $(\mathcal{F}u_m)(\xi)$ is the Fourier transform (see (2.1.3)) of u_m extended by zero onto the whole $\mathbb{R}^d$. The first equality in (3.3.16) follows from the variational principle, and the second one from Plancherel's theorem.

Denote

$$(3.3.17) \quad f(\xi) := \frac{\mathrm{Vol}_d(\Omega)}{(2\pi)^d} |\xi|^2, \qquad g(\xi) := \frac{(2\pi)^d}{\mathrm{Vol}_d(\Omega)} \sum_{m=1}^{k} |(\mathcal{F}u_m)(\xi)|^2 \geq 0.$$

Then (3.3.16) may be rewritten as

$$(3.3.18) \qquad \sum_{m=1}^{k} \lambda_m = \int_{\mathbb{R}^d} f(\xi) g(\xi) \, \mathrm{d}\xi.$$

We want to estimate the integral in the right-hand side of (3.3.18) using (3.3.14), but we need to show first that the function $g(\xi)$ defined by (3.3.17) satisfies the conditions of Proposition 3.3.25. By the definition of the Fourier transform and using the fact that $\{u_m\}$ is an orthonormal basis in $L^2(\Omega)$, we have

$$g(\xi) := \frac{(2\pi)^d}{\mathrm{Vol}_d(\Omega)} \sum_{m=1}^{k} \left| \left((2\pi)^{-\frac{d}{2}} e^{-\mathrm{i}\langle x,\xi\rangle}, u_m \right)_{L^2(\Omega)} \right|^2$$
$$\leq \frac{1}{\mathrm{Vol}_d(\Omega)} \left\| e^{-\mathrm{i}\langle x,\xi\rangle} \right\|_{L^2(\Omega)}^2 = 1$$

by Bessel's inequality, so that (3.3.14) is indeed applicable. By Plancherel's theorem $\|\mathcal{F}u_m\|_{L^2(\mathbb{R}^d)}^2 = \|u_m\|_{L^2(\Omega)}^2 = 1$, and we therefore have

$$A = \int_{\mathbb{R}^d} g(\xi) \, \mathrm{d}\xi = \frac{(2\pi)^d}{\mathrm{Vol}_d(\Omega)} \sum_{m=1}^{k} \int_{\mathbb{R}^d} |(\mathcal{F}u_m)(\xi)|^2 \, \mathrm{d}\xi = \frac{(2\pi)^d k}{\mathrm{Vol}_d(\Omega)}.$$

Further on, since $\mathcal{U}_f(t)$ for $t > 0$ is the ball of radius $r_t = \left(\frac{(2\pi)^d t}{\mathrm{Vol}_d(\Omega)} \right)^{\frac{1}{2}}$, the constant s appearing in (3.3.13) satisfies

$$\mathrm{Vol}_d \left(B_{r_s}^d \right) = (r_s)^d \omega_d = A,$$

with ω_d given by (B.1.1), from where

$$(3.3.19) \qquad r_s = 2\pi \left(\frac{k}{\omega_d \, \mathrm{Vol}_d(\Omega)} \right)^{\frac{1}{d}}.$$

We now apply (3.3.14) to the right-hand side of (3.3.18):

$$\int_{\mathbb{R}^d} f(\xi)g(\xi)\, d\xi \geq \frac{\mathrm{Vol}_d(\Omega)}{(2\pi)^d}\int_{B^d_{r_s}}|\xi|^2\, d\xi = \frac{\mathrm{Vol}_d(\Omega)r_s^{d+2}}{(2\pi)^d}\int_{\mathbb{B}^d}|\xi'|^2\, d\xi'$$

(3.3.20)
$$= \frac{\mathrm{Vol}_d(\Omega)r_s^{d+2}}{(2\pi)^d}\int_{\mathbb{S}^{d-1}}\int_0^1 \rho^{1+d}\, d\rho\, d\kappa$$

$$= \frac{\mathrm{Vol}_d(\Omega)r_s^{d+2}}{(2\pi)^d}\frac{\sigma_{d-1}}{d+2},$$

where we have used the changes of variables $\xi = r_s\xi'$ and $\xi' = \rho\kappa$, $\rho \in [0,1)$, $\kappa \in \mathbb{S}^{d-1}$, and have used σ_{d-1} to denote the volume of $\mathbb{S}^{d-1}$; see (B.1.2). Substituting (3.3.19) into (3.3.20) and taking into account $\frac{\sigma_{d-1}}{\omega_d} = d$, we obtain

$$\int_{\mathbb{R}^d} f(\xi)g(\xi)\, d\xi \geq \frac{4\pi^2 k^{1+\frac{2}{d}}}{(\mathrm{Vol}_d(\Omega)\omega_d)^{\frac{2}{d}}}\cdot\frac{d}{d+2}.$$

Finally, recalling the definition (3.3.2) of the Weyl constant C_d and using (3.3.18), we rewrite the last inequality as (3.3.12). □

Remark 3.3.26. The approach of Theorem 3.3.21 can be adapted to prove similar inequalities for the eigenvalues of the Neumann Laplacian; see [**Krö92**]. In this case

$$\sum_{m=1}^{k}\lambda_m^{\mathrm{N}}(\Omega) \leq \frac{d}{d+2}\frac{k^{1+\frac{2}{d}}}{(C_d\,\mathrm{Vol}_d(\Omega))^{\frac{2}{d}}},$$

and

$$\lambda_{k+1}^{\mathrm{N}}(\Omega) \leq \left(\frac{d+2}{2}\right)^{\frac{2}{d}}\left(\frac{k}{C_d\,\mathrm{Vol}_d(\Omega)}\right)^{\frac{2}{d}}, \qquad k \in \mathbb{N}.$$

For further details and other applications of Berezin–Li–Yau inequalities, including their relation to the *Lieb–Thirring inequalities* and to the asymptotics of the *Riesz means*, see [**Lap97**], [**Lie73**], [**LapSaf96**], [**LapWei00**], and [**FraLapWei23**].

Chapter 4

Nodal geometry
of eigenfunctions

Marie-Sophie
Germain
(1776–1831)

In this chapter, we present the nodal geometry of eigenfunctions. We prove Courant's nodal domain theorem and show that the nodal set of an eigenfunction is dense on the wave-length scale. We also obtain a lower bound for the size of the nodal set in dimension two and give an overview of results concerning Yau's conjecture on the volume of nodal sets of Laplace–Beltrami eigenfunctions. In particular, we discuss Donnelly–Fefferman's estimate on the doubling index of eigenfunctions and its relation to the nodal volume. We also outline the proof, following the work of Logunov and Malinnikova, of a polynomial upper bound on the nodal volume.

Richard
Courant
(1888–1972)

4.1. Courant's nodal domain theorem

4.1.1. Nodal domains and nodal sets. Let $\Omega \subset \mathbb{R}^d$ be a bounded domain, and let u be an eigenfunction of either the Dirichlet or Neumann Laplacian. Consider the set

$$\mathcal{Z}_u := \{x \in \Omega : u(x) = 0\},$$

called the *nodal set* of u. A connected component of $\Omega \setminus \mathcal{Z}_u$ is called a *nodal domain* of u. Similarly, one defines the nodal domains and nodal sets for Laplace–Beltrami eigenfunctions on a Riemannian manifold. For an illustration, the nodal set and the nodal domains of some particular

Dirichlet and Neumann eigenfunctions of a unit square are shown in Figure 4.1. See also Figure 4.2 for the nodal set and the nodal domains of a Laplace–Beltrami eigenfunction on the sphere.

Figure 4.1. The nodal sets and the nodal domains of the eigenfunction $u^D = \frac{1}{\sqrt{5}}(\sin(2\pi x)\sin(9\pi y) - \sin(9\pi x)\sin(2\pi y) - \sin(6\pi x)\sin(7\pi y) + 2\sin(7\pi x)\sin(6\pi y))$ corresponding to the Dirichlet eigenvalue $\lambda^D = 85\pi^2$ of the unit square $[0,1]^2$ (left; cf. Figure 1.2) and of the eigenfunction $u^N = \frac{1}{\sqrt{5}}(\cos(6\pi x)\cos(43\pi y) - \cos(11\pi x)\cos(42\pi y) + \cos(38\pi x)\sin(21\pi y) + 2\cos(27\pi x)\sin(34\pi y))$ corresponding to the Neumann eigenvalue $\lambda^N = 1885\pi^2$ (right).

Figure 4.2. The nodal set and the nodal domains of an eigenfunction of the Laplace–Beltrami operator $-\Delta_{\mathbb{S}^2}$ on the round sphere corresponding to the eigenvalue $\lambda = 17 \times 18$.

⊡ **Numerical Exercise 4.1.1.** Plot your own analogue of Figure 4.1 for some eigenfunctions of a Laplacian, computed either using separation of variables or numerically, on a domain of your choice.

The nodal sets and the nodal domains are important geometric characteristics which can be used to measure "complexity" of eigenfunctions. Their investigation goes back to E. Chladni's experiments with vibrating plates at the end of the 18th century to the beginning of the 19th century (while Chladni's figures do not exactly correspond to nodal sets of Laplace eigenfunctions, they illustrate the same phenomenon).

Ernst Florens Friedrich **Chladni**
(1756–1827)

We refer to [**Stö07**] for a fascinating story about Chladni's work, his meeting with Napoleon, and a prize won by Sophie Germain; see also Figure 4.3.

In what follows we assume for simplicity that Ω is a Euclidean domain, though essentially all the results hold for Riemannian manifolds, either closed or with boundary. Where necessary we will indicate adjustments that are needed in the Riemannian case.

4.1.2. Courant's theorem. Let us start with the following simple one-dimensional example.

Example 4.1.2. Consider the Dirichlet problem on $\Omega = (0,\ell)$. Its eigenfunctions are given by

$$u_k(x) = \sin\frac{\pi kx}{\ell},$$

with eigenvalues $\lambda_k = \frac{\pi^2 k^2}{\ell^2}$, for $k \in \mathbb{N}$. Therefore, $\mathcal{Z}_{u_k}$ consists of $k-1$ zeros equidistributed on $(0,\ell)$, and u_k has k nodal domains.

Figure 4.3. The 19th century engraving showing how vibrations are excited in a Chladni plate with a violin bow to create the sand figures of nodal lines.

■ **Exercise 4.1.3.** Consider the Sturm–Liouville eigenvalue problem on the interval $(a, b) \subset \mathbb{R}$,

$$-(pu')' + qu = \lambda wu \qquad \text{in } (a, b),$$
$$u(a) = u(b) = 0,$$

where $p, q, w \in C^2([a, b])$ and p, w are positive functions. The eigenvalues form a sequence $\lambda_1 \leq \lambda_2 \leq \cdots \nearrow +\infty$. Using the Sturm oscillation theorem, prove that the number of nodal domains of an eigenfunction u_k corresponding to the eigenvalue λ_k is equal to k. For a solution, see [**Shu20**, Chapter 3].

Example 4.1.2 shows that in one dimension, the kth eigenfunction has precisely k nodal domains. One can easily check using Exercise 1.1.9 that this is no longer true for the square. However, the following fundamental theorem due to R. Courant [**Cou23**] holds in all dimensions.

Theorem 4.1.4 (Courant's nodal domain theorem). Let $\Omega \subset \mathbb{R}^d$ be a bounded domain. Suppose that u is a Dirichlet eigenfunction on Ω corresponding to an eigenvalue λ_k. Then u has at most k nodal domains.

Remark 4.1.5. We state Courant's theorem for the Dirichlet boundary conditions for the sake of simplicity. Under additional assumptions on the regularity of $\partial\Omega$, the argument presented below can be generalised to other selfadjoint boundary conditions, such as Neumann, Robin, or Zaremba.

4.1.3. Restriction of an eigenfunction to a nodal domain. A non-trivial technical step in the proof of Courant's theorem is

Theorem 4.1.6. Let $u \in H_0^1(\Omega) \cap C(\Omega)$, and let $\Omega_1 \subset \Omega$ be a nodal domain of u. Then, $u|_{\Omega_1} \in H_0^1(\Omega_1)$.

Theorem 4.1.6 immediately follows from Lemma 4.1.7 below under the additional assumptions that $u \in C(\overline{\Omega}) \cap C^1(\Omega)$ and $u = 0$ on $\partial\Omega$. Note that these assumptions are satisfied on Euclidean domains with Lipschitz boundaries by Theorem 2.2.1, part (iv), and on closed manifolds by Theorem 2.2.17.

Lemma 4.1.7. Let $\Omega \subset \mathbb{R}^d$ be a bounded domain. Suppose that $u \in C(\overline{\Omega}) \cap C^1(\Omega)$ and $u = 0$ on $\partial\Omega$. Then $u \in H_0^1(\Omega)$.

Proof. We follow the argument in [**Buh16**]. Let $h : \mathbb{R} \to \mathbb{R}$ be a smooth monotone function such that $h(t) = 0$ on $(-1, 1)$ and $h(t) = t$ if $|t| > 2$. Set $h_\varepsilon(t) := \varepsilon h(t/\varepsilon)$. The function $v_\varepsilon := h_\varepsilon \circ u$ is an element of $C_0^1(\Omega)$, due to the assumptions on u. We leave it as an exercise for the reader to show that $v_\varepsilon \to u$ in $H^1(\Omega)$ as ε tends to zero, which implies $u \in H_0^1(\Omega)$. $\square$

The proof of Theorem 4.1.6 in the general case uses some fine properties of Sobolev spaces which are discussed below.[2] First, let us recall the following notions.

Definition 4.1.8 (Capacity). Let $E \subset \mathbb{R}^d$. The *capacity* of E is the number
$$\mathrm{cap}(E) := \inf_{u \in A(E)} \|u\|_{H^1(\mathbb{R}^d)}^2,$$
where
$$A(E) = \{u \in C_0^1(\mathbb{R}^d) \mid u \geq 1 \text{ in a neighbourhood of } E\}.$$

---

[2]We thank Dorin Bucur for outlining this argument.

The capacity is an outer measure and it may be used to refine the notion of zero measure, since $\mathrm{cap}(E) = 0$ implies that the Lebesgue measure of E vanishes. Note that in $\mathbb{R}^2$, a point has both zero measure and zero capacity, whereas a segment has zero measure and a positive capacity; cf. Remark 3.2.15.

Definition 4.1.9 (Quasi-everywhere). A property P holds *quasi-everywhere* in $X \subset \mathbb{R}^d$ if there exists $E \subset X$ such that $\mathrm{cap}(E) = 0$ and P holds in $X \setminus E$.

Definition 4.1.10 (Quasi-continuity). A function $u : \mathbb{R}^d \to \mathbb{R}$ is *quasi-continuous* if for all $\varepsilon > 0$ there exists $E \subset \mathbb{R}^d$ such that $\mathrm{cap}(E) < \varepsilon$ and the restriction $u|_{\mathbb{R}^d \setminus E}$ is a continuous function.

One can show that any function from the Sobolev space $H^1(\mathbb{R}^d)$ has a quasi-continuous representative. Further, the above notions are useful for characterising the space $H_0^1(\Omega)$ for an open subset $\Omega \subset \mathbb{R}^d$ or for defining the restriction of an $H^1(\mathbb{R}^d)$ function on an arbitrary subset of $\mathbb{R}^d$.

Theorem 4.1.11 ([**HeiKilMar93**, Theorem 4.5], [**Kin21**, Corollary 4.31]; see also [**Hed81**] and references therein). Let Ω be an open subset of $\mathbb{R}^d$. Then the function u belongs to the Sobolev space $H_0^1(\Omega)$ if and only if there exists a quasi-continuous function $v \in H^1(\mathbb{R}^d)$ such that $v(x) = 0$ quasi-everywhere outside Ω and $v(x) = u(x)$ almost everywhere in Ω.

The quasi-continuous representatives are unique in the following sense.

Theorem 4.1.12 ([**HeiKilMar93**, Theorem 4.12], [**Kin21**, Theorem 4.23]). Let $U \subset \mathbb{R}^d$ be open. Let v_1, v_2 be quasi-continuous functions defined in U. If $v_1 = v_2$ almost everywhere, then $v_1 = v_2$ quasi-everywhere.

We can now apply these notions in order to prove Theorem 4.1.6.

Proof of Theorem 4.1.6. Since $u \in H_0^1(\Omega)$, we can find v as in Theorem 4.1.11. Let

$$F = \{x \in \Omega^c : v(x) \neq 0\} \cup \{x \in \Omega : v(x) \neq u(x)\},$$

where $\Omega^c := \mathbb{R}^d \setminus \Omega$. Since $u \in C(\Omega)$ we can deduce from Theorem 4.1.12 that $\mathrm{cap}(F) = 0$. Let

$$w(x) := \begin{cases} v(x), & \text{if } x \in \Omega_1, \\ 0, & \text{if } x \notin \Omega_1. \end{cases}$$

We will show that w is quasi-continuous.

Let $\varepsilon > 0$ be given. There exists a set E such that $\mathrm{cap}(E) < \varepsilon$ and $v|_{E^c}$ is a continuous function. Consider the function w restricted to $(E \cup F)^c$. We pick an arbitrary converging sequence $x_k \to x_0$, where x_k and x_0 are points in $(E \cup F)^c$. Consider the possible cases:

- If $x_k \in \Omega_1 \cap (E \cup F)^c$ and $x_0 \in \Omega_1 \cap (E \cup F)^c$, then $w(x_k) = v(x_k)$, $w(x_0) = v(x_0)$, and the convergence $w(x_k) \to w(x_0)$ follows from the continuity of $v|_{E^c}$.

- If $x_k \in \Omega_1 \cap (E \cup F)^c$ and $x_0 \in (\partial \Omega_1) \cap \Omega \cap (E \cup F)^c$, then $w(x_k) = v(x_k)$, $w(x_0) = 0$, and $v(x_0) = u(x_0) = 0$. Thus, the continuity of $v|_{E^c}$ implies the convergence $w(x_k) \to w(x_0)$.

- If $x_k \in \Omega_1 \cap (E \cup F)^c$ and $x_0 \in (\partial \Omega_1) \cap (\partial \Omega) \cap (E \cup F)^c$, then $w(x_k) = v(x_k)$, $w(x_0) = 0$, and $v(x_0) = 0$. Again, we have as above $w(x_k) \to w(x_0)$.

- If $x_k \in \Omega_1^c \cap (E \cup F)^c$ and $x_0 \in \Omega_1^c \cap (E \cup F)^c$, then $w(x_k) = 0$ and $w(x_0) = 0$, and trivially $w(x_k) \to w(x_0)$.

It follows that $w|_{(E \cup F)^c}$ is continuous. We have found a quasi-continuous function w such that $w = 0$ everywhere in Ω_1^c and $w = u$ almost everywhere in Ω_1. Hence, by Theorem 4.1.11, $u \in H_0^1(\Omega_1)$.   □

4.1.4. Proof of Courant's theorem.
Below we give two slightly different proofs of Courant's theorem; one uses the strict domain monotonicity (see Proposition 3.2.2) and the other one directly relies on the unique continuation property of eigenfunctions (see also Remark 4.1.14). Since the latter is needed for the proof of the strict domain monotonicity, in the end the two arguments use the same set of ideas.

First proof of Theorem 4.1.4. Let u be an eigenfunction corresponding to an eigenvalue $\lambda = \lambda(\Omega)$ and suppose it has at least $k + 1$ nodal domains $\Omega_1, \ldots, \Omega_k, \Omega_{k+1}, \ldots$. To prove the theorem, if suffices to show that $\lambda > \lambda_k$. Set

$$\psi_i(x) = \begin{cases} u(x) & \text{if } x \in \Omega_i, \\ 0 & \text{otherwise.} \end{cases}$$

By Theorem 4.1.6, ψ_i is an element of $H_0^1(\Omega_i)$. Let $\mathcal{L} = \mathrm{Span}\{\psi_1, \ldots, \psi_k\}$. Since $\psi \in H_0^1(\Omega_i)$ and $-\Delta \psi_i = \lambda \psi_i$ in Ω_i, we deduce that ψ_i is a Dirichlet eigenfunction in Ω_i with the eigenvalue λ. Therefore,

$$(4.1.1) \qquad R[\psi_i] = \frac{\|\nabla \psi_i\|_{L^2(\Omega)}^2}{\|\psi_i\|_{L^2(\Omega)}^2} = \lambda.$$

Set

$$\widetilde{\Omega} = \bigcup_{i=1}^{k} \Omega_i;$$

then for any linear combination $\psi = \sum_{i=1}^{k} c_i \psi_i \in \mathcal{L}$, we have

$$\|\nabla \psi\|_{L^2(\widetilde{\Omega})}^2 = \sum_{i=1}^{k} |c_i|^2 \|\nabla \psi_i\|_{L^2(\Omega_i)}^2 = \lambda \sum_{i=1}^{k} |c_i|^2 \|\psi_i\|_{L^2(\Omega_i)}^2 = \lambda \|\psi\|_{L^2(\widetilde{\Omega})}^2.$$

Hence, $R_{\widetilde{\Omega}}[\psi] = \lambda$, and by the variational principle $\lambda \geq \lambda_k(\widetilde{\Omega})$. However, since there are at least $k+1$ nodal domains, $\Omega \setminus \widetilde{\Omega}$ contains a nonempty open set Ω_{k+1}, and thus by strict domain monotonicity (Proposition 3.2.2),

$$\lambda \geq \lambda_k(\widetilde{\Omega}) > \lambda_k(\Omega),$$

which completes the first proof of Theorem 4.1.4. $\qquad\qquad\qquad\square$

Remark 4.1.13. We recall that since the variational principle for the Dirichlet Laplacian can be applied without any assumptions on the regularity of the boundary and since $\psi_i \in H_0^1(\Omega_i)$, we do not need to impose any smoothness conditions on $\partial \Omega_i$ for the validity of (4.1.1).

Second proof of Theorem 4.1.4. This proof is due to Å. Pleijel [**Ple56**]. We argue essentially in the same way as above until the last step. As before, with $\mathcal{L} = \mathrm{Span}\{\psi_1, \dots, \psi_k\} \subset H_0^1(\Omega)$, we have that for all $f \in \mathcal{L}$, $R[f] = \lambda$. Assume that $\lambda = \lambda_k$ and that u has at least $k+1$ nodal domains. Since $\dim \mathcal{L} = k$, we can choose $f \in \mathcal{L}$ such that f is orthogonal in $L^2(\Omega)$ to the first $k-1$ Dirichlet eigenfunctions $u_1, \dots, u_{k-1}$. Then (see Remark 3.1.21)

$$\lambda_k = R[f] = \frac{\sum_{i=k}^{\infty} \lambda_i f_i^2}{\sum_{i=k}^{\infty} f_i^2},$$

where $f_i = (f, u_i)_{L^2(\Omega)}$ are the coefficients in the expansion of f in the basis $\{u_i\}$. By Theorem 3.1.9, this equality implies that f is an eigenfunction corresponding to the eigenvalue λ_k. Hence, f is real analytic in Ω. But $f|_{\Omega_{k+1}} \equiv 0$ by construction. It follows that $f \equiv 0$ on all Ω, and we get a contradiction. Therefore, an eigenfunction corresponding to λ_k can have at most k nodal domains, which completes the second proof of Courant's theorem. $\qquad\qquad\qquad\square$

Remark 4.1.14. Some changes are needed in the above argument in order to prove Courant's theorem on Riemannian manifolds. Note that Laplace eigenfunctions on smooth Riemannian manifolds are smooth but not necessarily real analytic. In this case, in the last step of the proof above one should use N. Aronszajn's unique continuation principle; see [**Aro57**]. It implies that eigenfunctions of elliptic operators with smooth coefficients may vanish at a given point only to a finite order and, as a consequence, cannot vanish on an open set. Later on we will also discuss a quantitative version of the unique continuation principle; see Theorem 4.3.7 and Remark 4.3.19.

∎ **Exercise 4.1.15.** Deduce from the second proof of Theorem 4.1.4 that without using the unique continuation property one can prove a weaker version of Courant's bound with k replaced by $k + m(\lambda_k) - 1$, where $m(\lambda_k)$ is the multiplicity of the eigenvalue λ_k.

Let us also make a few historical remarks. The proof of Courant's theorem in the Riemannian setting appeared first in an influential paper by S.-Y. Cheng [**Che75**]. The argument relied on a claim regarding the regularity of the nodal set that was used to justify the application of Green's formula; cf. Remark 4.1.13. However, as was pointed out by Y. Colin de Verdière, the proof of this claim was incomplete in dimensions three and higher. A corrected proof of Courant's theorem was presented several years later by P. Bérard and D. Meyer in [**BérMey82**]. Cheng's claim regarding the regularity of nodal sets has been finally proved in [**HarSim89**] by R. Hardt and L. Simon. For Laplace–Beltrami eigenfunctions, their result can be stated as follows.

Theorem 4.1.16. Let u be an eigenfunction of the Laplacian on a smooth Riemannian manifold of dimension d. Then its nodal set decomposes into a regular part $\mathcal{Z}_u \cap \{|\nabla u| > 0\}$, which is a smooth $(d-1)$-dimensional submanifold having a finite $(d-1)$-dimensional volume, and a singular part $\mathcal{Z}_u \cap \{|\nabla u| = 0\}$, which is a closed countably $(d-2)$-rectifiable subset (see [**Fed14**, §3.2.14] for the definition) of the manifold.

Remark 4.1.17. In general, there is no nontrivial lower bound for the number of nodal domains. Antonie Stern proved in 1925 that for a square and for a round sphere, there exist eigenfunctions with two nodal domains, corresponding to eigenvalues lying arbitrarily high in the spectrum. We refer to [**BérHel14**] for a recent exposition of these results.

4.1.5. Properties of subharmonic and harmonic functions. In order to deduce several important corollaries from Courant's theorem we need to review some properties of subharmonic and harmonic functions.

Definition 4.1.18 (Subharmonic and harmonic functions). Let Ω be an open set. A function $u \in C^2(\Omega)$ is called *subharmonic* in Ω if $\Delta u \geq 0$ in Ω. If $\Delta u = 0$ in Ω, we say that u is *harmonic* in Ω.

In fact, the notion of subharmonicity can be extended to continuous functions using the inequality (4.1.4) below; see [**AxlBouWad01**, p. 224].

Example 4.1.19. Let u be a Laplace eigenfunction on some domain, corresponding to an eigenvalue $\lambda \geq 0$, and let Ω be a nodal domain of u such that $u|_\Omega < 0$. Then u is subharmonic in Ω, since $-\Delta u = \lambda u \leq 0$ in Ω.

■ **Exercise 4.1.20.** Prove that if h is a harmonic function, then $|h|^2$ is subharmonic.

Subharmonic and harmonic functions satisfy a mean value property and a maximum principle that we discuss below. Given $x \in \mathbb{R}^d$, let, as before, $S_{x,r} = S_{x,r}^{d-1}$ and $B_{x,r} = B_{x,r}^d$ be the sphere and the open ball of radius r centred at x.

Definition 4.1.21 (Means over spheres and balls). The *spherical mean* of a locally integrable function u at the point $x \in \mathbb{R}^d$ is the function

$$M_{u,x}(r) = \fint_{S_{x,r}} u := \frac{1}{\mathrm{Vol}_{d-1}(S_{x,r})} \int_{S_{x,r}} u(x) \, dS_r.$$

We will consider also the mean over a ball

$$A_{u,x}(r) = \fint_{B_{x,r}} u := \frac{1}{\mathrm{Vol}_d(B_{x,r})} \int_{B_{x,r}} u(x) \, dx.$$

For a function u defined on a domain $\Omega \subset \mathbb{R}^d$, we assume in Definition 4.1.21 that $x \in \Omega$ and that r is chosen small enough for $\overline{B_{x,r}} \subset \Omega$.

Lemma 4.1.22. The derivative of a spherical mean is given by

$$(4.1.2) \qquad M'_{u,x}(r) = \frac{1}{\mathrm{Vol}_{d-1}(S_{x,r})} \int_{B_{x,r}} \Delta u(y) \, dy.$$

Proof. This is a standard result, and we follow the proof of [**Shu20**, Theorem 6.1]. Let us rewrite the spherical mean as an average over a unit sphere. Let σ_{d-1} be the volume of a unit sphere given by (B.1.2), and set $z = \frac{y-x}{r}$. Then switching to the variable z yields

$$M_{u,x}(r) = \frac{1}{\sigma_{d-1} r^{d-1}} \int_{S_{x,r}} u(y) \, dS_r(y) = \frac{1}{\sigma_{d-1}} \int_{S_{0,1}} u(x + rz) \, dS_1(z).$$

Therefore,

$$M'_{u,x}(r) = \frac{1}{\sigma_{d-1}} \int\limits_{S_{0,1}} \frac{\mathrm{d}}{\mathrm{d}r} u(x+rz) \, \mathrm{d}S_1(z) = \frac{1}{\sigma_{d-1} r^{d-1}} \int\limits_{S_{x,r}} \partial_n u(y) \, \mathrm{d}S_r(y),$$

where ∂_n is the outward normal derivative. Here we used that z is the unit normal at $y \in S_{x,r}$ and made a reverse change of variables. Taking now $v \equiv 1$ and $\Omega = B_{x,r}$ in Green's formula (2.1.7), we get

$$M'_{u,x}(r) = \frac{1}{\mathrm{Vol}_{d-1}(S_{x,r})} \int\limits_{B_{x,r}} \Delta u(y) \, \mathrm{d}y,$$

which completes the proof of the lemma. $\qquad\square$

Corollary 4.1.23. Let u be a subharmonic function. Then $M_{u,x}(r)$ and $A_{u,x}(r)$ are monotone nondecreasing in r.

Proof. Indeed, by (4.1.2) and Definition 4.1.18 the derivative of $M_{u,x}(r)$ is nonnegative for a subharmonic u and vanishes if u is harmonic. Moreover, it is easy to check that $A_{u,x}$ is a weighted average of $M_{u,x}$; namely, in $\mathbb{R}^d$ we have

$$(4.1.3) \qquad\qquad A_{u,x}(r) = \frac{\sigma_{d-1}}{\omega_d} \int\limits_0^1 t^{d-1} M_{u,x}(tr) \, \mathrm{d}t,$$

where ω_d, σ_{d-1} are the volumes of the unit ball $\mathbb{B}^d$ and the unit sphere $\mathbb{S}^{d-1}$, respectively. Hence it follows that $A_{u,x}$ is monotone nondecreasing as well. $\qquad\square$

From Corollary 4.1.23, the fact that $M_{u,x}(r)$ tends to $u(x)$ as r tends to zero, and the identity (4.1.3), we readily deduce

Corollary 4.1.24 (The mean value inequality for subharmonic functions). Let u be a subharmonic function in B_R. Then

$$(4.1.4) \qquad\qquad u(x) \le A_{u,x}(r) \le M_{u,x}(r)$$

for all $0 < r < R$. Additionally, if u is harmonic, then the inequalities are replaced by equalities.

We are now in a position to prove

Theorem 4.1.25 (The maximum principle for subharmonic functions). Let $\Omega \subset \mathbb{R}^d$ be a domain and let $u \in C^2(\Omega)$ be a subharmonic function. Then u cannot attain a maximum in Ω unless it is constant.

Proof. Let $x_0 \in \Omega$ be such that $u(x_0) \geq u(x)$ for all $x \in \Omega$. Set $m = u(x_0)$ and consider the level set $Z := \mathcal{L}_u(m)$. We want to show that $Z = \Omega$. This follows from the fact that Z is both open and closed in Ω. Firstly, since u is continuous and $Z = u^{-1}(\{m\})$, it is immediate that Z is closed. Let us show that Z is also open. Indeed, let $y \in Z$ and choose $\rho > 0$ such that $B_{y,\rho} \subset \Omega$. Then, for all $0 < r < \rho$, we have that $M_{u,y}(r) \geq m$ by the mean value property. Therefore, $u|_{S_{y,r}} \equiv m$ for all $0 < r < \rho$ since m is the maximum. Thus, we get $u|_{B_{y,\rho}} \equiv m$, and so $B_{y,\rho} \subset Z$. It follows that Z is open, and since Ω is connected we have $Z = \Omega$. This completes the proof of the theorem. $\square$

Corollary 4.1.26. Let u satisfy $-\Delta u = \lambda u$ in a domain Ω, and let $x_0 \in \Omega$ be such that $u(x_0) = 0$. Then either u vanishes in a neighbourhood of x_0 or u attains both positive and negative values in every neighbourhood of x_0.

Proof. Suppose u does not change sign in a ball $B_{x_0,r}$. We can assume that u is nonpositive there. The function u is subharmonic in B (see Example 4.1.19). Then, by Theorem 4.1.25 u is identically zero in $B_{x_0,r}$. $\square$

Remark 4.1.27. The maximum principle holds for second-order elliptic operators in divergence form, in particular, for the Laplace–Beltrami operator on a Riemannian manifold. The proof of this fact uses Hopf's lemma; see [**Eva10**, §6.4.2].

The next theorem shows that for harmonic functions the L^2 and L^∞ norms are in a sense comparable. Such a comparison is also possible for solutions of other elliptic equations and can be viewed as part of elliptic regularity.

Theorem 4.1.28 (Comparison of L^2 and L^∞ norms). Let h be harmonic in a ball $B_{R(1+\delta)} \subset \mathbb{R}^d$, with $R, \delta > 0$. Then,

$$\fint_{B_R} |h|^2 \leq \sup_{x \in B_R} |h(x)|^2 \leq \left(1 + \frac{1}{\delta}\right)^d \fint_{B_{R(1+\delta)}} |h|^2.$$

Proof. The left inequality is trivially true for any function. Let $x_* \in \overline{B_R}$ be such that $|h(x_*)| = \sup_{x \in B_R} |h(x)|$. Then by the mean value property and the Cauchy–Schwartz inequality,

$$|h(x_*)|^2 = \left| \fint_{B_{x_*, \delta R}} h \right|^2 \leq \fint_{B_{x_*, \delta R}} |h|^2 \leq \frac{\mathrm{Vol}_d(B_{R(1+\delta)})}{\mathrm{Vol}_d(B_{x_*, \delta R})} \fint_{B_{R(1+\delta)}} |h|^2$$

$$= \left(1 + \frac{1}{\delta} \right)^d \fint_{B_{R(1+\delta)}} |h|^2. \qquad \square$$

We record the following important property of *positive* harmonic functions (see also [**GilTru01**, Theorem 2.5]).

Theorem 4.1.29 (Harnack's inequality in concentric balls). Let h be a positive harmonic function in a ball $B_{x_0, R} \subset \mathbb{R}^d$. Then for all $x \in B_{x_0, R/2}$ we have $h(x) \leq 2^d h(x_0)$.

Proof. By the mean value property of harmonic functions,

$$h(x) = A_{h,x}(R/2) \leq 2^d A_{h,x_0}(R) = 2^d h(x_0),$$

where we have used the fact that $B_{x, R/2} \subset B_{x_0, R}$ and the positivity to compare the integrals over these balls. $\qquad \square$

4.1.6. Corollaries of Courant's theorem. Using Courant's theorem one can show that the first eigenvalue and the first eigenfunctions have some special features.

Theorem 4.1.30. An eigenfunction corresponding to the eigenvalue $\lambda_1^{\mathrm{D}}(\Omega)$ does not vanish in Ω.

Proof. By Courant's theorem, the first eigenfunction has exactly one nodal domain; i.e., it does not change sign. The assertion of the theorem then follows from Corollary 4.1.26. $\qquad \square$

■ **Exercise 4.1.31.** Show that an eigenfunction of the Dirichlet Laplacian cannot have nonpositive values at local maxima or nonnegative values at local minima.

Corollary 4.1.32. The first eigenvalue λ_1^{D} is simple.

Proof. By contradiction, assume that u_1, u_2 are two linearly independent first eigenfunctions. We can choose $u_1 \perp u_2$ in $L^2(\Omega)$. But this is impossible since they do not vanish.   □

Corollary 4.1.33. The only Dirichlet eigenfunction that does not change sign is the first eigenfunction. In particular, if $\Omega' \subset \Omega$ is a nodal domain of an eigenfunction in Ω with eigenvalue λ, then $\lambda_1(\Omega') = \lambda$.

We leave the proof of Corollary 4.1.33 as an exercise for the reader.

Corollary 4.1.34. The second eigenfunction of the Dirichlet Laplacian has precisely two nodal domains.

Proof. Indeed, it cannot have one nodal domain by Corollary 4.1.33, and it cannot have more than two nodal domains by Courant's theorem.   □

Remark 4.1.35. An eigenvalue λ_k is called *Courant-sharp* if it has an eigenfunction with exactly k nodal domains. In one dimension, all eigenvalues are Courant-sharp. Furthermore, λ_1 and λ_2 are always Courant-sharp. How many Courant-sharp eigenvalues can there be? We will return to this question in Remark 5.1.23.

4.2. Density of nodal sets

4.2.1. Geometric features of nodal sets. In the previous section we focused on the properties of nodal domains of Laplace eigenfunctions. Let us now explore the geometric features of the nodal sets. Looking at Figure 4.1 we observe that the nodal lines become more dense as the eigenvalue grows. This is also seen from looking at the eigenfunction $u_{m,1}(x,y) = \sin mx \sin y$, corresponding to the eigenvalue $\lambda = \lambda_{m,1} = m^2 + 1$ of the square $(0,\pi)^2$; it has the nodal set composed of m equally spaced vertical lines. Let us investigate this phenomenon in more detail.

Definition 4.2.1.   Given a set X in a metric space, we say that X is ε-*dense* (or *dense at the scale* ε) for some $\varepsilon > 0$, if any open ball of radius bigger than ε intersects X.

Returning to the eigenfunctions of the square $(0,\pi)^2$, we see that $\mathcal{Z}_{u_{m,1}}$ is $\frac{1}{m}$-dense, and therefore the scale at which the nodal set is dense is approximately $\frac{1}{\sqrt{\lambda_{m,1}}}$.

Theorem 4.2.2. Let f be a solution of the equation $-\Delta f = \lambda f$ with $\lambda > 0$ in a domain $\Omega \subset \mathbb{R}^d$. Then the nodal set of f is $\frac{c_d}{\sqrt{\lambda}}$-dense where

(4.2.1) $$c_d = j_{\frac{d}{2}-1,1} = \sqrt{\lambda_1(\mathbb{B}^d)}.$$

The first proof. We follow the argument in [**BérMey82**, Appendix D]. Let $\Omega' \Subset \Omega$ be a smooth bounded subdomain such that f does not vanish in Ω'. Without loss of generality, suppose that $f > 0$ in Ω'. Let $u_1 > 0$ be the first Dirichlet eigenfunction in Ω', whose corresponding first eigenvalue is $\lambda_1^D(\Omega')$. By Green's formula (2.1.7) we have

$$\begin{aligned}
\left(\lambda_1^D(\Omega') - \lambda\right)(u_1, f)_{L^2(\Omega')} &= (-\Delta u_1, f)_{L^2(\Omega')} - (u_1, -\Delta f)_{L^2(\Omega')} \\
&= -\int_{\partial\Omega'} \left((\partial_n u_1)f - u_1(\partial_n f)\right) \mathrm{d}s \\
&= -\int_{\partial\Omega'} (\partial_n u_1)f\, \mathrm{d}s \geq 0.
\end{aligned}$$

Indeed, since $u_1|_{\partial\Omega'} = 0$ and $u_1 > 0$ in Ω', the exterior normal derivative satisfies $\partial_n u_1 \leq 0$, and $f|_{\partial\Omega'} \geq 0$ by continuity. Since $(u_1, f)_{L^2(\Omega')} > 0$, we find that

(4.2.2) $$\lambda_1^D(\Omega') \geq \lambda.$$

Taking Ω' to be a ball B_r and recalling that $\lambda_1^D(B_r) = c_d r^{-2}$, we conclude from (4.2.2) that $r \leq c_d \lambda^{-1/2}$. $\qquad\square$

The second proof. This proof is essentially based on [**BerNirVar94**]. Let $\Omega' \Subset \Omega$ be a smooth bounded domain, where f is positive on the closure of Ω'. Let u_1 be the first Dirichlet eigenfunction of Ω', so that $u_1 > 0$ in Ω'. Consider the quotient $g = \frac{u_1}{f}$. A direct computation shows that

$$-\Delta g = \left(\lambda_1^D(\Omega') - \lambda\right)g + 2\frac{\langle \nabla g, \nabla f\rangle}{f}.$$

The maximum of g on $\overline{\Omega'}$ is attained at an interior point $x_0 \in \Omega'$, since g vanishes on $\partial\Omega'$. Since Δ is the trace of the Hessian, one has $-\Delta g(x_0) \geq 0$, while $g(x_0) > 0$ and $\nabla g(x_0) = 0$. Hence we deduce (4.2.2) and conclude the argument as in the first proof. $\qquad\square$

The third proof. Consider the spherical mean (see Definition 4.1.21)

$$A(r) := A_{f,x}(r) = \fint_{\partial B_r} f.$$

By Lemma 4.1.22, or simply by superposition, the radial function A satisfies the equation

$$-\Delta A = \lambda A,$$

with

$$\Delta A(x) = A''(r) + \frac{d-1}{r} A'(r)$$

and $r = |x|$. Let $\tilde{J}(\rho) := A(r)$, with $\rho = r\sqrt{\lambda}$ as a dimensionless quantity. Then, $\tilde{J}$ satisfies the equation

$$\tilde{J}''(\rho) + \frac{d-1}{\rho} \tilde{J}'(\rho) + \tilde{J}(\rho) = 0.$$

Finally, we set $J(\rho) = \rho^{\frac{d}{2}-1} \tilde{J}(\rho)$, and we find that J satisfies the Bessel equation

$$J''(\rho) + \frac{1}{\rho} J'(\rho) + \left(1 - \frac{(d/2-1)^2}{\rho^2}\right) J(\rho) = 0.$$

We conclude that

$$A(r) = Cu(x_0) \left(r\sqrt{\lambda}\right)^{1-d/2} J_{d/2-1}\left(r\sqrt{\lambda}\right),$$

with $C = 2^{d/2-1}\Gamma(d/2)$. Hence, for

$$r_0 = \frac{j_{\frac{d}{2}-1,1}}{\sqrt{\lambda}}$$

we have $A(r_0) = 0$, and it follows that f must vanish at a point on the circle $\left\{x : |x| = \frac{c_d}{\sqrt{\lambda}}\right\}$. □

The fourth proof. The following elegant proof based on Harnack's inequality and lifting to harmonic functions (cf. Exercise 4.3.17) is due to T. Colding and W. Minicozzi [**ColMin11**]. This argument gives a density result without the sharp constant. Assume that f is positive in a ball $B_{x_0,r}$. Consider a harmonic function $h(x,t) := f(x)\cosh\left(t\sqrt{\lambda}\right)$ in the $(d+1)$-dimensional ball $B^{d+1}_{(x_0,0),r}$. Since h is positive there, by Harnack's inequality (Theorem 4.1.29)

$$h(x_0, r/2) \leq 2^{d+1} h(x_0, 0) = 2^{d+1} f(x_0).$$

It follows that

$$f(x_0)\cosh\left(r\sqrt{\lambda}/2\right) \leq 2^{d+1} f(x_0).$$

Equivalently, $\cosh(r\sqrt{\lambda}/2) \leq 2^{d+1}$, or $r \leq (2\operatorname{arccosh} 2^{d+1})/\sqrt{\lambda}$. □

Given a bounded domain Ω, let ρ_Ω denote its inradius. The following result is an immediate corollary of the density of nodal sets. In view of Corollary 4.1.33, it also easily follows from the domain monotonicity for the first Dirichlet eigenvalue.

Proposition 4.2.3 ([**PólSze51**, p. 98]). Let $\Omega \subset \mathbb{R}^d$ be a bounded domain, and let u_λ be a Dirichlet eigenfunction corresponding to an eigenvalue λ. Let $\Omega_\lambda \subset \Omega$ be a nodal domain of u_λ. Then

$$\rho_{\Omega_\lambda} \leq \frac{c_d}{\sqrt{\lambda}}.$$

We refer to §5.2.3 for further results relating the inradius and the first Dirichlet eigenvalue.

■ **Exercise 4.2.4.** Prove the analogue of Proposition 4.2.3 for compact Riemannian manifolds (if the boundary is nonempty, assume Dirichlet boundary conditions). *Hint:* Use the fact that any Riemannian metric is locally close to Euclidean. A complete proof can be found in [**Man08**].

4.2.2. A lower bound on the size of the nodal set in dimension two. Let us prove the following lower bound on the size of the nodal sets for Dirichlet eigenfunctions of planar domains.

Theorem 4.2.5 ([**BrüGro72**]). Let $\Omega \subset \mathbb{R}^2$ be a bounded domain, and let u_λ be an eigenfunction of $-\Delta_\Omega^{\mathrm{D}}$ corresponding to an eigenvalue $\lambda > \lambda_1$. Then, the total length of the nodal set satisfies $L(\mathcal{Z}_{u_\lambda}) \geq C\sqrt{\lambda}$, where C is a positive constant independent of λ.

Proof. Let c_2 be defined by (4.2.1), and let us partition the domain Ω using a square grid of size

$$(4.2.3) \qquad\qquad h := \frac{2c_2}{\sqrt{\lambda}}.$$

Choose a grid square $Q \subset \Omega$ and consider the bigger square $3Q$ of side length $3h$ formed by Q and all its neighbours (see Figure 4.4); we assume that Q is such that $3Q \subset \Omega$. By Theorem 4.2.2, there exists $p \in Q \cap \mathcal{Z}_{u_\lambda}$. If u is identically zero in a neighbourhood of p, the theorem is trivially true (in fact, this situation is impossible since the eigenfunctions are real analytic). Otherwise, consider a nodal line passing through the point p. There are two possibilities.

If this nodal line leaves $3Q$, then its length is at least h.

If the nodal line stays in $3Q$, by Corollary 4.1.26 there exists a nodal domain Ω' such that $p \in \partial\Omega'$ and $\Omega' \subset 3Q$. Let D_r be a disk of minimal radius r which contains Ω'. By the domain monotonicity, Corollary 4.1.33, and (4.2.3),

$$\frac{c_2^2}{r^2} = \lambda_1^{\mathrm{D}}(D_r) \leq \lambda_1^{\mathrm{D}}(\Omega') = \lambda = \frac{4c_2^2}{h^2};$$

thus $r \geq \frac{h}{2}$. Therefore,

$$\text{Length}(\partial\Omega' \cap 3Q) > 2\text{diam}(\Omega') \geq 2r \geq h.$$

In either case, we get that the size of the nodal set contained in each square $3Q$ is at least $h = \frac{2c_2}{\sqrt{\lambda}}$. Since for large λ there are $O(\lambda)$ such squares inside Ω, there exists $C > 0$ such that $L(\mathscr{Z}_{u_\lambda}) \geq C\sqrt{\lambda}$. $\qquad\square$

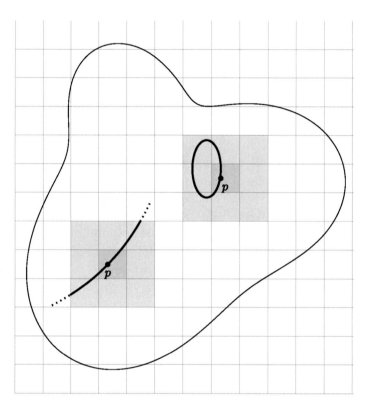

Figure 4.4. Grid squares Q (darker shading) and $3Q$ (lighter shading) inside a planar domain, with $p \in Q$, and a nodal line passing through p and existing $3Q$ on the left, or staying closed in $3Q$ on the right.

Remark 4.2.6. For Euclidean domains with Neumann boundary conditions the proof of Theorem 4.2.5 can be repeated essentially verbatim. In order to generalise it for surfaces with a Riemannian metric, some further observations are required. Note that all the measurements in the proof of Theorem 4.2.5 are made in small neighbourhoods of size $O\left(\frac{1}{\sqrt{\lambda}}\right)$. Due to the existence of local isothermal coordinates on a surface, we may assume that in each neighbourhood the Riemannian metric has the form $\mathrm{d}s^2 = h(x,y)\left(\mathrm{d}x^2 + \mathrm{d}y^2\right)$ with $\frac{1}{K} \leq h(x,y) \leq K$ for some $K > 0$. Then the Riemannian lengths and their Euclidean counterparts are comparable;

i.e., they differ by at most a factor which is controlled by K. Moreover, as follows from the variational principle and the conformal equivalence of the Dirichlet energy in two dimensions (3.1.14), the eigenvalues of the Laplacian in the Riemannian metric ds^2 are comparable to the corresponding eigenvalues of the Euclidean Laplacian. Hence, the proof of Theorem 4.2.5 could be adapted to the Riemannian case. This result was obtained in [**Brü78**] and independently by S.-T. Yau [**Yau82**, problem 74].

Interestingly enough, the analogue of Theorem 4.2.5 for surfaces can be proved with an explicit universal constant. The following result is due to A. Savo [**Sav01**].

Theorem 4.2.7. Let M be a compact Riemannian surface without boundary. Then

$$(4.2.4) \qquad L\left(\mathcal{Z}_{u_\lambda}\right) > \frac{1}{11}\,\mathrm{Area}(M)\sqrt{\lambda}$$

for sufficiently large λ.

It is a challenging open question to find the optimal constant in inequality (4.2.4). It is suggested in [**Sav01**] that the possible answer is $\frac{1}{\pi}$ with equality attained by the eigenfunctions $u_m(x,y) = \sin mx$, $m \to \infty$, on a flat square torus.

4.3. Yau's conjecture on the volume of nodal sets

4.3.1. Nodal volume and doubling index. In higher dimensions, the method of the proof of Theorem 4.2.5 fails for the following reason. It is easy to see that the above argument does not rule out "needle-like" nodal sets, for which the diameter is large, but the volume could be made arbitrarily small. Still, in 1982, S.-T. Yau [**Yau82**] made a conjecture that the following two-sided inequality holds for an arbitrary closed d-dimensional Riemannian manifold M:

$$(4.3.1) \qquad C_1\sqrt{\lambda} \le \mathcal{H}^{d-1}(\mathcal{Z}_{u_\lambda}) \le C_2\sqrt{\lambda},$$

with some constants $C_1, C_2 > 0$ depending only on the metric. Here $\mathcal{H}^{d-1}(\cdot)$ denotes the $(d-1)$-dimensional Hausdorff measure, which is a generalisation of the notion of the $(d-1)$-dimensional volume (see [**Fed14**, Introduction and §3.2.46] for the definition). Yau's conjecture has attracted a lot of attention in the past decades. In 1988, the following fundamental result was proved by H. Donnelly and C. Fefferman.

Theorem 4.3.1 ([**DonFef88**]). Assume that the Riemannian metric on M is real analytic. Then Yau's conjecture (4.3.1) holds.

In particular, this proves the upper bound in Yau's conjecture for the standard two-dimensional sphere and both upper and lower bounds for higher-dimensional spheres, all previously unknown cases.

The approach of Donnelly–Fefferman has been recently significantly developed by A. Logunov and E. Malinnikova (see [**LogMal18b**] and references therein), who obtained several breakthrough results for smooth manifolds.

> **Theorem 4.3.2** ([**Log18a**, **Log18b**]). Let M be a closed d-dimensional Riemannian manifold endowed with a smooth Riemannian metric. Then
>
> $$(4.3.2) \qquad C_1\sqrt{\lambda} \le \mathcal{H}^{d-1}(\mathcal{Z}_{u_\lambda}) \le C_2\lambda^S,$$
>
> where $S = S(M)$ is a positive constant.

In particular, the lower bound in Yau's conjecture holds. The polynomial upper bound in (4.3.2) is a breakthrough compared with the Hardt–Simon exponential estimate $O\left(\lambda^{c\sqrt{\lambda}}\right)$ that has been known earlier [**HarSim89**, Theorem 5.3]. Note that the upper bound $O\left(\lambda^{1/2}\right)$ in (4.3.1) is still not proved even in two dimensions. In the planar case, the best-known exponent is $\frac{3}{4} - \varepsilon$ for a certain small $\varepsilon > 0$ [**LogMal18a**]. H. Donnelly and C. Fefferman [**DonFef90**] and R.-T. Dong [**Don92**] have previously proved a two-dimensional upper bound with the exponent $\frac{3}{4}$.

The goal of this section is to explain some ideas behind the proofs of Theorems 4.3.1 and 4.3.2, with a particular focus on the upper bound in (4.3.2) which we discuss in detail. One of the key observations is that in order to estimate the nodal volume one needs to understand well the growth properties of the eigenfunctions; see Remark 4.3.9 below. Recall that a *geodesic ball* $B := B_{x,r} \subset M$ is the image of the Euclidean ball $B_{0,r} \subset T_xM$ under the exponential map (see [**Bur98**, §3.3]), where $r > 0$ is small enough so that this map is a diffeomorphism. Similarly to the Euclidean balls, we write $cB := B_{x,cr}$.

Definition 4.3.3 (The doubling index). Let $B \subset M$ be a geodesic ball such that $2B \subset M$ is also a geodesic ball, and assume that $f \in C(\overline{2B})$ is not the zero function. The L^∞-*doubling index* of f (or simply its *doubling index*) is the number

$$\beta(f, B) := \log_2\left(\frac{\sup\limits_{x \in 2B}|f(x)|}{\sup\limits_{x \in B}|f(x)|}\right) = \log_2\left(\frac{\|f\|_{L^\infty(2B)}}{\|f\|_{L^\infty(B)}}\right).$$

Example 4.3.4. If P_n is a homogeneous polynomial of degree n in d variables, then
$$\beta\left(P_n, B_{0,r}\right) = n.$$

The doubling index is closely related to the *vanishing order* of a smooth function. The vanishing order $\mathrm{ord}_x\left(f\right)$ of a function f at the point x is defined as the maximal integer k such that all the derivatives of f of order smaller than k vanish at x. If no such k exists, we say that f vanishes to infinite order at x. For instance, $\mathrm{ord}_x\left(f\right) = 0$ if $f(x) \neq 0$, $\mathrm{ord}_x\left(f\right) = 1$ if x is a simple zero of f, and $f(x) = e^{-1/x^2}$ vanishes to infinite order at $x = 0$.

■ **Exercise 4.3.5** (Doubling index and vanishing order). Let f be a smooth function.

(i) Show that if f has a finite vanishing order at x, then
$$\mathrm{ord}_x\left(f\right) = \lim_{r \to 0} \beta\left(f, B_{x,r}\right).$$

(ii) Show that if there exists a constant $C > 0$ such that $\beta(f, B_{x,r}) \leq C$ for all small enough $r > 0$, then $\mathrm{ord}_x\left(f\right) \leq C$.

The following important fact of independent interest established in [**DonFef88**] is heavily used in the proofs of both Theorems 4.3.1 and 4.3.2. Roughly speaking, it says that eigenfunctions grow like polynomials of degree $\sqrt{\lambda}$, similarly to the spherical harmonics.

Theorem 4.3.6 (The Donnelly–Fefferman growth bound). Let u_λ be a Laplace eigenfunction on a Riemannian manifold M. Then for any geodesic ball $B \subset M$ such that $2B$ is also a geodesic ball in M,
$$\beta(u_\lambda, B) \leq C_M \sqrt{\lambda},$$
where C_M is a constant depending only on the geometry of M.

We review the main ideas involved in its proof in §4.3.2. In view of the second part of Exercise 4.3.5, Theorem 4.3.6 immediately implies

Theorem 4.3.7. Let u_λ be a Laplace eigenfunction on a smooth Riemannian manifold M. Then
$$\mathrm{ord}_x\left(u_\lambda\right) \leq C_M \sqrt{\lambda}$$
at any point $x \in M$.

Theorem 4.3.7 could be viewed as a quantitative version of the Aronszajn's unique continuation result for Laplace eigenfunctions; see Remark 4.1.14.

■ **Exercise 4.3.8.** Use spherical harmonics to show that the bounds in Theorems 4.3.6 and 4.3.7 are sharp.

Remark 4.3.9 (The doubling index and nodal volume). There is a natural link between the degree of a polynomial and the size of its zero set. In one real dimension, a polynomial of degree d has at most d zeros; in higher dimensions, Milnor's bound on the number of connected components of the zero set in terms of the degree yields an estimate on the nodal volume [**HarSim89**, Theorem 2.1]. In one complex dimension the number of zeros, counted with multiplicities, equals the degree. For a holomorphic function in $\mathbb{C}$, the number of zeros is bounded by its growth (Jensen's formula); see, e.g., [**LogMal18b**, §4.2]. This result and the Crofton formula play an important role in the proof of the upper bound in Yau's conjecture in the real analytic case. In the smooth case, one needs to develop other methods which connect the growth of a harmonic function to the size of its nodal set. For solutions of second-order elliptic equations with smooth coefficients one has an important result of R. Hardt and L. Simon [**HarSim89**, Theorem 1.7], which together with Theorem 4.3.6 implies the *existence of an upper bound* on the size of the nodal set. In particular, if h is a solution of such an equation (e.g., a harmonic function or an eigenfunction of a Laplacian) in a ball $2B$, then

$$(4.3.3) \qquad \mathcal{H}^{d-1}(\mathscr{Z}_h \cap B) \leq C\beta(h,B)^{C\beta(h,B)},$$

with some constant $C > 0$ independent of h.

4.3.2. The Donnelly–Fefferman growth bound: A sketch of the proof. In this section we prove Theorem 4.3.6 under the simplifying assumption that M is endowed with a locally Euclidean metric, which allows us to consider only harmonic functions. The proof we give illustrates the main ideas and is adaptable to general smooth Riemannian manifolds using the standard techniques of elliptic theory, since our arguments do not rely on the real analyticity of harmonic functions.

The proof of Theorem 4.3.6 is based on a monotonicity property of the doubling index of a harmonic function, which goes back to T. Carleman [**Car33**], S. Agmon [**Agm65**], and F. J. Almgren [**Alm00**]. The monotonicity in the context of general elliptic equations of second order and related applications are due to N. Garofalo and F.-H. Lin [**GarLin86**].

Definition 4.3.10 (The height and frequency functions). Consider a continuous function f defined in a ball $B_{x_0,R} \subset \mathbb{R}^d$. The *height function* of f is given by

$$H_f(r) = H_f(x_0, r) := \fint_{\partial B_{x_0,r}} f^2, \qquad r \in (0, R)$$

(see Definition 4.1.21).

The *frequency function* of f is defined by

$$(4.3.4) \qquad N_f(r) = N_f(x_0, r) := \frac{r H'_f(x_0, r)}{2 H_f(x_0, r)}, \qquad r \in (0, R).$$

The height function of a harmonic function h is monotonically nondecreasing, since h^2 is subharmonic (see Lemma 4.1.22). Recall that a function is called *logarithmically convex* if its logarithm is a convex function. The following result holds.

Theorem 4.3.11 (Monotonicity of the frequency function). Let h be a harmonic function defined in a Euclidean ball B_R. The function $t \mapsto H_h(e^t)$ defined in $\mathbb{R}$ is logarithmically convex. Equivalently, the frequency function $N_h(r)$ is monotonically nondecreasing.

Proof. In dimension two, working in polar coordinates (r, θ) one easily verifies the convexity of $t \mapsto \log H_{h_m}(e^t)$ for $h_m(r, \theta) := r^{|m|} e^{im\theta}$. Indeed, in this case $\log H_{h_m}(e^t) = 2|m|t$ is just linear. Then, the orthogonal decomposition

$$h(r, \theta) = \sum_{m \in \mathbb{Z}} a_m h_m(r, \theta)$$

shows that

$$H_h(e^t) = \sum_{m \in \mathbb{Z}} |a_m|^2 e^{2|m|t},$$

the logarithm of which is convex (see Exercise 4.3.13 below).

Similarly, for a ball $B_R \subset \mathbb{R}^{d+1}$, $d \geq 2$, one uses the expansion into spherical harmonics to write

$$h(x) = \sum_{k=0}^{\infty} \sum_{\widetilde{P}_{k,j} \in \widetilde{\mathcal{H}}_k} c_{k,j} |x|^k \widetilde{P}_{k,j}\left(\frac{x}{|x|}\right),$$

where $\widetilde{\mathcal{H}}_k$ is a space of homogeneous harmonic polynomials (or spherical harmonics) of degree k whose elements $\{\widetilde{P}_{k,j}\}$ are chosen to be orthonormal in $L^2(\mathbb{S}^d)$; the dimension of $\widetilde{\mathcal{H}}_k$ is given in Theorem 1.2.16. The height function $H_{h_{k,j}}(e^t)$ for each term $h_{k,j} = c_{k,j} |x|^k \widetilde{P}_{k,j}\left(\frac{x}{|x|}\right)$ in this expansion is equal to $|c_{k,j}|^2 e^{2kt}$ and hence is logarithmically convex. Therefore, as above, $H_h(e^t)$ is also logarithmically convex. The equivalence of this property to the monotonicity of the frequency function follows immediately by noting that

$$\frac{\mathrm{d}}{\mathrm{d}t} \log H_h(e^t) = 2N_h(e^t). \qquad \square$$

Remark 4.3.12. Theorem 4.3.11 may be proved using integration by parts, which is adaptable to general manifolds, where no orthogonal decomposition is available; see, e.g., [Agm65], [LogMal20].

■ Exercise 4.3.13. Show that if f_1 and f_2 are positive functions in some open interval $I \subset \mathbb{R}$ and if $\log f_1, \log f_2$ are convex, then $\log(f_1 + f_2)$ is also convex. *Hint:* Use the geometric-arithmetic mean inequality.

■ Exercise 4.3.14. Show that the frequency function can be expressed as

$$N_h(x_0, r) = \frac{r \int_{B(x_0,r)} |\nabla h|^2 \,\mathrm{d}x}{\int_{\partial B(x_0,r)} |h|^2 \,\mathrm{d}S_r}.$$

In what follows we often use a shortcut notation $B_r := B_{x_0,r}$ for concentric balls provided the centre x_0 can be an arbitrary fixed point.

Theorem 4.3.15. Let $R > 0$, and let h be a harmonic function in $\Omega \supset B_{cR}$ for some fixed $c > 1$. Then, the quantity

$$(4.3.5) \qquad N(h, B_r, c) := \frac{1}{2} \log_2 \frac{\oint_{\partial B_{cr}} |h|^2}{\oint_{\partial B_r} |h|^2} = \frac{1}{2} \log_2 \frac{H_h(cr)}{H_h(r)}$$

is monotonically nondecreasing in r for $r \in (0, R)$.

Proof. Using the definition (4.3.4), we have

$$\int_1^c \frac{N_h(tr)}{t} \,\mathrm{d}t = \int_1^c \frac{rH_h'(tr)}{2H_h(tr)} \,\mathrm{d}t = \frac{1}{2} \int_1^c \frac{\mathrm{d}}{\mathrm{d}t} (\log H_h(tr)) \,\mathrm{d}t$$

$$= \frac{1}{2} (\log H_h(cr) - \log H_h(cr)) = (\log 2)N(h, B_r, c),$$

and therefore

$$(4.3.6) \qquad N(h, B_r, c) = \frac{1}{\log 2} \int_1^c \frac{N_h(tr)}{t} \,\mathrm{d}t.$$

Since by Theorem 4.3.11 the frequency function N_h is monotone, (4.3.6) shows that $N(h, B_r, c)$ is monotone in r. $\qquad \square$

In view of (4.3.6), we call $N(h, B_r, c)$ the *integrated frequency*. It can also be viewed as an L^2 analogue of the doubling index.

■ **Exercise 4.3.16.** One may define versions of the height and frequency functions H_h, N_h for balls as

$$H_h^b(r) := \fint_B |h|^2, \quad N_h^b(r) = \frac{r\,(H_h^b)'\,(r)}{2H_h^b(r)}.$$

Show using essentially the same arguments as above that

$$N^b(h, B_r, c) := \frac{1}{2}\log_2 \frac{\fint_{B_{cr}}|h|^2}{\fint_{B_r}|h|^2}$$

is monotonically nondecreasing in r. Show also that

$$H_h^b(r) \le H_h(r), \quad N_h^b(r) \le N(r), \quad N^b(h, B_r, c) \le N(h, B_r, c).$$

Hint: Observe that

$$\frac{H_h^b(r)}{H_h(r)} = \frac{1}{\mathrm{Vol}(B_1)}\int_0^1 \frac{H_h(tr)}{H_h(r)}\,\mathrm{Vol}(\partial B_t)\,\mathrm{d}t,$$

where the integrand is monotonically nonincreasing in r by Theorem 4.3.15.

To prove Theorem 4.3.6 we will apply the following lifting trick to reduce it to the case of harmonic functions (cf. proof of Theorem 2.2.1, part (ii)).

■ **Exercise 4.3.17** (Lifting trick). Consider an open product Riemannian manifold $M \times I$, where $I \subset \mathbb{R}$, and let u_λ be an eigenfunction of the Laplace–Beltrami operator on M corresponding to an eigenvalue λ. Show that the function

$$h(x,t) := u_\lambda(x)\cosh\left(\sqrt{\lambda}t\right), \qquad (x,t) \in M \times (-1,1),$$

is harmonic in $M \times I$.

Theorem 4.3.18 (A local version of the Donnelly–Fefferman growth bound). Let u satisfy the equation $-\Delta u = \lambda u$ in a ball B_R. Then the following statements hold.

  (i) For all $0 < r \le \frac{2}{3}s < \frac{R}{3}$ one has

$$\beta\,(u, B_r) \le C(\beta\,(u, B_s) + s\sqrt{\lambda} + 1).$$

  (ii) The following three-ball inequality holds for all $0 < r < \frac{R}{4}$:

$$\sup_{x \in B_{2r}} |u(x)| \le Ce^{Cr\sqrt{\lambda}}\left(\sup_{x \in B_r}|u(x)|\right)^\alpha\left(\sup_{x \in B_{4r}}|u(x)|\right)^{1-\alpha},$$

with some $\alpha \in (0,1)$ independent of λ and u.

Here, $C > 0$ denotes some constants (possibly different) which are independent of λ and u.

Remark 4.3.19 (Aronszajn's unique continuation principle). In particular, by fixing s and letting r tend to zero, it follows from the corresponding version of Theorem 4.3.18 for smooth manifolds and Exercise 4.3.5 that if $-\Delta_g u = \lambda u$ and u has a zero of infinite order, then u is identically zero (see also [**Aro57**]).

Proof of Theorem 4.3.18. We follow [**Man13**]. Let $B = B_{x_0,r}$, and let B_r^{d+1} be the $(d+1)$-dimensional ball with centre $(x_0, 0)$ and radius r. Lift the eigenfunction $u(x)$ to a harmonic function $h(x,t)$ on B_r^{d+1} as in Exercise 4.3.17.

We observe that for any r, δ with $r(1+\delta) \in (0, R)$,

$$\sup_{x \in B_r} |u(x)|^2 \leq \sup_{(x,t) \in B_r^{d+1}} |h(x,t)|^2$$

(4.3.7)
$$\overset{\text{Theorem 4.1.28, Corollary 4.1.24}}{\leq} \left(1 + \frac{1}{\delta}\right)^{d+1} \fint_{\partial B_{r(1+\delta)}^{d+1}} |h|^2,$$

while

(4.3.8) $$\fint_{\partial B_r^{d+1}} |h|^2 \leq \sup_{\partial B_r^{d+1}} |h|^2 \leq \sup_{B_r} |u|^2 \cdot (\cosh(r\sqrt{\lambda}))^2 \leq \sup_{B_r} |u|^2 \cdot e^{2r\sqrt{\lambda}}.$$

Applying Theorem 4.3.15 to h,

$$\frac{\sup_{x \in B_{2r}} |u(x)|^2}{\sup_{x \in B_r} |u(x)|^2} \overset{(4.3.7),(4.3.8)}{\leq} 3^{d+1} e^{2r\sqrt{\lambda}} \frac{\fint_{\partial B_{3r}^{d+1}} |h(x,t)|^2}{\fint_{\partial B_r^{d+1}} |h(x,t)|^2}$$

$$= 3^{d+1} e^{2r\sqrt{\lambda}} \frac{\fint_{\partial B_{3r}^{d+1}} |h(x,t)|^2}{\fint_{\partial B_{\sqrt{3}r}^{d+1}} |h(x,t)|^2} \cdot \frac{\fint_{\partial B_{\sqrt{3}r}^{d+1}} |h(x,t)|^2}{\fint_{\partial B_r^{d+1}} |h(x,t)|^2}$$

$$\overset{(4.3.5)}{=} 3^{d+1} e^{2r\sqrt{\lambda}} \frac{H_h(3r)}{H_h(\sqrt{3}r)} \cdot \frac{H_h(\sqrt{3}r)}{H_h(r)}$$

$$\overset{\text{Theorem 4.3.15}}{\leq} 3^{d+1} e^{2r\sqrt{\lambda}} \left(\frac{H_h(3r)}{H_h(\sqrt{3}r)}\right)^2$$

$$\overset{\text{Theorem 4.3.15}}{\leq} 3^{d+1} e^{4s\sqrt{\lambda}/3} \left(\frac{H_h(2s)}{H_h(2s/\sqrt{3})}\right)^2$$

$$\overset{(4.3.8),(4.3.7)}{\leq} C e^{28s\sqrt{\lambda}/3} \left(\frac{\sup_{x \in B_{2s}} |u(x)|^2}{\sup_{x \in B_s} |u(x)|^2}\right)^2,$$

where C denotes a positive constant independent of λ. Taking logarithms on both sides we obtain the first part of the theorem. By substituting $s = 2r$ in the preceding inequality we arrive at

$$\sup_{x \in B_{2r}} |u(x)| \le C \mathrm{e}^{19r\sqrt{\lambda}} \left(\sup_{x \in B_r} |u(x)| \right)^{1/3} \left(\sup_{x \in B_{4r}} |u_\lambda(x)| \right)^{2/3},$$

thus establishing the statement in part (ii) with $\alpha = \frac{1}{3}$. $\qquad\square$

Finally, we apply Theorem 4.3.18 together with the fact that the manifold M is compact in order to prove Theorem 4.3.6.

Proof of Theorem 4.3.6. Fix $r_0 > 0$ such that every ball of radius $3r_0$ is geodesic; i.e., $3r_0$ is smaller than the *injectivity radius* of M; see [**Cha84**, p. 118]. We first show that Theorem 4.3.6 holds for any ball $B_{p,r}$ of radius $r \ge r_0$. Normalise u_λ so that $\sup_M |u_\lambda| = 1$. Let $x_* \in M$ be a point where $|u_\lambda(x_*)| = 1$. Let $x_0, x_1, \ldots, x_N$ be a sequence of points such that $x_0 = p, x_N = x_*$, $d(x_j, x_{j+1}) < r_0$ and such that N depends on the geometry of M and r_0 only. Observe that $B_{x_j, 2r_0} \supset B_{x_{j+1}, r_0}$. The three-ball inequality of Theorem 4.3.18 with $\alpha = \frac{1}{3}$ gives, taking into account that $|u_\lambda| \le 1$,

$$\sup_{x \in B_{x_j, r_0}} |u_\lambda(x)| \ge C^{-1} \mathrm{e}^{-3Cr_0\sqrt{\lambda}} \left(\sup_{x \in B_{x_j, 2r_0}} |u_\lambda(x)| \right)^3$$

$$\ge C^{-1} \mathrm{e}^{-3Cr_0\sqrt{\lambda}} \left(\sup_{x \in B_{x_{j+1}, r_0}} |u_\lambda(x)| \right)^3.$$

Using this inequality recursively for $j = N - 1, N - 2, \ldots, 0$, we arrive at

$$\sup_{x \in B_{x_0, r_0}} |u_\lambda(x)| \ge C' \mathrm{e}^{-C''r_0\sqrt{\lambda}} \left(\sup_{x \in B_{x_*, r_0}} |u_\lambda(x)| \right)^{3^N} = C' \mathrm{e}^{-C''r_0\sqrt{\lambda}},$$

where

$$C' = C^{-(3^N-1)/2}, \qquad C'' = 3 \frac{3^N - 1}{2}.$$

The preceding inequality shows that for all $r \ge r_0$,

$$\sup_{x \in B_{x_0, r}} |u_\lambda(x)| \ge \sup_{x \in B_{x_0, r_0}} |u_\lambda(x)| \ge C' \mathrm{e}^{-C''r_0\sqrt{\lambda}}$$

$$\ge C' \mathrm{e}^{-C''r_0\sqrt{\lambda}} \sup_{x \in B_{x_0, 2r}} |u_\lambda(x)|.$$

Recalling Definition 4.3.3, we have proved, in other words, that for all $r \ge r_0$,

(4.3.9) $$\beta(u_\lambda, B_{p,r}) \le C_1 \sqrt{\lambda} + C_2,$$

where C_1, C_2 depend only on r_0 and the geometry of M. For $0 < r < r_0$ we apply part (i) of Theorem 4.3.18 with $s = \frac{3}{2}r_0$ and inequality (4.3.9) to get that for any ball $B = B(x, r) \subset M$

$$\beta(u_\lambda, B) \le C_3 \sqrt{\lambda} + C_4,$$

where C_3, C_4 depend only on M. Finally, since we may assume that $\lambda \ge \lambda_1(M) > 0$ (the case $\lambda = 0$ being trivial), we can absorb the additive constant C_4 in the multiplicative constant C_3.   □

4.3.3. Distribution of doubling indices: A combinatorial approach.
Spectacular recent progress on Yau's conjecture due to Logunov and Malinnikova [**LogMal18a**, **Log18a**, **Log18b**] is based on a better understanding of the distribution of doubling indices. Theorem 4.3.6 gives a worst-case scenario, but in reality, in most of the balls the doubling index is much smaller. In view of Remark 4.3.9 this should lead to better nodal estimates. We note that this observation in various forms is also key for the proof of Theorem 4.3.1, as well as the lower bound in Theorem 4.3.2. For instance, in dimension two, the upper bound in Yau's conjecture (4.3.1) is equivalent to showing that the doubling indices of an eigenfunction u_λ on balls of radii $C/\sqrt{\lambda}$ are bounded on average; see [**NazPolSod05**, **RoF15**].

Below we survey some of the important insights on the distribution of doubling indices that have led to the proof of the polynomial upper bound in Theorem 4.3.2. Remarkably, a key idea discussed in this subsection is purely combinatorial.

In what follows, we work with cubes rather than with balls; it does not make an essential difference and is more convenient for combinatorial purposes. However, minor technical issues appear. We denote by Q a cube in $\mathbb{R}^d$ and by αQ a concentric cube with parallel sides of length $\alpha s(Q)$, where $s(Q)$ is the side length of Q.

Slightly abusing notation, given a continuous function $f : \ell Q \to \mathbb{R}$, we define the *doubling index of a cube* Q by

$$\beta(f, Q) = \log_2 \frac{\|f\|_{L^\infty(\ell Q)}}{\|f\|_{L^\infty(Q)}},$$

where ℓ is a fixed large odd integer depending on dimension (one can take $\ell > 2\sqrt{d}$). The integer ℓ appears in order to allow the comparison of $\beta(f, Q)$ with relevant quantities of the inscribed and circumscribed balls.

Lemma 4.3.20 (Combinatorial lemma for an arbitrary function). Let f be a continuous function in $\ell Q \subset \mathbb{R}^d$. Subdivide ℓQ into $(\ell K)^d$ equal subcubes of side length $\frac{1}{K}s(Q)$. Assume that $\beta(h, q) > \beta_0$ for each subcube q with $\ell q \subset \ell Q$. Then $\beta(h, Q) > K\beta_0$.

Proof. Find a subcube q_0 of Q and a point $x_0 \in q_0$ such that $|f(x_0)| = \max_{x \in Q} |f(x)|$. Since $\beta(h, q_0) > \beta_0$ we can find a point $x_1 \in \ell q_0$ such that $|f(x_1)| > 2^{\beta_0} |f(x_0)|$, and we can find a subcube q_1 such that $x_1 \in q_1$. Observe that if $K > 1$, then $\ell q_1 \subset \ell Q$ and $\beta(h, q_1) > \beta_0$. At the $(j+1)$th step, as long as $j < K$ we find a point $x_{j+1} \in \ell q_j$ such that $|f(x_{j+1})| > 2^{\beta_0} |f(x_j)|$ and a subcube q_{j+1} such that $x_{j+1} \in q_{j+1}$. For $j = K - 1$ we get $|f(x_K)| > 2^{K\beta_0} |f(x_0)|$; see Figure 4.5. $\qquad\square$

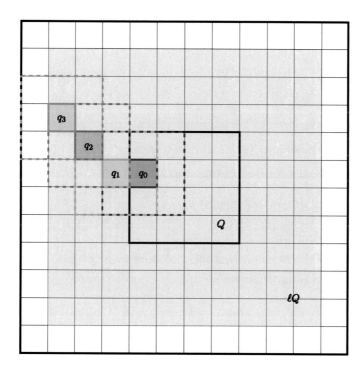

Figure 4.5. Cubes Q and ℓQ, the latter subdivided into $(\ell K)^d$ equal subcubes, shown here for $d = 2$, $\ell = 3$, and $K = 4$. The subcubes q satisfying $\ell q \subset \ell Q$ are shaded grey. Also, this gives an example of the sequence q_j of subcubes appearing in the proof of Lemma 4.3.20; the corresponding cubes ℓq_j are shown by dashed lines.

In order to iterate Lemma 4.3.20 one has to know that an upper bound on the doubling index does not grow after a subdivision. In general this is obviously false, but for harmonic functions it is essentially the content of the monotonicity theorem (Theorem 4.3.15) after replacing the L^2 estimates by L^∞ estimates (with the same arguments in the proofs of Theorems 4.3.18 and 4.3.6). One obtains Lemma 4.3.21 stated below, which provides such a monotonicity result when the cubes are not concentric and when the inner cube is far from the boundary of the exterior one (cf. the case $\lambda = 0$ of

Theorem 4.3.18). We have fixed ℓ to be larger than $2\sqrt{d}$ above exactly in order for this lemma to hold.

Lemma 4.3.21 ([**LogMal18a**, **Hal22**]). There exist a positive constant C_0 and a positive odd integer T such that for any harmonic function h in a cube $\ell Q \subset \mathbb{R}^d$ and for any subcube $q \subset \frac{1}{T}Q$,

$$\beta(h, q) \leq C_0 \beta(h, Q) + C_0.$$

Given a harmonic function $h : \ell Q \to \mathbb{R}$, we introduce the notation

$$\beta^{\mathrm{sup}}(h, Q) := \sup_{q \subset Q} \beta(h, q).$$

The quantity β^{sup} is convenient since it is monotonic with respect to the inclusion of cubes. Lemma 4.3.21 implies that

(4.3.10) $$\beta^{\mathrm{sup}}(h, q) \leq 2C_0 \max\{\beta(h, Q), 1\}$$

for a function h harmonic in ℓQ and any $q \subset \frac{1}{T}Q$.

Set $\beta_0 = 2C_0$. Iterating Lemma 4.3.20 we get

Lemma 4.3.22 (Combinatorial lemma for harmonic functions). Let a cube $Q^0 \subset \mathbb{R}^d$ be subdivided into A^{md} equal subcubes Q^m, where $m \in \mathbb{N}$ and $A \in \mathbb{N}$ is greater than some constant A_0. For any harmonic function h in ℓQ^0, one can regroup the subcubes Q^m into $m+1$ disjoint subsets $G_0^m, \ldots, G_m^m$ such that

(4.3.11) $$\beta^{\mathrm{sup}}(h, Q_j^m) \leq \max\left\{ \frac{\beta^{\mathrm{sup}}(h, Q^0)}{2^j}, \beta_0 \right\} \qquad \text{for all } Q_j^m \in G_j^m$$

and

$$\#G_j^m = \binom{m}{j} \cdot \left(A^d - 1 \right)^{m-j}.$$

Proof. Set $\beta := \beta^{\mathrm{sup}}(h, Q^0)$ and $s_0 = s(Q^0)$. We argue by induction. For $m = 0$ there is nothing to prove. Suppose $G_0^m, \ldots, G_m^m$ are defined and satisfy the required properties. Partition each subcube $Q_j^m \in G_j^m$ of side length $A^{-m}s_0$ into $(\ell K)^d$ equal sized subcubes $q_{j,k}^m$ with side length $\frac{s_0}{A^m \ell K}$ as in Lemma 4.3.20. Since $\beta\left(h, \frac{1}{\ell}Q_j^m\right) \leq \beta^{\mathrm{sup}}(h, Q_j^m) \leq \max\{2^{-j}\beta, \beta_0\}$ by (4.3.11), we can apply the contrapositive of Lemma 4.3.20 to the cube $\frac{1}{\ell}Q_j^m$ of side length $\frac{s_0}{A^m \ell}$, and therefore we can find a subcube q_{j,k_0}^m of Q_j^m such that

$$\beta\left(h, q_{j,k_0}^m\right) \leq \frac{1}{K} \max\{2^{-j}\beta, \beta_0\}$$

and $\ell q_{j,k_0}^m \subset Q_j^m$. Consider the cube $\frac{1}{T} q_{j,k_0}^m$ of side length $\frac{s_0}{A^m T \ell K}$. Applying (4.3.10) with $Q = q_{j,k_0}^m$ and $q = \frac{1}{T} q_{j,k_0}^m$, we have, given that $\beta_0 = 2C_0$ and choosing $K > 4C_0$,

(4.3.12)
$$\beta^{\text{sup}}\left(h, \frac{1}{T} q_{j,k_0}^m\right) \leq 2C_0 \max\left\{\beta\left(h, q_{j,k_0}^m\right), 1\right\}$$
$$\leq \max\left\{\frac{2C_0}{K} 2^{-j}\beta, \frac{2C_0}{K}\beta_0, 2C_0\right\} \leq \max\left\{2^{-j-1}\beta, \beta_0\right\}.$$

We have proved that if we partition Q_j^m into $(T\ell K_0)^d$ equal subcubes with $K_0 := \lceil 4C_0 \rceil$, there exists at least one such subcube $\frac{1}{T} q_{j,k_0}^m$ for which (4.3.12) holds (here we used the fact that T was chosen to be odd in Lemma 4.3.21).

Let us now repartition Q_j^m into A^d equal subcubes q with

$$A \geq A_0 := 3T\ell K_0 = 3T\ell \lceil 4C_0 \rceil$$

(which corresponds to partitioning the original cube Q^0 into $A^{(m+1)d}$ subcubes). Then $\frac{1}{T} q_{j,k_0}^m$ contains at least one such subcube q, and from (4.3.12) and the monotonicity of β^{sup} we have $\beta^{\text{sup}}(h, q) \leq \max\{2^{-j-1}\beta, \beta_0\}$.

We add q to G_{j+1}^{m+1}. We add the other $A^d - 1$ remaining subcubes of Q_j^m to G_j^{m+1}. Counting the contributions of G_j^m and G_{j+1}^m to G_{j+1}^{m+1}, we arrive at the following recursion:

$$\#G_{j+1}^{m+1} = \#G_j^m + (A^d - 1) \cdot \#G_{j+1}^m.$$

This is a classical recursion of a weighted Pascal triangle with initial condition $\#G_0^0 = 1$. Its solution is $\#G_j^m = \binom{m}{j} \cdot (A^d - 1)^{m-j}$; see Exercise 4.3.23 below. This completes the proof of the theorem. $\qquad\square$

■ **Exercise 4.3.23** (Weighted Pascal triangle). Let $g(m, j)$ be a function defined for all pairs of nonnegative integers (m, j) such that $0 \leq j \leq m$. Assume that it satisfies the recursion $g(m, j) = ag(m-1, j) + bg(m-1, j-1)$, where $g(m, j)$ is interpreted as 0 when $j < 0$ or $j > m$ and where $a, b > 0$ are fixed constants. In addition assume that $g(0, 0) = 1$. Prove that $g(m, j) = \binom{m}{j} a^{m-j} b^j$.

4.3.4. A polynomial upper bound: An overview.
The proof of the polynomial upper bound on the size of the nodal set is based on further improvements of Lemma 4.3.22. Fix a number A. When m is large enough (independently of the harmonic function h) one observes that $\#G_0$, the number of subcubes for which the doubling index is greater than β_0 and is not guaranteed to decrease by Lemma 4.3.22, is arbitrarily small compared to the total number of subcubes in the subdivision of $Q = Q^0$. While the

total number of subcubes is A^{md}, at most $\#G_0 = (A^d - 1)^m < 0.01A^{md}$ subcubes $q := Q_0^m$ satisfy $\beta^{\sup}(h, q) > \max\{\beta^{\sup}(h, Q)/2, \beta_0\}$. It turns out that the number of these "bad" cubes is even smaller by an order of magnitude; i.e., it can be compared to the number of subcubes on ∂Q.

Theorem 4.3.24 ([**Log18a**]). Subdivide a cube $Q \subset \mathbb{R}^d$ into A^d equal subcubes q, where A is greater than some constant A_1. Let h be a harmonic function in ℓQ. Then, the number of subcubes such that $\beta^{\sup}(h, q) > \max\{\beta^{\sup}(h, Q)/2, \beta_0\}$ is at most $0.9A^{d-1}$.

The proof of Theorem 4.3.24 is based on the following two ideas. For a harmonic function one can improve Lemma 4.3.20 as follows: if there exist only a few (say, $d+1$) bad subcubes (i.e., where the doubling index is large) that are well distributed in ℓQ, then one can still deduce that the doubling index of Q is even bigger (the simplex lemma [**Log18a**, §2]). Accordingly, if the doubling index of Q is small, the bad subcubes should be spread along a hyperplane. This idea can be applied to deduce that the number of bad subcubes is at most A^{d-1}. In turn, the cubes along a hyperplane cannot be all bad, since, otherwise, a quantitative version of the Cauchy data uniqueness theorem can be applied to show the doubling index of Q would be too big (the hyperplane lemma [**Log18a**, §4]). As a consequence one shows that at most $0.9A^{d-1}$ of the subcubes are bad.

We have now all the required ingredients to complete the overview of the polynomial upper bound in Theorem 4.3.2 provided the metric on M is flat. While the proof in the general case is more technical, the argument in the flat case highlights essentially all the conceptual ideas.

Proof of the upper bound in Theorem 4.3.2 for flat metrics. We start by applying the lifting trick to an eigenfunction. Let $Q_1 \subset \mathbb{R}^d$ be a unit cube. In view of Theorem 4.3.6, it is sufficient to prove that for any harmonic function $h : \ell Q_1 \to \mathbb{R}$ for which $\beta^{\sup}(h, Q_1) \leq \beta$, we have

$$\mathcal{H}^{d-1}(\mathcal{Z}_h \cap Q_1) \leq C\beta^{2S}$$

for some $S > 0$ independently of h (although Theorem 4.3.6 refers to the doubling indices on balls, the doubling indices on cubes are essentially equivalent, as mentioned earlier). Let

$$F(\beta) := \sup_{h \in H_\beta} \mathcal{H}^{d-1}(\mathcal{Z}_h \cap Q_1),$$

where

$$H_\beta := \{h : \ell Q_1 \to \mathbb{R}, h \text{ is harmonic with } \beta^{\sup}(h, Q_1) < \beta\}.$$

It follows from (4.3.3) that $F(\beta)$ is finite. Note that for any cube q and a harmonic function $h : \ell q \to \mathbb{R}$ one has

$$\mathcal{H}^{d-1}(\mathcal{Z}_h \cap Q_1) \leq F\left(\beta^{\mathrm{sup}}(h, q)\right) s(q)^{d-1}.$$

Let $h_o : \ell Q_1 \to \mathbb{R}$ be harmonic, where $\beta^{\mathrm{sup}}(h_o, Q_1) \leq \beta$ optimises $F(\beta)$ up to a small positive ε; i.e., it satisfies

(4.3.13) $$\mathcal{H}^{d-1}(\mathcal{Z}_{h_o} \cap Q_1) \leq F(\beta) < \mathcal{H}^{d-1}(\mathcal{Z}_{h_o} \cap Q_1) + \varepsilon.$$

To give a bound on the size of its nodal set, subdivide Q_1 into A^d equal sub-cubes q where A is large as in Theorem 4.3.24. Collecting the contributions to the nodal set from all subcubes q, it is clear that

(4.3.14) $$\mathcal{H}^{d-1}(\mathcal{Z}_{h_o} \cap Q_1) \leq A^d F(\beta/2) \frac{1}{A^{d-1}} + 0.9 A^{d-1} F(\beta) \frac{1}{A^{d-1}},$$

and we derive from (4.3.13) and (4.3.14) that

$$F(\beta) < 10 A F(\beta/2) + 10\varepsilon.$$

Since $\varepsilon > 0$ is arbitrary, we get

$$F(\beta) \leq 10 A F(\beta/2).$$

This inequality implies (see Exercise 4.3.25) that F is bounded by a polynomial in β of degree $\log_2(10A) =: 2S$, which completes the proof of the theorem. $\square$

■ **Exercise 4.3.25.** Let $f : [1, \infty) \to \mathbb{R}$ be a nonnegative monotonically nondecreasing function. Suppose that $f(2x) < Af(x)$ for all $x \geq 1$ and some $A > 1$. Prove that $f(x) < Cx^S$ for all $x > 1$, where C, S are positive constants depending on A only.

4.4. Nodal sets on surfaces and eigenvalue multiplicity bounds

4.4.1. Local structure of the nodal set. Let u be a smooth function in a neighbourhood of the origin in $\mathbb{R}^d$, and suppose that it has vanishing order $N \in \mathbb{N}$ at $x = 0$. Then, by Taylor's theorem,

(4.4.1) $$u(x) = P_N(x) + o\left(|x|^N\right) \qquad \text{as } x \to 0,$$

where P_N is a nonzero homogeneous polynomial of degree N. If u is a solution of a linear partial differential equation, we have the following simple result.

Theorem 4.4.1. Let $\mathcal{A}$ be a linear differential operator with C^∞ smooth coefficients in a neighbourhood $0 \ni W \subset \mathbb{R}^d$, and let $\mathcal{A}_0$ be its principal part with the coefficients fixed at $x = 0$. Suppose that $u \in C^\infty(W)$ is a solution of the equation $\mathcal{A}u = 0$ that has vanishing order $N \in \mathbb{N}$ at $x = 0$. Then

(4.4.2) $\mathcal{A}_0 P_N = 0,$

where P_N is defined in (4.4.1).

Proof. We follow the argument in [**Alb71**, Theorem 2.12]. Let m be the degree of $\mathcal{A}$, and let us represent $u(x)$ in the form $u(x) = P_N(x) + R(x)$. Write $\mathcal{A} = \mathcal{A}_0 + \mathcal{A}_1 + \mathcal{A}_2$ where $\mathcal{A}_0 + \mathcal{A}_1$ is the principal part of $\mathcal{A}$ and $\mathcal{A}_2$ is of a smaller degree. Then

$$0 = \mathcal{A}u = \mathcal{A}_0 P_N + (\mathcal{A}_1 + \mathcal{A}_2) P_N + \mathcal{A}R.$$

Since $\mathcal{A}_0 P_N$ is the Taylor polynomial of $\mathcal{A}u$ of degree $N-m$, we may conclude by Taylor's theorem that $(\mathcal{A}_1 + \mathcal{A}_2) P_N + \mathcal{A}R = o\left(|x|^{N-m}\right)$. It follows that $0 = \mathcal{A}_0 P_N + o\left(|x|^{N-m}\right)$. This is possible only if $\mathcal{A}_0 P_N = 0$.   □

Remark 4.4.2 (Bers's theorem). Theorem 4.4.1 can be viewed as an elementary version of the celebrated Bers's theorem [**Ber55**], which guarantees that any solution u of an *elliptic* equation with *Hölder* coefficients has a polynomial asymptotics (4.4.1) near its zero set, as if u were a smooth function. In addition (4.4.2) is also satisfied.

We can now prove the following result which in a way is a two-dimensional version of Theorem 4.1.16.

Theorem 4.4.3 ([**Che75**]). Let M be a compact Riemannian surface. The nodal set of a Laplace eigenfunction on M consists of C^1 immersed circles. The nodal critical points of an eigenfunction (i.e., the zeros of its gradient lying on the nodal set) are isolated, and at each such point the nodal lines divide the angle 2π equally.

Proof. Let us apply Theorem 4.4.1 to a Laplace eigenfunction $u(x)$ with eigenvalue λ on a Riemannian manifold (with $\mathcal{A} = \Delta_g + \lambda$). Note that by elliptic regularity (see Theorem 2.2.17) $u(x)$ is smooth and by Theorem 4.3.7 it has a finite vanishing order N. Choose coordinates at a neighbourhood of a point x_0 in which the Riemannian metric $g_{ij}(x_0) = \delta_{ij}$. The principal part of Δ_g at the point x_0 is the Euclidean Laplacian. Then, P_N is a harmonic homogeneous polynomial of degree N. Note that in two dimensions, homogeneous harmonic polynomials of degree N have a particularly simple form $\operatorname{Re} Az^N$, where $z = x_1 + ix_2$ and $A \in \mathbb{C}$. The nodal set of such a harmonic polynomial is a union of straight lines going through the origin and dividing the unit disk into $2N$ congruent sectors.

It was shown in [**Che75**] (see also [**BérMey82**, Appendix E] and the discussion before Theorem 4.1.16) that in two dimensions there exists a C^1-diffeomorphism f near x such that $u(x) = P_N(f(x))$. Hence, the nodal set $\mathcal{Z}_u$ is locally diffeomorphic to the nodal set of a harmonic polynomial, and the first statement follows. Note also that f maps nodal critical points of u to nodal critical points of P_N, which are isolated, and the second statement follows.

It remains to prove the equiangular property of nodal lines. Take a path $(r(t)\cos\varphi(t), r(t)\sin\varphi(t))$ lying in the nodal set $\mathcal{Z}_u$ and starting at a critical point $r(0) = 0$. We can write

$$0 = u(r(t)\cos\varphi(t), r(t)\sin\varphi(t))$$
$$= A_1 r(t)^N \cos N\varphi(t) + A_2 r(t)^N \sin N\varphi(t) + R\left(r(t)\cos\varphi(t), r(t)\sin\varphi(t)\right)$$
$$= A r(t)^N \sin\left(N\varphi(t) + \alpha\right) + R\left(r(t)\cos\varphi(t), r(t)\sin\varphi(t)\right),$$

where A_1, A_2, A, and α are some constants and $R(r(t)\cos\varphi, r(t)\sin\varphi) = o\left(r(t)^N\right)$ as $t \to 0$. It follows that

$$\lim_{t\to 0} \sin(N\varphi(t) + \alpha) = 0,$$

from which one concludes that $\varphi(0)$ can take only values of the form $(k\pi - \alpha)/N$ for some $k \in \mathbb{Z}$. Recall that the nodal set of the harmonic polynomial P_N consists of $2N$ rays emanating from the critical point. Since f is a C^1–diffeomorphism, the images of different rays under f cannot yield the same value of $\varphi(0)$ mod 2π, and the equiangular property follows. $\square$

4.4.2. Multiplicity bounds. We have the following.

Lemma 4.4.4 ([**Nad87**], [**KarKokPol14**]). Let M be a Riemannian surface, and let $u_1, \ldots, u_{2n}$ be a collection of linearly independent eigenfunctions corresponding to some eigenvalue λ. Then, for a given point $x \in M$, there exists a nontrivial linear combination $\sum_{i=1}^{2n} \alpha_i u_i$ with vanishing order at x of at least n.

Proof. Let $V = \text{Span}\{u_1, \ldots, u_{2n}\}$, and let V_i be the subspace of elements $u \in V$ such that $\text{ord}_x(u) \geq i$. Clearly, $V_{i+1} \subset V_i$. We need to show that $V_n \neq \{0\}$. Suppose the contrary. Let us calculate $\dim V$. We have

$$\dim V = \sum_{j=0}^{n-1} \dim(V_j/V_{j+1}).$$

As follows from the proof of Theorem 4.4.3, V_j/V_{j+1} can be identified with a subspace of the space of harmonic homogeneous polynomials of degree j. In turn, the latter space is of dimension one for $j = 0$ and of dimension two for $j \geq 1$. Therefore, $\dim V \leq 1 + 2(n-1) < 2n$, which is a contradiction. $\square$

It is useful to think about the nodal set of an eigenfunction on a Riemannian surface as a graph with edges being the arcs of the nodal lines and the vertices being the critical points. If there is a closed nodal line without critical points on it, we may introduce an artificial vertex with the edge being a cycle. The graph constructed this way is called the *nodal graph* of an eigenfunction.

Let us recall some general facts about graphs on surfaces. Given a graph Γ, let the *degree* of a vertex x, denoted $\deg_\Gamma x$, be the number of edges incident to x; if there is an edge that starts and ends at x, it is counted twice. Let e be the number of edges in the graph. Then

$$2e = \sum_x \deg_\Gamma x.$$

Let f be the number of faces of Γ, i.e., the number of connected components of $M \setminus \Gamma$. Euler's inequality states that

$$v - e + f \geq \chi(M),$$

where $\chi(M)$ is the Euler characteristic of M. It becomes an equality (the well-known Euler formula) if all the faces are topological disks. The following theorem is due to N. Nadirashvili [**Nad87**]); weaker versions were earlier obtained by G. Besson [**Bes80**] and S.-Y. Cheng [**Che75**].

Theorem 4.4.5. The multiplicity $m(\lambda_k)$ of the eigenvalue λ_k on a Riemannian surface M satisfies the inequality

(4.4.3) $m(\lambda_k) \leq 2k - 2\chi(M) + 5.$

Proof. Suppose the contrary. Then there exist $2k - 2\chi(M) + 6$ linearly independent λ_k-eigenfunctions. By Lemma 4.4.4, there exists an eigenfunction with the vanishing order $k - \chi(M) + 3$ at some point x_0. Consider the nodal graph of this eigenfunction. The number of faces of this graph is the number of nodal domains. Therefore, by Courant's theorem, since we number our eigenvalues as $0 = \lambda_0 < \lambda_1 \leq \lambda_2 \leq \cdots$, we have

$$k + 1 \geq f \geq \chi(M) + e - v.$$

At the same time, $e = \frac{1}{2} \sum_x \deg_\Gamma(x)$, and the degree of each vertex is at least two. Hence, in order to obtain a lower bound on the right-hand side we can assume that x_0 is the only vertex. Since $\deg_\Gamma(x_0) = 2(k - \chi(M) + 3)$, we get

$$k + 1 \geq \chi(M) + k - \chi(M) + 3 - 1 = k + 2,$$

which is a contradiction.   □

In some cases, further refinements of the bound (4.4.3) can be obtained using a careful analysis of the structure of the nodal graph. Multiplicity estimates could also be proved in a similar way for the Dirichlet and Neumann eigenvalues on surfaces with boundary; see [**KarKokPol14**, §6] for details.

The estimate (4.4.3) in general is not sharp.

■ **Exercise 4.4.6.** Deduce from Weyl's law (see Theorem 3.3.4) that the multiplicity $m(\lambda)$ on a d-dimensional manifold satisfies

$$(4.4.4) \qquad m(\lambda_k) = o(k) \qquad \text{as } k \to \infty.$$

It follows from (4.4.4) that the estimate (4.4.3) is not of the correct order in k asymptotically. Yet, in a few cases it yields sharp multiplicity bounds.

Corollary 4.4.7. On the sphere, $m\left(\lambda_1(\mathbb{S}^2)\right) \leq 3$, which is sharp and is attained by the round metric. On the projective plane, $m\left(\lambda_1(\mathbb{RP}^2)\right) \leq 5$, which is again sharp and is attained by the round metric.

We leave the proof of Corollary 4.4.7 as an exercise for the reader.

Chapter 5

Eigenvalue inequalities

Georg **Faber**
(1877–1966)

In this chapter, we prove various geometric eigenvalue inequalities, in particular, due to Faber–Krahn, Cheeger, and Szegő–Weinberger. We also present the results of Hersch and Yang–Yau, as well as other isoperimetric inequalities for Laplace–Beltrami eigenvalues on surfaces. Furthermore, we discuss universal inequalities for Dirichlet eigenvalues on Euclidean domains and related commutator identities.

Edgar **Krahn**
(1894–1961)

5.1. The Faber–Krahn inequality

Throughout the chapter, we will use

Definition 5.1.1. Let $\Omega \subset \mathbb{R}^d$ be a measurable set of finite volume. Its *symmetric rearrangement* is an open ball $\Omega^* = B_{R^*_\Omega}^d$, where the radius $R^* = R^*_\Omega$ is determined by the condition $\operatorname{Vol}_d(\Omega^*) = \operatorname{Vol}_d(\Omega)$. Therefore

$$R^*_\Omega = \left(\operatorname{Vol}_d(\Omega)\omega_d^{-1} \right)^{\frac{1}{d}},$$

where ω_d is the volume of the unit ball $\mathbb{B}^d$; see (B.1.1).

We will also use Notation 3.3.23 for level and superlevel sets of a function and the volume of a superlevel set.

5.1.1. Motivation.
The Faber–Krahn inequality states that among all Euclidean domains of given volume, the first Dirichlet eigenvalue is minimal for the ball.

Theorem 5.1.2 (Faber–Krahn inequality). Let $\Omega \subset \mathbb{R}^d$ be a bounded domain. Then

$$(5.1.1) \qquad\qquad \lambda_1^{\mathrm{D}}(\Omega) \geq \lambda_1^{\mathrm{D}}(\Omega^*).$$

Inequality (5.1.1) was conjectured in 1877 by Lord Rayleigh in his famous book on the theory of sound [**Ray77**]. Moreover, he proved, using perturbation theory, that a ball is a local minimiser for $\lambda_1 = \lambda_1^{\mathrm{D}}$ among all domains of a given volume. A complete proof of (5.1.1) was obtained independently by G. Faber and E. Krahn [**Fab23**], [**Kra25**].

Remark 5.1.3. In view of Definition 5.1.1 and Exercise 1.2.21, Theorem 5.1.2 can be reformulated as follows: for any bounded domain $\Omega \subset \mathbb{R}^d$,

$$\lambda_1(\Omega)\,\mathrm{Vol}(\Omega)^{2/d} \geq \omega_d^{2/d}\, j_{\frac{d}{2}-1,1}^2,$$

where $j_{\frac{d}{2}-1,1}$ is the first zero of the Bessel function of the first kind of order $\frac{d}{2} - 1$. In particular, for $d = 2$,

$$\lambda_1(\Omega)\,\mathrm{Area}(\Omega) \geq \pi j_{0,1}^2 \approx 5.76\pi.$$

Note that this estimate confirms Pólya's conjecture (Conjecture 3.3.14) for the first Dirichlet eigenvalue of a planar domain, which in this case reads $\lambda_1(\Omega)\,\mathrm{Area}(\Omega) \geq 4\pi$.

In order to get some physical intuition, it is instructive to look at the Faber–Krahn inequality from the viewpoint of the heat equation on a bounded domain $\Omega \in \mathbb{R}^d$:

$$\begin{cases} \frac{\partial u(t,x)}{\partial t} = \Delta u(t,x) & \text{for } (t,x) \in (0,\infty) \times \Omega, \\ u = 0 & \text{on } \partial\Omega, \\ u(0,x) = u_0(x). \end{cases}$$

Here $u(t,x)$ is the temperature at the point $x \in \Omega$ at the time $t > 0$, and $u_0(x)$ is the initial temperature distribution. Using the Fourier method, we obtain

$$u(t,x) = \sum_{k=1}^{\infty} c_k \mathrm{e}^{-\lambda_k t} u_k(x),$$

where λ_k and u_k are the Dirichlet eigenvalues and eigenfunctions, respectively, and the coefficients c_k are determined by the initial condition u_0. Consider the *heat content* of Ω,

$$(5.1.2) \qquad\qquad Q_\Omega(t) := \int_\Omega u(t,x)\,\mathrm{d}x,$$

and the *rate of the relative heat loss*,

$$\alpha_\Omega(t) := -\frac{Q'_\Omega(t)}{Q_\Omega(t)}.$$

Clearly,

$$\lim_{t\to\infty} \alpha_\Omega(t) = \lambda_1(\Omega).$$

In other words, the smaller λ_1 is, the smaller the long-term heat loss is. At the same time, it is natural to assume that in order to minimise the heat loss due to the fact that the boundary is kept at the zero temperature, one needs to minimise the boundary surface of Ω. This leads to the isoperimetric problem: given the fixed interior volume, minimise the $(d-1)$-dimensional volume of the boundary. It is well known that the solution of this problem is a ball, which is in agreement with the Faber–Krahn inequality.

Interestingly enough, while the argument above is in no way rigorous, the isoperimetric inequality indeed plays the key role in the proof of the Faber–Krahn inequality. We present the details below.

5.1.2. The co-area formula. One of the technical tools used in the proof of the Faber–Krahn inequality is an important result from geometric measure theory called the *co-area formula* (see, for instance, [**Maz85**, §1.2.4]).

Theorem 5.1.4 (The co-area formula). Let $\Omega \subset \mathbb{R}^d$ be a domain, let $h : \Omega \to \mathbb{R}$ be an integrable function, and let $F : \Omega \to [a,b] \subset \mathbb{R}$ be a smooth function. Then

$$(5.1.3) \qquad \int_\Omega h(x)\,|\nabla F(x)|\,\mathrm{d}x = \int_a^b \int_{\mathcal{L}_F(t)} h(x)\,\mathrm{d}\Sigma_t\,\mathrm{d}t,$$

where the sets $\mathcal{L}_F(t)$ are the level sets of F, see Notation 3.3.23, and $\mathrm{d}\Sigma_t$ is the surface measure on $\mathcal{L}_F(t)$.

Note that since F is smooth, the set of critical values has measure zero by Sard's theorem. Therefore, by the implicit function theorem, the level sets $\mathcal{L}_F(t)$ are smooth hypersurfaces for almost all t. One can also check that the interior integral on the right is an integrable function of t, and hence the iterated integral is well-defined.

Remark 5.1.5. The smoothness assumption on F in Theorem 5.1.4 can be relaxed. In particular, the co-area formula holds if F is Lipschitz or if it is a function of bounded variation; see [**EvaGar15**].

The co-area formula can be viewed as a kind of a "curvilinear Fubini theorem", as the following example shows.

Example 5.1.6. Let $\Omega = B_R \subset \mathbb{R}^d$ and $F(x) = |x|$. Then $\nabla F(x) = \frac{x}{|x|}$ and $|\nabla F(x)| = 1$ for all x. In view of Remark 5.1.5 one can apply the co-area formula. It follows that $\mathcal{L}_F(r) = S_r := \{x \in \mathbb{R}^d : |x| = r\}$, and thus

$$\int_{B_R} h(x) \, dx = \int_0^R \int_{S_r} h(x) \, dS_r \, dr,$$

which is the usual integration formula in spherical coordinates.

Suppose that the set of critical points

$$\mathcal{C}_F := \{x \in \Omega : \nabla F = 0\}$$

of a function F has measure zero. Substituting formally $h(x) = \frac{1}{|\nabla F(x)|}$ into (5.1.3) we obtain

$$(5.1.4) \qquad \mathrm{Vol}(\Omega) = \int_a^b \int_{\mathcal{L}_F(t)} \frac{1}{|\nabla F(x)|} \, d\Sigma_t \, dt.$$

To justify this result (see, for instance, [**Dan11**]), take $\varepsilon > 0$ and set

$$h_\varepsilon(x) = \frac{1}{|\nabla F(x)| + \varepsilon}.$$

Applying (5.1.3) to h_ε we get

$$(5.1.5) \qquad \int_{\Omega \backslash \mathcal{C}_F} h_\varepsilon(x) \, |\nabla F(x)| \, dx = \int_{[a,b] \backslash F(\mathcal{C}_F)} \int_{\mathcal{L}_F(t)} h_\varepsilon(x) \, d\Sigma_t \, dt.$$

Using the monotone convergence theorem as $\varepsilon \to 0$ and taking into account that $\mathcal{C}_F$ has measure zero, we obtain (5.1.4).

Remark 5.1.7. As was pointed out in [**CadFar18**], the assumption that the set of critical points of F has measure zero has been often neglected in the literature, though it is necessary for the validity of (5.1.4).

5.1.3. Symmetric decreasing rearrangement. The proof of the Faber–Krahn inequality also uses the notion of *symmetric decreasing rearrangement* of a function. There are several equivalent ways to define it; we essentially follow the approach of [**LieLos97**]. First, given a set $A \subset \mathbb{R}^d$ of finite volume, we define the symmetric rearrangement of its characteristic function by $\chi_A^* := \chi_{A^*}$.

Definition 5.1.8 (Symmetric decreasing rearrangement). Let $u : \Omega \to \mathbb{R}$ be a measurable nonnegative function on an open bounded set $\Omega \subset \mathbb{R}^d$. The *symmetric decreasing rearrangement* of u is a function $u^* : \Omega^* \to \mathbb{R}$ defined by the relation

$$(5.1.6) \qquad u^*(x) = \int_0^{+\infty} \chi^*_{\mathcal{V}_u(t)}(x)\, \mathrm{d}t,$$

where $\mathcal{V}_u(t)$ are the superlevel sets of u; see Notation 3.3.23.

■ **Exercise 5.1.9.** Show that $u^*(x)$ is a lower semi-continuous radially symmetric function which is nonincreasing in $|x|$.

Recall the "layer cake representation" formula (see [**LieLos97**, Theorem 1.13]):

$$(5.1.7) \qquad u(x) = \int_0^{+\infty} \chi_{\mathcal{V}_u(t)}(x)\, \mathrm{d}t.$$

Comparing the two formulas above, we observe that $u^*(x)$ is obtained from $u(x)$ by symmetrisation of its superlevel sets. It then easily follows that the functions u and u^* are *equimeasurable*; i.e., $V_u(t) = V_{u^*}(t)$ for any $t \in \mathbb{R}$; see Notation 3.3.23.

■ **Exercise 5.1.10.** Symmetric decreasing rearrangement of a function u is sometimes alternatively defined as

$$u^*(x) := \sup \{ t : x \in (\mathcal{V}_u(t))^* \}.$$

Show that the two definitions are equivalent.

Integrating both sides of the layer cake representation (5.1.7) over Ω, applying Fubini's theorem, and making a change of variables $t = s^p$ yields

$$(5.1.8) \qquad \int_\Omega u(x)^p\, \mathrm{d}x = p \int_0^{+\infty} s^{p-1} V_u(s)\, \mathrm{d}s$$

for any $p \geq 1$. Since u and u^* are equimeasurable, (5.1.8) implies

$$(5.1.9) \qquad \|u\|_{L^p(\Omega)} = \|u^*\|_{L^p(\Omega^*)}.$$

5.1.4. Proof of the Faber–Krahn inequality. We follow the argument that essentially goes back to E. Krahn [**Kra25**]; see also [**Dan11**].

We will first prove

Proposition 5.1.11 (The Pólya–Szegő principle). *Let $\Omega \subset \mathbb{R}^d$ be a bounded domain, and let u be the first Dirichlet eigenfunction on Ω. Then*

$$(5.1.10) \qquad \|\nabla u\|_{L^2(\Omega)} \geq \|\nabla u^*\|_{L^2(\Omega^*)}.$$

Proof of Proposition 5.1.11. Without loss of generality, we can assume $u > 0$ in Ω. Let $\Omega^* = B_{R^*}$ be the symmetric rearrangement of Ω, and let $v(x) := u^*(x)$ be the symmetric decreasing rearrangement of u.

Since u is real analytic by Theorem 2.2.1(ii), the set of its critical points $\mathcal{C}_u$ has measure zero, and therefore by (5.1.4) we have

$$(5.1.11) \qquad V_u(t) = \int\limits_{\mathcal{V}_u(t)} \mathrm{d}x = \int\limits_{t}^{\max_{x \in \Omega} u(x)} \int\limits_{\mathcal{L}_u(s)} \frac{1}{|\nabla u|} \, \mathrm{d}\Sigma_s \, \mathrm{d}s.$$

Since $V_u(t)$ is a nonincreasing function of t, it is differentiable almost everywhere. In view of (5.1.11), its derivative is given by

$$V_u'(t) = -\int\limits_{\mathcal{L}_u(t)} \frac{1}{|\nabla u|} \, \mathrm{d}\Sigma_t.$$

Note that the integral on the right is well-defined for almost all t since $V_u(t) \leq \mathrm{Vol}(\Omega) < \infty$; this also follows from Sard's theorem, implying that $|\nabla u| > 0$ on the level set $\mathcal{L}_u(t)$ for almost all t.

We would like to obtain an analogue of (5.1.11) for v. However, we cannot apply (5.1.4) to the function v directly, since a priori the set $\mathcal{C}_v$ of the critical points of v may have a positive measure. Since v is radially decreasing and hence of bounded variation, one can apply the co-area formula. Arguing as in (5.1.5), we obtain

$$(5.1.12) \quad V_v(t) = \rho_v(t) + \int\limits_{\mathcal{V}_v(t) \setminus \mathcal{C}_v} \mathrm{d}x = \rho_v(t) + \int\limits_{t}^{\max_{x \in \Omega^*} v(x)} \int\limits_{\mathcal{L}_v(s)} \frac{1}{|\nabla v|} \, \mathrm{d}\Sigma_s \, \mathrm{d}s,$$

where

$$\rho_v(t) := \mathrm{Vol}_d \left(\mathcal{V}_v(t) \cap \mathcal{C}_v \right).$$

By [**CiaFus02**, Lemma 2.4] it follows that $\rho'_v(t) = 0$ for almost all t. Differentiating both sides of (5.1.12) with respect to t we get

$$(5.1.13) \qquad V'_v(t) = - \int_{\mathcal{L}_v(t)} \frac{1}{|\nabla v|} \, d\Sigma_t$$

for almost all t, as in (5.1.11).

Since $V_u(t) = V_v(t)$ for all t, their derivatives must coincide provided they are well-defined. Hence, $V'_u(t) = V'_v(t)$ for almost all t, which implies

$$(5.1.14) \qquad \int_{\mathcal{L}_u(t)} \frac{1}{|\nabla u|} \, d\Sigma_t = \int_{\mathcal{L}_v(t)} \frac{1}{|\nabla v|} \, d\Sigma_t.$$

Let us show that

$$(5.1.15) \qquad \int_{\mathcal{L}_u(t)} |\nabla u| \, d\Sigma_t \geq \int_{\mathcal{L}_v(t)} |\nabla v| \, d\Sigma_t$$

for almost all t. Indeed, by the Cauchy–Schwarz inequality,

$$(5.1.16) \qquad \left(\int_{\mathcal{L}_u(t)} \frac{1}{|\nabla u|} \, d\Sigma_t \right) \left(\int_{\mathcal{L}_u(t)} |\nabla u| \, d\Sigma_t \right)$$
$$\geq \left(\int_{\mathcal{L}_u(t)} d\Sigma_t \right)^2 = \left(\mathrm{Vol}_{d-1} \left(\mathcal{L}_u(t) \right) \right)^2.$$

However, by the isoperimetric inequality, $\mathrm{Vol}_{d-1} \left(\mathcal{L}_u(t) \right) \geq \mathrm{Vol}_{d-1} \left(\mathcal{L}_v(t) \right)$, since, by the definition of the symmetric decreasing rearrangement, the sets $\mathcal{L}_u(t)$ and $\mathcal{L}_v(t)$ bound the same volume, and $\mathcal{L}_v(t)$ is a sphere because v is a radial function. Furthermore, for the same reason, $|\nabla v|$ is constant on the spheres $\mathcal{L}_v(t)$, which leads to the case of equality in the Cauchy–Schwarz inequality analogous to (5.1.16),

$$(5.1.17) \qquad \left(\mathrm{Vol}_{d-1} \left(\mathcal{L}_v(t) \right) \right)^2 = \left(\int_{\mathcal{L}_v(t)} \frac{1}{|\nabla v|} \, d\Sigma_t \right) \left(\int_{\mathcal{L}_v(t)} |\nabla v| \, d\Sigma_t \right).$$

Hence, (5.1.15) follows from (5.1.14) combined with (5.1.16) and (5.1.17).

Applying the co-area formula once again and taking into account that $\max_{x \in \Omega} u(x) = \max_{x \in \Omega^*} v(x)$ we get

$$\int_{\Omega} |\nabla u|^2 \, \mathrm{d}x = \int_0^{\max_{x \in \Omega} u(x)} \int_{\mathcal{L}_u(t)} |\nabla u| \, \mathrm{d}\Sigma_t \, \mathrm{d}t$$

$$\geq \int_0^{\max_{x \in \Omega} u(x)} \int_{\mathcal{L}_v(t)} |\nabla v| \, \mathrm{d}\Sigma_t \, \mathrm{d}t = \int_{\Omega^*} |\nabla v|^2 \, \mathrm{d}x,$$

where the inequality follows from (5.1.15). This completes the proof of the Pólya–Szegő principle (5.1.10). □

Remark 5.1.12. The justification of (5.1.13) in the proof of the Pólya–Szegő principle follows the approach of [**Fus08**, formula (3.14)]. It is omitted in most available proofs of the Faber–Krahn equality; cf. Remark 5.1.7.

Remark 5.1.13. Given a nonnegative measurable function $u : \mathbb{R}^d \to \mathbb{R}$ of compact support one can define its symmetric rearrangement $u^* : \mathbb{R}^d \to \mathbb{R}$ by formula (5.1.6). A more general version of the Pólya–Szegő principle holds: for any $p \geq 1$ and any nonnegative $u \in W^{1,p}(\mathbb{R}^d)$ of compact support, one has

$$\int_{\mathbb{R}^d} |\nabla u|^p \, \mathrm{d}x \geq \int_{\mathbb{R}^d} |\nabla u^*|^p \, \mathrm{d}x.$$

We refer to [**Fus08**, Theorem 3.1] and [**Kaw85**, Remark 2.16] for details.

We are now in the position to prove Theorem 5.1.2. The inequality (5.1.10) together with the equality (5.1.9) for L^2 norms yields the inequality

$$R_{\Omega}[u] \geq R_{\Omega^*}[v]$$

for the Rayleigh quotients of u and v. Note that for $x \in \partial\Omega^* = S_{R^*}$, we have

$$v(x) = \int_0^{+\infty} \chi^*_{\mathcal{V}_u(t)}(x) \, \mathrm{d}t = 0$$

since $\mathcal{V}_u(t)^*(x) \subset B_{R^*}$ for any $t > 0$. It remains to show that $v \in H_0^1(\Omega^*)$. Let us extend $u \in H_0^1(\Omega)$ by zero to the whole $\mathbb{R}^d$ and apply the Pólya–Szegő principle to this extension (cf. Remark 5.1.13). The resulting function is the extension of v by zero and it lies in $H^1(\mathbb{R}^d)$. Given that v is radially decreasing, it follows that it is continuous up to the boundary $\partial\Omega^*$ where it vanishes, and hence it belongs to $H_0^1(\Omega^*)$. Therefore, one can use v as a test function for the first eigenvalue of the Dirichlet problem on the ball Ω^*. Hence,

$$\lambda_1(\Omega) = R_\Omega[u] \geq R_{\Omega^*}[v] \geq \lambda_1(\Omega^*),$$

which proves the Faber–Krahn inequality.

Remark 5.1.14 (Equality in the Faber–Krahn inequality). Let us inspect the proof of Theorem 5.1.2 in order to characterise the case of equality in the Faber–Krahn inequality. Note that the case of equality in the geometric isoperimetric inequality implies that the domain is a ball up to a set of measure zero (see [**Fus04**, Theorem 4.11]). Therefore, it follows from (5.1.16) that if the equality in the Faber–Krahn inequality is attained, $\mathcal{V}_u(t)$ are balls up to sets of measure zero for almost all t. At the same time, since $\Omega = \bigcup_{t>0} \mathcal{V}_u(t)$, it follows that Ω is a ball up to a set of measure zero. In particular, if Ω is sufficiently regular (for example, Lipschitz), it has to be a ball.

In fact, with some extra work one can prove an even more precise characterisation: the equality in the Faber–Krahn inequality implies that Ω is a ball up to a set of zero capacity; cf. Remark 3.2.15. Indeed, suppose Ω is an open set achieving the equality. As was shown above, it is equal to a ball B up to a set of zero measure. Therefore, $\Omega \subset B$, and the result follows from the following characterisation of domain monotonicity for Dirichlet eigenvalues proved in [**AreMon95**, Theorem 3.1]: $\lambda_1(\Omega) = \lambda_1(B)$ if and only if the capacity of $B \setminus \Omega$ is equal to zero. We refer to [**Dan11**, Remark 5.1] for more details.

We want to address the stability of the Faber–Krahn inequality. Namely, suppose that $\lambda_1(\Omega)$ is close to $\lambda_1(\Omega^*)$ for an open set Ω. Does it imply that Ω is in some sense close (up to rigid motions) to the ball Ω^*? The answer to this question is positive. In order to state it properly, we need the following.

Definition 5.1.15 (The Fraenkel asymmetry). The *Fraenkel asymmetry* of a set Ω is defined by

$$(5.1.18) \qquad \mathcal{A}(\Omega) = \inf_{y \in \mathbb{R}^d} \frac{\mathrm{Vol}_d\left(\Omega \Delta B_{y,R^*}\right)}{\mathrm{Vol}_d(\Omega)},$$

where R^* is the radius of the symmetric rearrangement Ω^* of Ω and $\Omega \Delta B := (\Omega \setminus B) \cup (B \setminus \Omega)$ denotes the *symmetric difference* of sets Ω and B.

The following result had been conjectured independently by N. Nadirashvili [**Nad97**, p. 200] and by T. Bhattacharya and A. Weitsman [**BhaWei99**, §8] in the late 1990s, recently proved in [**BraDePVel15**].

Theorem 5.1.16. There exists $a_d > 0$ such that for any bounded domain $\Omega \subset \mathbb{R}^d$,

$$(5.1.19) \qquad a_d \mathcal{A}(\Omega)^2 \leq \mathrm{Vol}\,(\Omega)^{2/d} \left(\lambda_1(\Omega) - \lambda_1(\Omega^*) \right).$$

Moreover, one can check that the power two in the left-hand side of (5.1.19) is the smallest possible.

Remark 5.1.17 (Torsional rigidity). Given a bounded domain Ω, the quantity

$$T(\Omega) := \sup_{u \in H_0^1(\Omega) \setminus \{0\}} \frac{\left(\int_\Omega u \, \mathrm{d}x \right)^2}{\int_\Omega |\nabla u|^2 \, \mathrm{d}x}$$

is called the *torsional rigidity* of Ω (see [**PólSze51**]). Its physical meaning is as follows: $T(\Omega)$ measures the amount of resistance of a beam with a cross-section Ω against torsional deformation. The celebrated inequality of A. de Saint-Venant, proved by G. Pólya [**Pól49**], states that the ball has the maximal torsional rigidity among all domains of given volume, that is,

$$T(\Omega) \leq T(\Omega^*).$$

■ **Exercise 5.1.18.** Prove the Saint-Venant inequality using an adaptation of the proof of the Faber–Krahn inequality.

Apart from the symmetric rearrangement, there exist other symmetrisation techniques which are used to prove eigenvalue inequalities. Probably the most important one is the Steiner symmetrisation of a set, which is a symmetrisation with respect to a hyperplane; see [**PólSze51**, Chapter 1]. The corresponding Steiner rearrangement of a function shares the essential features with the symmetric decreasing rearrangement; in particular, it preserves the L^2 norm of a function and does not increase the Dirichlet energy. Therefore, the Steiner rearrangement does not increase the fundamental tone. Motivated by this approach, in 1951 G. Pólya and G. Szegő made the following well-known conjecture (see [**PólSze51**, page 159]), which they proved for $n = 3$ and $n = 4$.

Conjecture 5.1.19 (Pólya–Szegő conjecture). Among all polygons with n sides and a given area, λ_1 is minimised by a regular n-gon.

For $n = 3$, one can show that given any triangle, there exists a sequence of Steiner symmetrisations under which it converges to an equilateral triangle. For $n = 4$ the argument is even easier: any quadrilateral can be transformed into a rectangle using a sequence of not more than three symmetrisations. We refer to [**Hen06**, §3.3.2] for details of the proof in these two cases. However, this method no longer works for a higher number of vertices n of a polygon; indeed, it is easy to check that in this case a Steiner symmetrisation may increase the number of sides of an n-gon. Therefore new ideas will be required to prove Conjecture 5.1.19 for $n \geq 5$. See also §A.1.2 for a discussion about the asymptotics of the first eigenvalue of the regular n-gon as $n \to \infty$.

5.1.5. Applications of the Faber–Krahn inequality. Faber–Krahn inequality combined with the Courant nodal domain theorem implies a sharp isoperimetric inequality for the second Dirichlet eigenvalue. It was proved by E. Krahn in [**Kra26**] and later rediscovered independently by P. Szego (the son of G. Szegő) and I. Hong; see [**Hen06**, §4.1]

Theorem 5.1.20 (Krahn–Szego inequality). Among all (possibly disconnected) Euclidean domains of a given volume, the second Dirichlet eigenvalue is minimised by the disjoint union of two identical balls.

Proof. Let Ω be a connected bounded domain of given volume, which by rescaling we may assume to be equal to one. By Corollary 4.1.34 of the Courant nodal domain theorem, the second Dirichlet eigenfunction has precisely two nodal domains Ω_1 and Ω_2. Let Ω_1^* and Ω_2^* be the symmetric decreasing rearrangements of Ω_1 and Ω_2, respectively. Applying the Faber–Krahn inequality one obtains

$$\lambda_2(\Omega) = \lambda_1(\Omega_1) = \lambda_1(\Omega_2) \geq \max\left\{\lambda_1(\Omega_1^*), \lambda_1(\Omega_2^*)\right\}.$$

At the same time, since $\mathrm{Vol}(\Omega_1^*) + \mathrm{Vol}(\Omega_2^*) = \mathrm{Vol}(\Omega)$, one has

$$\max\left(\lambda_1(\Omega_1^*), \lambda_1(\Omega_2^*)\right) \geq \lambda_1(B_R) = \lambda_2\left(B_R \sqcup B_R'\right),$$

where B_R, B_R' are identical balls such that $\mathrm{Vol}(B_R) = \mathrm{Vol}(B_R') = \frac{1}{2}\mathrm{Vol}(\Omega)$. Here the first step follows from rescaling and the last step uses the fact that the spectrum of a disjoint union of domains is a union of their spectra. This completes the proof of the Krahn–Szego inequality for connected domains.

If $\Omega = \Omega_1' \sqcup \Omega_2'$ is not connected, we can modify the above argument as follows. In this case

$$\lambda_2(\Omega) = \max\left\{\lambda_1(\Omega_1), \lambda_1(\Omega_2)\right\}$$

for some disjoint sets $\Omega_1 \sqcup \Omega_2 \subset \Omega$, which are either connected components of Ω or nodal domains of the second eigenfunction of a connected component. Applying the Faber–Krahn inequality and rescaling if necessary, we again arrive at the conclusion that the minimum of λ_2 is attained by a domain which is a disjoint union of two identical balls of volume $\frac{1}{2}\mathrm{Vol}(\Omega)$. This completes the proof of the theorem. $\qquad\square$

Remark 5.1.21. In view of Remark 5.1.14, it follows from the proof of Theorem 5.1.20 that the minimum of the second Dirichlet eigenvalue is attained if and only if the domain is equal to a disjoint union of two identical balls up to a set of zero capacity. In particular, the minimum of λ_2 is not attained in the class of connected domains.

Let us now discuss an application of the Faber–Krahn inequality to the nodal geometry. The following result is due to Å. Pleijel [**Ple56**] and could be viewed as an asymptotic refinement of Courant's nodal domain theorem.

> **Theorem 5.1.22** (Pleijel's nodal domain theorem). Let $\Omega \subset \mathbb{R}^d$ be a bounded domain and let u_k be an orthogonal basis of Dirichlet eigenfunctions corresponding to eigenvalues λ_k^D. Let η_k be the number of nodal domains of u_k. Then
>
> $$(5.1.20) \qquad \limsup_{k \to \infty} \frac{\eta_k}{k} < 1.$$

Proof. For simplicity, assume $d = 2$; the proof in higher dimensions is analogous (see [**BérMey82**, Lemme 9] for the last step). Let $\Omega \subset \mathbb{R}^2$, and let $\Omega_l \subset \Omega$, $l = 1, \ldots, \eta_k$, be the nodal domains of an eigenfunction u_k. Then, $\lambda_k^D(\Omega) = \lambda_1^D(\Omega_l)$ for all l. At the same time, by the Faber–Krahn inequality,

$$(5.1.21) \qquad \frac{\mathrm{Area}(\Omega_l)}{\pi j_{0,1}^2} \geq \frac{1}{\lambda_1^D(\Omega_l)} = \frac{1}{\lambda_k^D(\Omega)}.$$

Summing up the inequalities (5.1.21) over $l = 1, \ldots, \eta_k$, we get

$$\frac{\mathrm{Area}(\Omega)}{\pi j_{0,1}^2} \geq \frac{\eta_k}{\lambda_k^D(\Omega)}.$$

By Weyl's law,

$$\lim_{k \to \infty} \frac{\lambda_k^D(\Omega)}{k} = \frac{4\pi}{\mathrm{Area}(\Omega)},$$

and hence

$$(5.1.22) \qquad \limsup_{k \to \infty} \frac{\eta_k}{k} \leq \frac{4}{j_{0,1}^2} \simeq 0.691 < 1. \qquad \square$$

Remark 5.1.23 (Courant-sharp eigenvalues and nodal deficiency). Pleijel's theorem implies that in dimension $d \geq 2$, only finitely many eigenvalues λ_k^D admit eigenfunctions satisfying $\eta_k = k$. We recall that such eigenvalues are called *Courant-sharp*; see Remark 4.1.35. The nonnegative quantity $k - \eta_k$ is called the *nodal deficiency* of an eigenfunction and it admits interesting interpretations in terms of the Morse indices of certain functionals and operators; see [**BerKucSmi12, CoxJonMar17, BerCHS22**] and references therein.

Remark 5.1.24 (Pleijel's nodal domain theorem in other settings). Inequality (5.1.20) also holds in the case of Neumann boundary conditions. The main difficulty in the Neumann case is to handle the nodal domains which touch the boundary. On those nodal domains the corresponding Neumann eigenvalue is equal to the first eigenvalue of a mixed Dirichlet–Neumann problem, and therefore the Faber–Krahn inequality cannot be applied. Pleijel's theorem for Neumann boundary conditions was first established in [**Pol09**] for piecewise analytic planar domains. The result was later extended in [**Lén19**] to arbitrary dimensions and more general Robin boundary conditions for domains with $C^{1,1}$ boundary. Analogues of Pleijel's theorem exist in other settings as well, in particular, for compact Riemannian manifolds [**BérMey82**], and for certain Schrödinger operators in $\mathbb{R}^d$ [**Cha18, ChaHelHoO18**].

Remark 5.1.25 (Optimal Pleijel constant). One may wonder whether *Pleijel's constant* 0.691 in the right-hand side of (5.1.22) is close to being optimal for planar domains (a similar question could also be asked in arbitrary dimension). By taking separable eigenfunctions on rectangles, it is easy to check that Pleijel's constant is not smaller than $\frac{2}{\pi} \simeq 0.636$. It was conjectured in [**Pol09**] that this value is optimal for planar domains with either Dirichlet or Neumann boundary conditions. Slight improvements of the constant in (5.1.22) were obtained in [**Bou15, Ste14**] by using quantitative stability results for the Faber–Krahn inequality and estimates on the packing density by disks.

5.2. Cheeger's inequality and its applications

5.2.1. Cheeger's inequality. By the variational principle, in order to estimate the first eigenvalue from above it is sufficient to find an appropriate test function. Estimating eigenvalues from *below* is, a priori, a more difficult task. The importance of Cheeger's inequality [**Che71**] is that it provides a rather simple geometric lower bound for the first eigenvalue. In order to state the result, we need to introduce the following definition.

Definition 5.2.1 (The Dirichlet Cheeger constant). Let Ω be a compact Riemannian manifold with boundary or a bounded Euclidean domain, of dimension d. The *Dirichlet Cheeger constant* is defined by

$$(5.2.1) \qquad h_{\mathrm{D}} := h_D(\Omega) = \inf_A \frac{\mathrm{Vol}_{d-1}(\partial A)}{\mathrm{Vol}_d(A)},$$

where the infimum is taken over all compactly embedded open subsets $A \Subset \Omega$ with smooth boundary.

The subsets A appearing in (5.2.1) are not assumed to be connected. Let us also remark that the smoothness assumption on ∂A is not restrictive, since any set of bounded perimeter can be approximated by sets with smooth boundary. We note as well that the Dirichlet Cheeger constant is somewhat reminiscent of the isoperimetric constant

$$\frac{\mathrm{Vol}_{d-1}(\partial A)^{\frac{d}{d-1}}}{\mathrm{Vol}_d(A)};$$

however, unlike the latter it is not scaling invariant.

Theorem 5.2.2 (Dirichlet Cheeger inequality [**Che71**]). Let Ω be a compact Riemannian manifold with boundary or a bounded Euclidean domain. Then the first Dirichlet eigenvalue of Ω satisfies Cheeger's inequality

(5.2.2) $$\lambda_1^{\mathrm{D}}(\Omega) \geq \frac{1}{4}h_{\mathrm{D}}^2(\Omega).$$

In order to prove Theorem 5.2.2 we will need

Lemma 5.2.3. Let $\varphi \geq 0$ be a smooth function such that $\varphi|_{\partial\Omega} = 0$. Then

$$\int_{\Omega} |\nabla\varphi|\,\mathrm{d}x \geq h_{\mathrm{D}}(\Omega)\int_{\Omega} \varphi\,\mathrm{d}x.$$

Proof of Lemma 5.2.3. Applying a version of the co-area formula (5.1.3) for Riemannian manifolds (see, for instance, [**CadFar18**, §2.3]), we obtain

$$\int_{\Omega} |\nabla\varphi|\,\mathrm{d}x = \int_0^{\infty}\int_{\mathcal{L}_\varphi(t)} \mathrm{d}\Sigma_t\,\mathrm{d}t = \int_0^{\infty} \mathrm{Vol}_{d-1}(\mathcal{L}_\varphi(t))\,\mathrm{d}t$$

$$\geq h_{\mathrm{D}}\int_0^{\infty} V_\varphi(t)\,\mathrm{d}t = h_{\mathrm{D}}\int_{\Omega} \varphi\,\mathrm{d}x,$$

where the last equality follows from the layer-cake representation (5.1.7). $\square$

Proof of Theorem 5.2.2. We follow the argument given in [**SchYau94**, §III.1]; see also [**Bus80**]. Let u be the first Dirichlet eigenfunction of Ω. Then

(5.2.3) $$\int_{\Omega} |\nabla(u^2)|\,\mathrm{d}x = 2\int_{\Omega} |u|\,|\nabla u|\,\mathrm{d}x \leq 2\|u\|_{L^2(\Omega)}\|\nabla u\|_{L^2(\Omega)}$$

$$= 2\sqrt{\lambda_1^{\mathrm{D}}(\Omega)}\|u\|_{L^2(\Omega)}^2.$$

Here the first inequality is simply the Cauchy–Schwarz inequality, and the last equality holds since $\lambda_1^{\mathrm{D}}(\Omega) = R[u]$, where $R[u]$ is the Rayleigh quotient of u.

We now use Lemma 5.2.3 with $\varphi := u^2$, which implies

$$\int_\Omega \left| \nabla \left(u^2 \right) \right| \, \mathrm{d}x \ge h_{\mathrm{D}} \|u\|_{L^2(\Omega)}^2.$$

Combining this with the inequality (5.2.3), we get $2\sqrt{\lambda_1^{\mathrm{D}}(\Omega)} \ge h_D$, and hence (5.2.2). □

Consider now the case of the Neumann boundary conditions.

Definition 5.2.4 (The Neumann Cheeger constant). Let Ω be a compact Riemannian manifold (with or without boundary) or a bounded Euclidean domain, of dimension d. The *Neumann Cheeger constant* is defined by

$$(5.2.4) \qquad h_{\mathrm{N}} := h_{\mathrm{N}}(\Omega) = \inf_\Gamma \frac{\mathrm{Vol}_{d-1}(\Gamma)}{\min\left\{\mathrm{Vol}_d(\Omega_1), \mathrm{Vol}_d(\Omega_2)\right\}},$$

where the infimum is taken over all smooth hypersurfaces Γ (not necessarily connected) separating Ω into two open sets Ω_1 and Ω_2; see Figure 5.1.

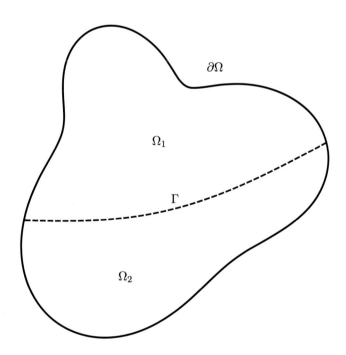

Figure 5.1. A hypersurface Γ splitting a domain Ω into two open sets Ω_1 and Ω_2.

Theorem 5.2.5 (Neumann Cheeger inequality). Let Ω be a compact Riemannian manifold with boundary or a bounded Lipschitz domain, of dimension d, and let $\lambda_2^N(\Omega)$ be its first nonzero eigenvalue of the Neumann Laplacian $-\Delta_\Omega^N$. Then

$$(5.2.5) \qquad\qquad \lambda_2^N(\Omega) \geq \frac{1}{4}h_N^2(\Omega).$$

Proof. By Corollary 4.1.34 of Courant's nodal domain theorem, an eigenfunction u corresponding to the first nonzero eigenvalue has exactly two nodal domains Ω_+ and Ω_- separated by the nodal set $\mathcal{Z}_u$. Without loss of generality, assume that $\mathrm{Vol}_d(\Omega_+) \leq \mathrm{Vol}_d(\Omega_-)$. The function u satisfies mixed boundary conditions on $\partial\Omega_+$: the Dirichlet one on $\mathcal{Z}_u$ and the Neumann one on $\partial\Omega_+ \setminus \mathcal{Z}_u = \partial\Omega_+ \cap \partial\Omega$ (if this part of $\partial\Omega_+$ is nonempty). The first eigenvalue of this mixed (Zaremba) problem satisfies $\lambda_1^Z(\Omega_+, \mathcal{Z}_u) = \lambda_2^N(\Omega)$. Let us define the *mixed Cheeger constant* (cf. [**Bus82**])

$$h_{\mathrm{DN}}(\Omega_+) = \inf_A \frac{\mathrm{Vol}_{d-1}(\partial A \cap \Omega_+)}{\mathrm{Vol}_d(A)},$$

where the infimum is taken over all open sets $A \subset \Omega_+$ with smooth boundary such that $\partial A \cap \mathcal{Z}_u = \emptyset$. Arguing as in the proof of Theorem 5.2.2 we obtain

$$\lambda_1^Z(\Omega_+, \mathcal{Z}_u) \geq \frac{1}{4}h_{\mathrm{DN}}^2(\Omega_+).$$

At the same time,

$$h_{\mathrm{DN}}(\Omega_+) \geq h_N(\Omega).$$

Indeed, the volume of any subdomain $A \subset \Omega_+$ is smaller than $\mathrm{Vol}_d(\Omega_+) \leq \mathrm{Vol}_d(\Omega_-)$, and $\Gamma := \partial A$ can be taken as a separating hypersurface for Ω in (5.2.4). Therefore,

$$\lambda_2^N(\Omega) = \lambda_1^Z(\Omega_+, \mathcal{Z}_u) \geq \frac{1}{4}h_{\mathrm{DN}}^2(\Omega_+) \geq \frac{1}{4}h_N^2(\Omega),$$

which completes the proof of (5.2.5). $\qquad\qquad\qquad\qquad\qquad\qquad \square$

An exact analogue of Theorem 5.2.5 holds for closed Riemannian manifolds.

Theorem 5.2.6 (Cheeger's inequality for closed manifolds). Let M be a closed Riemannian manifold of dimension d, and let $\lambda_1(\Omega)$ be the first nonzero eigenvalue of the Laplace–Beltrami operator $-\Delta_M$ (we recall that for a closed connected manifold, in our Notation 2.1.41, $0 = \lambda_0 < \lambda_1$). Then

$$\lambda_1(M) \geq \frac{1}{4}h_N^2(M),$$

where $h_N(M)$ is the Neumann Cheeger constant (5.2.4).

The proof of Theorem 5.2.6 is almost identical to that of Theorem 5.2.5, the only difference being that instead of the mixed problem on Ω_+ we have a pure Dirichlet problem and should use the Cheeger constant $h_D(\Omega_+)$ instead of $h_{DN}(\Omega_+)$ in the intermediate bounds.

5.2.2. Examples and further results. The following example shows that in the Riemannian setting Cheeger's inequality is sharp in any dimension.

Example 5.2.7. Let $\mathbb{H}^d$ be the hyperbolic space of constant sectional curvature -1 (see [**Cha84**, §2.5] and [**Bur98**, §3.4] for the definitions). Let $B_r \subset \mathbb{H}^d$ be a geodesic ball of radius r. An explicit computation shows that

$$\frac{\mathrm{Vol}_{d-1}(\partial B_r)}{\mathrm{Vol}_d(B_r)} > d - 1$$

for any $r > 0$. Moreover, the isoperimetric inequality for the hyperbolic space (see [**Oss78**, formula (4.23)]) states that a geodesic ball has the minimal volume of the boundary among all smooth domains of given volume. Therefore, $h_D(B_r) > d - 1$ for any $r > 0$. At the same time, another computation yields

$$\lambda_1^D(B_r) = \frac{(d-1)^2}{4} + O\left(\frac{1}{r^2}\right) \qquad \text{as } r \to \infty.$$

Therefore, the inequality (5.2.2) is sharp, with the equality attained in the limit as the radius of the geodesic ball in the hyperbolic space tends to infinity. We refer to [**Bus80**] for further details on this example, as well as its generalisation to the case of closed manifolds.

■ **Exercise 5.2.8.** Using the isoperimetric inequality for the sphere, show that

$$h_N\left(\mathbb{S}^d\right) = \frac{2}{B\left(\frac{d}{2}, \frac{1}{2}\right)},$$

where B is the Euler beta function [**DLMF22**, §5.12]. In particular, show that $h_N\left(\mathbb{S}^2\right) = 1$.

Example 5.2.7 admits the following important extension ([**Yau75**]; see also [**Cha84**]). Let M be a complete simply connected d-dimensional manifold with all sectional curvatures bounded above by some $-\kappa < 0$. Using comparison theorems, one can generalise the isoperimetric inequality mentioned above to manifolds of variable negative curvature. For any bounded domain $\Omega \subset M$ we have

$$\frac{\mathrm{Vol}_{d-1}(\partial\Omega)}{\mathrm{Vol}_d(\Omega)} > (d-1)\sqrt{\kappa}.$$

Cheeger's inequality then implies *McKean's inequality* [**McK70**],

$$\lambda_1^{\mathrm{D}}(\Omega) > \frac{(d-1)^2\kappa}{4},$$

for any bounded domain $\Omega \subset M$. Note that this inequality has no analogue in the Euclidean space: there exists no nontrivial uniform bound for the first Dirichlet eigenvalue of a bounded domain in $\mathbb{R}^d$.

It follows from Cheeger's inequality that if the first eigenvalue is small, the Cheeger constant is small as well. In fact, for closed manifolds the converse is also true. Recall that the Ricci curvature Ric of a Riemannian manifold (M,g) is a 2-tensor which is the trace of the Riemann curvature tensor (see [**Bur98**, §4.1.1]). We write $\mathrm{Ric} \geq -\kappa$ if $\mathrm{Ric}(\xi,\xi) \geq -\kappa\,|\xi|_g^2$ for any $\xi \in TM$.

Theorem 5.2.9 (Buser's inequality [**Bus82**]). Let M be a closed Riemannian manifold of dimension d, with Ricci curvature $\mathrm{Ric} \geq -\kappa$, $\kappa \geq 0$. Then

$$(5.2.6) \qquad \lambda_1(M) \leq 2\sqrt{(d-1)\kappa}\,h_{\mathrm{N}}(M) + 10h_{\mathrm{N}}^2(M).$$

As indicated in [**Bus82**], there is no direct analogue of Theorem 5.2.9 for manifolds with boundary.

Example 5.2.10 (Cheeger's dumbbell). Buser's inequality can be illustrated by the following example. Let M_ε be a surface, obtained by taking two identical round spheres and smoothly attaching them to each other by a thin cylinder of length one and radius $\varepsilon > 0$; see Figure 5.2. Take a test function which is equal to 1 on one sphere, -1 on the other, and is changing linearly along the cylinder in such a way that it is orthogonal to constants on M_ε. It then follows from the variational principle that $\lambda_1(M_\varepsilon) = O(\varepsilon)$ as $\varepsilon \to 0$. Moreover, with some extra work one can show that $\lambda_1(M_\varepsilon) \neq o(\varepsilon)$; see [**JimMor92**]. At the same time, it is clear that $h_{\mathrm{N}}(M_\varepsilon) = O(\varepsilon)$. This explains the presence of the first term (which is linear in h_{N}) in Buser's inequality.

■ **Exercise 5.2.11.** Use Exercise 5.2.8 to show that the term containing h_{N}^2 is essential in Buser's inequality (5.2.6). Another way to verify this is to consider a sequence of flat square tori $\mathbb{T}_n = \mathbb{R}^2/(n\mathbb{Z})^2$ as $n \to \infty$. See [**Ben15**, Example 3.6] for further details on Cheeger's constants of the flat tori and the Klein bottles.

Remark 5.2.12. The Ricci curvature assumption in Buser's inequality is also necessary: there exists a sequence of metrics on a torus with Ricci curvature unbounded from below, such that their Cheeger constants h_{N} tend to zero,

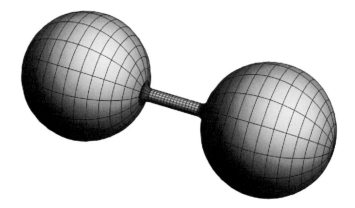

Figure 5.2. Cheeger's dumbbell.

and the first nonzero eigenvalues are uniformly bounded from below. We refer to [**Col17**, Example 23] for details. Let us also note that a lower bound on the Ricci curvature often arises as an assumption in spectral inequalities; see [**HasKokPol16**]. At the same time, the dependence on the dimension in the first term of (5.2.6) can be removed; as was shown in [**Led04**, Theorem 5.2], see also [**DePMon21**, formula (7)],

$$\lambda_1(M) \leq \max\{6\sqrt{\kappa}h_{\mathrm{N}}, 36h_{\mathrm{N}}^2\}.$$

5.2.3. The first eigenvalue and the inradius of planar domains. Let us present another application of Cheeger's inequality. Our exposition closely follows [**Gri06**].

Let Ω be a simply connected planar domain, and let ρ_Ω be its inradius, i.e., the radius of the largest disk contained inside Ω. Define the *reduced inradius*

$$\widetilde{\rho}_\Omega = \frac{\rho_\Omega}{1 + \frac{\pi\rho_\Omega^2}{|\Omega|}},$$

where $|\Omega| = \mathrm{Area}(\Omega)$. Clearly, $0 < \frac{\pi\rho_\Omega^2}{|\Omega|} \leq 1$ and hence $\frac{\rho_\Omega}{2} < \widetilde{\rho}_\Omega \leq \rho_\Omega$.

Theorem 5.2.13 ([**Gri06**]). *The first Dirichlet eigenvalue of simply connected $\Omega \subset \mathbb{R}^2$ satisfies*

$$(5.2.7) \qquad \lambda_1^{\mathrm{D}}(\Omega) \geq \frac{1}{4\widetilde{\rho}_\Omega^2}.$$

Remark 5.2.14. Note that by domain monotonicity, $\lambda_1^{\mathrm{D}}(\Omega) \leq \frac{j_{0,1}^2}{\rho_\Omega^2}$, where the right-hand side is the first Dirichlet eigenvalue of a disk of radius ρ_Ω; cf. Proposition 4.2.3. Together with (5.2.7), it means that $\lambda_1(\Omega)\rho_\Omega^2$ is uniformly bounded away from both zero and infinity for all simply connected planar domains. Earlier versions of (5.2.7) were obtained in [**Mak65**] and [**Hay78**]; see also an improvement in [**BañCar94**]. In [**Oss77**], the bound was extended to nonsimply connected planar domains, for which the constant on the right-hand side depends on the connectivity. In higher dimensions, a straightforward generalisation of (5.2.7) is false. Indeed, take a unit cube, split it into small cubes with the side length $\frac{1}{n}$, and remove all the vertices of those cubes. The remaining open set is simply connected, its inradius tends to zero as $n \to \infty$, while the first Dirichlet eigenvalue remains unchanged, since a point has capacity zero in $\mathbb{R}^3$ (see Remark 3.2.15). We refer to [**MazShu05**] for a more delicate higher-dimensional generalisation of (5.2.7).

Theorem 5.2.13 immediately follows from Cheeger's inequality combined with

Proposition 5.2.15. Let $\Omega \subset \mathbb{R}^2$ be a simply connected domain. Then

$$(5.2.8) \qquad\qquad h_{\mathrm{D}}(\Omega) \geq \frac{1}{\widetilde{\rho}_\Omega}.$$

The proof of the proposition is based on a Bonnesen-type isoperimetric inequality originally due to A. Besicovitch, which could be viewed as a strengthening of the usual isoperimetric inequality for planar domains.

Theorem 5.2.16 ([**Gri06**], [**Oss78**])**.** Let $\Omega \subset \mathbb{R}^2$ be simply connected. Then

$$(5.2.9) \qquad\qquad |\partial\Omega|^2 - 4\pi\,|\Omega| \geq (|\partial\Omega| - 2\pi\rho_\Omega)^2\,,$$

where $|\Omega| := \mathrm{Vol}_2(\Omega)$ denotes the area of Ω and $|\partial A| = \mathrm{Vol}_1(\partial A)$ is its perimeter.

Proof of Proposition 5.2.15. Recall the definition of the Dirichlet Cheeger constant in the planar setting,

$$h_{\mathrm{D}}(\Omega) = \inf\left\{ \frac{|\partial A|}{|A|} : A \Subset \Omega \text{ smooth} \right\}.$$

Since Ω is simply connected, it suffices to consider only simply connected A. Indeed, if A is not simply connected, filling in the holes increases the area and decreases the perimeter, while keeping the set inside Ω.

Let us now show that $A \subset \Omega$ implies that $\widetilde{\rho}_A \leq \widetilde{\rho}_\Omega$. Indeed, consider a function

$$f_a(\rho) := \frac{\rho}{1 + \frac{\pi\rho^2}{a}}.$$

It is easy to check that $f_a'(\rho) \geq 0$ if $\pi\rho^2 \leq a$. Hence, for these values of ρ, $f_a(\rho)$ in increasing. Since $|A| \leq |\Omega| := \mathrm{Vol}_2(\Omega)$,

$$\widetilde{\rho}_A = f_{|A|}(\rho_A) \leq f_{|\Omega|}(\rho_A) \leq f_{|\Omega|}(\rho_\Omega) = \widetilde{\rho}_\Omega.$$

Now, applying (5.2.9) to A, we get

$$|\partial A|^2 - 4\pi |A| \geq (|\partial A| - 2\pi\rho_A)^2.$$

Therefore,

$$\rho_A |\partial A| \geq |A| + \pi\rho_A^2,$$

and hence

$$\frac{|\partial A|}{|A|} \geq \frac{1 + \frac{\pi\rho_A^2}{|A|}}{\rho_A} = \frac{1}{\widetilde{\rho}_A} \geq \frac{1}{\widetilde{\rho}_\Omega}.$$

Since $A \Subset \Omega$ is arbitrary, this completes the proof of the proposition.   □

Let us show that Proposition 5.2.15 gives a sharp estimate. We claim that

$$h_{\mathrm{D}}(\mathbb{D}) = 2 = \frac{1}{\widetilde{\rho}_{\mathbb{D}}}$$

for the unit disk. One can compute the Cheeger constant $h_{\mathrm{D}}(\mathbb{D})$ for the unit disk using the isoperimetric inequality. The following useful lemma provides a more elementary way to do this.

Lemma 5.2.17 ([**Gri06**, Proposition 1]). Let $\Omega \subset \mathbb{R}^d$, let V be a smooth vector field on Ω, and let $h \geq 0$. Assume that $|V(x)| \leq 1$ and $\mathrm{div}\, V(x) \geq h$ for all $x \in \Omega$. Then $h_{\mathrm{D}}(\Omega) \geq h$.

Proof. Let $A \Subset \Omega$ be an open set with smooth boundary. Then

$$\mathrm{Vol}_d(\partial A) \geq \int_{\partial A} \langle V(x), n \rangle \, \mathrm{d}s = \int_A \mathrm{div}\, V(x) \, \mathrm{d}x \geq h \, \mathrm{Vol}_d(A),$$

where the equality in the middle holds by the divergence theorem and the inequalities follow from the assumptions. The result then immediately follows from (5.2.2).   □

Example 5.2.18 ([**Gri06**]). Let $\mathbb{B}^d \subset \mathbb{R}^d$ be the unit ball. Choosing $A \Subset \mathbb{B}^d$ arbitrarily close to $\mathbb{B}^d$ in (5.2.2) and taking into account that $\frac{\mathrm{Vol}_{d-1}(\mathbb{S}^d)}{\mathrm{Vol}_d(\mathbb{B}^d)} = d$, we find that $h_{\mathrm{D}}(\mathbb{B}^d) \leq d$. At the same time, applying the lemma above to the vector field $V(x) = x$ we get $h_{\mathrm{D}}(\mathbb{B}^d) \geq \mathrm{div}\, x = d$. Therefore, $h_D(\mathbb{B}^d) = d$.

The following two remarks give some more information on the optimality of the constant in Cheeger's inequality.[3]

Remark 5.2.19. Recall that by Exercise 1.2.21, $\lambda_1(\mathbb{B}^d) = j^2_{\frac{d}{2}-1,1}$, i.e., the square of the first zero of the Bessel function $J_{\frac{d}{2}-1}$. Using the asymptotics of the first Bessel zero as the order of the Bessel function tends to infinity [**Wat95**, §15.81], [**DLMF22**, Eq. 10.21.40], we observe that $\lambda_1(\mathbb{B}^d) = \frac{d^2}{4}(1 + o(1))$ as $d \to \infty$. Since $h_D(\mathbb{B}^d) = d$, this shows that the constant $1/4$ in Cheeger's inequality (5.2.2) is asymptotically sharp for Euclidean domains as the dimension $d \to \infty$. We refer to [**Fto21**, **BriButPri22**] for further discussion and related results.

Remark 5.2.20. It would be interesting to understand whether the constant $\frac{1}{4}$ in Cheeger's inequality (5.2.2) admits an improvement for Euclidean domains of a given dimension. For convex planar domans, such a result was obtained in [**Par17**]. In the same paper, a nice way to unify the inequalities (5.2.2), (5.2.7), and (5.2.8) is presented. Indeed, all these inequalities can be viewed as relations between the first Dirichlet eigenvalues of the p-Laplacians $-\Delta_p$, which are nonlinear operators defined by

$$\Delta_p u = \operatorname{div}\left(|\nabla u|^{p-2}\nabla u\right).$$

Note that for $p = 2$ it is the usual Laplace operator. Moreover, one can show that $\lambda_1(-\Delta_p, \Omega) \to h_D(\Omega)$ as $p \to 1$ and $\lambda_1(-\Delta_p, \Omega) \to \frac{1}{\rho_\Omega}$ as $p \to \infty$.

In conclusion, let us note that we have covered just a few aspects of Cheeger's inequality. In particular, aside from its significance in analysis and geometry, it has important applications to probability and graph theory. For further reading on this topic see, for instance, [**Chu97**].

5.3. Upper bounds for Laplace eigenvalues

5.3.1. The Szegő–Weinberger inequality. The Faber–Krahn inequality has stimulated further research on isoperimetric inequalities for Laplace eigenvalues in various settings. Let us start with the Neumann problem for bounded Euclidean domains Ω. The Neumann spectrum $0 = \mu_1 < \mu_2 \leq \cdots$, $\mu_j = \mu_j(\Omega) = \lambda_j^N(\Omega)$, always starts with the zero eigenvalue, and therefore the Neumann analogue of the fundamental tone is the second (i.e., first nonzero) Neumann eigenvalue μ_2. Recall that by formula (3.1.11), the first nonzero Neumann eigenvalue is given by

$$(5.3.1) \qquad \mu_2(\Omega) = \min_{\substack{u \in H^1(\Omega)\setminus\{0\} \\ u \perp 1}} \frac{\|\nabla u\|^2_{L^2(\Omega)}}{\|u\|^2_{L^2(\Omega)}}.$$

---

[3]We thank Dorin Bucur and Dmitry Faifman for useful discussions on Remarks 5.2.19 and 5.2.20.

Remark 5.3.1 (Physical interpretation of μ_2). Recall that the Neumann boundary conditions for the heat equation correspond to a perfectly insulated boundary. Therefore, as the time $t \to \infty$, the temperature distribution becomes constant at each point of the domain; mathematically, this follows from the fact that the first Neumann eigenvalue μ_1 is equal to zero. The first nonzero Neumann eigenvalue μ_2 defines the exponential rate of convergence to this constant distribution.

As follows from Exercise 1.1.15, Neumann (respectively, Dirichlet) eigenvalues do not admit nontrivial lower (respectively, upper) bounds under the volume constraint. Therefore, while in the Dirichlet case we were looking for a minimum of the first eigenvalue, in the Neumann case we should be looking for a maximum. The following theorem was first proved by G. Szegő [**Sze54**] for simply connected planar domains, and it was later generalised by H. F. Weinberger [**Wei56**] to arbitrary domains in any dimension.

Gábor **Szegő**

(1895–1985)

Theorem 5.3.2 (Szegő–Weinberger inequality). Let $\Omega \subset \mathbb{R}^d$ be a Lipschitz domain, and let $\Omega^* \subset \mathbb{R}^d$ be a ball of the same volume. Then $\mu_2(\Omega) \le \mu_2(\Omega^*)$ with equality attained if and only if Ω is a ball.

The following exercise forms one of the crucial steps in the proof of Theorem 5.3.2.

■ **Exercise 5.3.3.** Consider the ball $B_R^d \subset \mathbb{R}^d$ of radius R. Using your solution of Exercise 1.2.21 or directly by separation of variables in spherical coordinates, show that

$$(5.3.2) \qquad \mu_2\left(B_R^d\right) = \left(\frac{p_{d,1,1}'}{R}\right)^2,$$

where $p'_{d,1,1}$ is the first zero of the derivative of an ultraspherical Bessel function $U_{d,1}(r) := r^{1-\frac{d}{2}} J_{\frac{d}{2}}(r)$. Moreover, show that the multiplicity of the eigenvalue $\mu_2\left(B_R^d\right)$ is equal to d and that the corresponding eigenfunctions are given by $u_i(x) = \frac{g(r)x_i}{r}$, $i = 1, \ldots, d$, where x_i are the coordinate functions and

$$(5.3.3) \qquad\qquad g(r) = r^{1-\frac{d}{2}} J_{\frac{d}{2}}\left(\frac{p'_{d,1,1}r}{R}\right).$$

Proof of Theorem 5.3.2. Let R be the radius of the ball Ω^*, and let $\mu_2^* := \mu_2(\Omega^*) = \mu_2(B_R^d)$ be its first nonzero eigenvalue, given by (5.3.2).

It is easy to show using Bessel equation (1.1.16) that the function $g(r)$ defined by (5.3.3) satisfies the equation

$$(5.3.4) \qquad g''(r) + \frac{d-1}{r}g'(r) + \left(\mu_2^* - \frac{d-1}{r^2}\right)g(r) = 0.$$

In particular, $r = R$ is the first zero of $g'(r)$, and it follows from (1.1.17) that $g(r)$ is monotone increasing and positive for $0 < r < R$. Let us define an extension of $g(r)$:

$$G(R) := \begin{cases} g(r) & \text{if } r \le R, \\ g(R) & \text{if } r > R. \end{cases}$$

It is clear that $G(r) \in C([0, +\infty))$ and it follows from (1.1.17) that $\frac{G(r)}{r}$ has bounded derivatives as $r \to 0^+$. Therefore, the functions $f_i(x) := \frac{G(r)x_i}{r}$, $i = 1, \ldots, d$, are in $H^1(\Omega)$.

We will use the following.

Lemma 5.3.4 (The "centre of mass" lemma). There exists a choice of the origin O of the coordinate system such that

$$\int_\Omega f_i(x)\,\mathrm{d}x = \int_\Omega \frac{G(r)x_i}{r}\,\mathrm{d}x = 0 \qquad \text{for all } i = 1, \ldots, d.$$

This lemma is proved using a topological argument. In fact, an argument of this kind appears also in the proof of Szegő, as well as in the proof of Hersch's inequality; see §5.3.2. Let us postpone the proof of Lemma 5.3.4 for later, and note that for the choice of the origin O given by this lemma, the functions $f_i(x)$, $i = 1, \ldots, d$, are orthogonal to constants and hence admissible for the variational characterisation (5.3.1).

Let us calculate the Rayleigh quotients $R[f_i]$. Taking into account that $\frac{\partial r}{\partial x_j} = \frac{x_j}{r}$, we have

$$\frac{\partial f_i(x)}{\partial x_j} = \frac{G'(r)x_ix_j}{r^2} - \frac{G(r)x_ix_j}{r^3} + \delta_{ij}\frac{G(r)}{r}, \qquad i,j = 1,\ldots,d.$$

Therefore, a direct computation yields

$$|\nabla f_i|^2 = \sum_{j=1}^{d}\left(\frac{\partial f_i}{\partial x_j}\right)^2 = \frac{G'(r)^2x_i^2}{r^2} + \frac{G(r)^2\left(1 - \frac{x_i^2}{r^2}\right)}{r^2}, \qquad i = 1,\ldots,d.$$

Thus by the variational principle we get

$$\left(\int_\Omega \frac{G(r)^2}{r^2}x_i^2\,\mathrm{d}x\right)\mu_2(\Omega) \le \int_\Omega \left(\frac{G'(r)^2x_i^2}{r^2} + \frac{G(r)^2\left(1 - \frac{x_i^2}{r^2}\right)}{r^2}\right)\mathrm{d}x.$$

Summing up for $i = 1,\ldots,d$, we obtain

$$(5.3.5) \qquad \mu_2(\Omega) \le \frac{\int_\Omega \left(G'(r)^2 + \frac{(d-1)G(r)^2}{r^2}\right)\mathrm{d}x}{\int_\Omega G(r)^2\,\mathrm{d}x}.$$

Let $\Omega_1 := \Omega \cap \Omega^*$ and $\Omega_2 := \Omega \setminus \Omega^*$, where we assume that $\Omega^* = B_{O,R}^d$ is now centred at O; see Figure 5.3.

Then

$$\int_\Omega G(r)^2\,\mathrm{d}x = \int_{\Omega_1} G(r)^2\,\mathrm{d}x + G(R)^2\int_{\Omega_2}\mathrm{d}x,$$

and, since $G(r)$ is nondecreasing,

$$(5.3.6) \qquad \begin{aligned}\int_{\Omega^*} G(r)^2\,\mathrm{d}x &= \int_{\Omega_1} G(r)^2\,\mathrm{d}x + \int_{\Omega^*\setminus\Omega_1} G(r)^2\,\mathrm{d}x\\ &\le \int_{\Omega_1} G(r)^2\,\mathrm{d}x + G(R)^2\int_{\Omega^*\setminus\Omega_1}\mathrm{d}x.\end{aligned}$$

Note that $\mathrm{Vol}(\Omega) = \mathrm{Vol}(\Omega^*)$, and hence $\mathrm{Vol}(\Omega_2) = \mathrm{Vol}(\Omega^* \setminus \Omega_1)$. Therefore,

$$\int_\Omega G(r)^2\,\mathrm{d}x \ge \int_{\Omega^*} G(r)^2\,\mathrm{d}x = \int_{\Omega^*} g(r)^2\,\mathrm{d}x.$$

Let us investigate the numerator in the Rayleigh quotient (5.3.5). Differentiating the integrand, we get

$$(5.3.7) \qquad \begin{aligned}\frac{\mathrm{d}}{\mathrm{d}r}&\left(G'(r)^2 + (d-1)\frac{G(r)^2}{r^2}\right)\\ &= 2G'(r)G''(r) + 2(d-1)\frac{rG(r)G'(r) - G(r)^2}{r^3}.\end{aligned}$$

For $r > R$ this expression is negative since $G(r)$ is constant. For $r \leq R$, we use the Bessel-type equation (5.3.4), which yields

$$g''(r) = -\frac{d-1}{r}g'(r) + \left(\frac{d-1}{r^2} - \mu_2^*\right)g(r).$$

Substituting it into (5.3.7) gives

$$\frac{\mathrm{d}}{\mathrm{d}r}\left(G'(r)^2 + (d-1)\frac{G(r)^2}{r^2}\right)$$

$$= -2\mu_2^* g(r)g'(r) - 2(d-1)\frac{(rg'(r) - g(r))^2}{r^3}$$

$$= -\mu_2^*(g(r)^2)' - 2(d-1)\frac{(rg'(r) - g(r))^2}{r^3} < 0,$$

since $g(r)^2$ is monotone increasing. Therefore, the integrand in the numerator in (5.3.5) is monotone decreasing for $r > 0$, and arguing as in (5.3.6), we get

$$(5.3.8) \quad \int_\Omega \left(G'(r)^2 + \frac{(d-1)G(r)^2}{r^2}\right)\mathrm{d}x \leq \int_{\Omega^*}\left(g'(r)^2 + \frac{(d-1)g(r)^2}{r^2}\right)\mathrm{d}x.$$

It follows that

$$(5.3.9) \quad \mu_2(\Omega) \leq \frac{\int_{\Omega^*}\left(g'(r)^2 + \frac{(d-1)g(r)^2}{r^2}\right)\mathrm{d}x}{\int_{\Omega^*} g(r)^2\,\mathrm{d}x}.$$

At the same time, since $\frac{g(r)x_i}{r}$ is an eigenfunction on Ω^* corresponding to the first nonzero eigenvalue μ_2^*, it realises the equality in the variational characterisation (5.3.1):

$$\mu_2^*\int_{\Omega^*}\frac{g(r)^2 x_i^2}{r^2}\,\mathrm{d}x = \int_{\Omega^*}\frac{g'(r)^2}{r^2}x_i^2 + \frac{g(r)^2\left(1 - \frac{x_i^2}{r^2}\right)}{r^2}\,\mathrm{d}x.$$

Summing up for $i = 1, \ldots, d$, we get

$$\mu_2^*\int_{\Omega^*} g(r)^2\,\mathrm{d}x = \int_{\Omega^*}\left(g'(r)^2 + \frac{(d-1)g(r)^2}{r^2}\right)\mathrm{d}x.$$

In view of (5.3.9), this implies

$$(5.3.10) \quad \mu_2(\Omega) \leq \mu_2^* = \mu_2(\Omega^*).$$

Moreover, it is easy to see that the equality in both (5.3.6) and (5.3.8) is attained if and only if $\Omega = \Omega^*$ up to a set of measure zero. Since by assumption Ω has Lipschitz boundary (which is a common assumption for the Neumann boundary value problem but in fact is not necessary for the validity of (5.3.10); see Remark 5.3.6), $\mu_2(\Omega) = \mu_2(\Omega^*)$ if and only if $\Omega = \Omega^*$. □

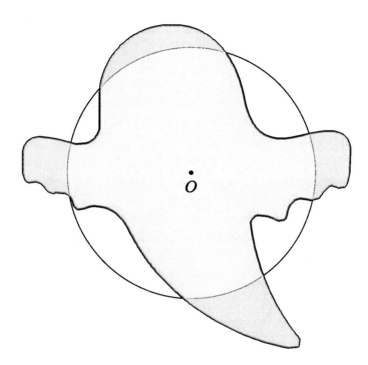

Figure 5.3. An example of a domain Ω and a ball $\Omega^* = B^d_{O,R}$. Ω is decomposed into $\Omega_1 := \Omega \cap \Omega^*$ (lighter shading) and $\Omega_2 := \Omega \setminus \Omega^*$ (darker shading).

It remains to prove the "centre of mass" lemma.

Proof of Lemma 5.3.4. Let $O_1 \in \mathbb{R}^d$ be the origin of some initial coordinate system. Consider a ball $B^d \supset \Omega$, and let $F = (F_1, \ldots, F_d) : B^d \to \mathbb{R}^d$ be a map defined by

$$F_i(y_1, \ldots, y_d) = \int_\Omega \frac{G(|x - y|)(x_i - y_i)}{|x - y|} \, \mathrm{d}x.$$

We want to show that there exists $y = (y_1, \ldots, y_d) \in B^d$ such that $F(y) = 0$. Indeed, if this is the case, choosing $O = y$ as the new origin of the coordinate system proves the result.

Clearly, F is continuous. Take $y \in \partial B^d$. The outward unit normal at y is given by $n = \frac{y}{|y|}$. Then,

$$\langle F(y), n \rangle = \sum_{i=1}^{d} F_i(y) \frac{y_i}{|y|} = \int_{\Omega} \frac{\langle x, y \rangle - |y|^2}{|x-y||y|} G(|x-y|) \, dx.$$

Since $\Omega \subset B^d$ and $y \in \partial B^d$, we have $|y| > |x|$ and hence $\langle x, y \rangle - |y|^2 < 0$. Therefore, $\langle F(y), n \rangle < 0$ for all $y \in \partial B^d$. Therefore, F is a continuous vector field on B^d which points inward on the boundary ∂B^d. Then there exists $\varepsilon > 0$ such that the continuous transformation $y \mapsto y + \varepsilon F(y)$ maps $\overline{B^d}$ into itself. Recall that by Brouwer's theorem (see, for example, [**Mil97**, §2]) such a transformation has a fixed point. Moreover, since F points inward on the boundary, there are no fixed points on ∂B^d. Hence, there exists $y \in B^d$ such that $F(y) = 0$. This completes the proof of the lemma. $\qquad \square$

Remark 5.3.5. For simply connected planar domains, G. Szegő [**Sze54**] proved Theorem 5.3.2 using the Riemann mapping theorem. While his method cannot be extended to higher dimensions, it has been generalised to other contexts, in particular, by R. Weinstock for the first nonzero Steklov eigenvalue (see §7.1.3), as well as by J. Hersch for the first nonzero Laplace eigenvalue on a sphere (see §5.3.2). Note also that Szegő's approach yields a stronger result (cf. Proposition 5.3.11):

$$\frac{1}{\mu_2(\Omega)} + \frac{1}{\mu_3(\Omega)} \geq \frac{2}{\mu_2(\Omega^*)},$$

for any simply connected planar domain Ω.

Remark 5.3.6 (Stability of the Szegő–Weinberger inequality). Similarly to the Faber–Krahn inequality, the Szegő–Weinberger inequality is stable; it was shown in [**BraPra12**, Theorem 4.1] (see also [**BraDeP17**]) that for any bounded open set $\Omega \subset \mathbb{R}^d$,

$$\mathrm{Vol}(\Omega^*)^{\frac{2}{d}} \mu_2(\Omega^*) - \mathrm{Vol}(\Omega)^{\frac{2}{d}} \mu_2(\Omega) \geq c_d \mathcal{A}(\Omega)^2,$$

where $\mathcal{A}(\Omega)$ is the Fraenkel asymmetry defined by (5.1.18); cf. Theorem 5.1.16. Moreover, the exponent 2 on the right-hand side is sharp. Note that the stability result, as well as the Szegő–Weinberger inequality itself, could be stated for arbitrary open bounded domains, with μ_2 defined by (5.3.1); assuming that the Neumann spectrum is discrete is not necessary.

Remark 5.3.7 (Higher eigenvalues). Among all Euclidean domains of fixed volume, the second nonzero Neumann eigenvalue μ_3 is maximised by a disjoint union of two identical balls. This result was proved in [**GirNadPol09**] for simply connected planar domains using an argument inspired by Szegő's proof, and it was extended in [**BucHen19**] to arbitrary Euclidean domains using an argument inspired by Weinberger's proof presented above. This result, together with the Szegő–Weinberger inequality, as well as with the Faber–Krahn and the Krahn–Szego inequalities, implies that Pólya's conjecture (3.3.8) is true for $k = 1, 2$. For higher eigenvalues, both for the Dirichlet and the Neumann boundary conditions, little is known apart from some numerics, showing that some peculiar shapes may arise as extremal geometries (see [**AntFre12**, Figures 1 and 2]).

5.3.2. Hersch's theorem for the first eigenvalue on the sphere. The Faber–Krahn and Szegő–Weinberger inequalities gave rise to a new direction in spectral geometry called *isoperimetric inequalities for Laplace eigenvalues*. In the following two subsections we are going to review some of the main results in this subject.

Let (M, g) be a closed d-dimensional Riemannian manifold, and let $\lambda_1(g) := \lambda_1(M, g)$ be the first nonzero eigenvalue of the Laplacian.[4] As was shown in (2.1.21), the quantity

$$\overline{\lambda}_1(M, g) := \lambda_1(g) \operatorname{Vol}(M, g)^{2/d}$$

is invariant under rescaling. Adapting the Cheeger dumbbell example (see Example 5.2.10) it is easy to see that $\inf_{g} \overline{\lambda}_1(M, g) = 0$ for any M, where the infimum is taken over all Riemannian metrics on M. Note that the eigenvalues of the closed eigenvalue problem satisfy the same variational principle as the Neumann eigenvalues, and therefore it is natural to consider the maximisation problem in this setting.

As it turns out, $\sup_{g} \overline{\lambda}_1(M, g) = +\infty$ on any compact Riemannian manifold M of dimension $d \geq 3$ [**ColDod94**]. We will therefore restrict ourselves to the case of surfaces. Note that if $d = 2$, $\overline{\lambda}_1(M, g) = \lambda_1(M, g) \operatorname{Area}(M, g)$. Let us start with the simplest surface, namely, the 2-sphere.

Theorem 5.3.8 (Hersch's theorem [**Her70**]). *Let* $(\mathbb{S}^2, g)$ *be a sphere endowed with a Riemannian metric* g. *Then*

(5.3.11) $$\overline{\lambda}_1(\mathbb{S}^2, g) \leq 8\pi,$$

with the equality attained if and only if g *is a round metric.*

Proof. We follow the argument given in [**SchYau94**]. Let g_0 be the standard round metric on $\mathbb{S}^2$. Then $\operatorname{Area}(\mathbb{S}^2, g_0) = 4\pi$ and, as was shown in §1.2.3, $\lambda_1(g_0) = 2$ with multiplicity three. The corresponding eigenspace is generated by the restriction to the sphere of the coordinate functions x_1, x_2, x_3; see Exercise 1.2.19. Let g be an arbitrary metric on $\mathbb{S}^2$ normalised in such

---

[4]We recall once more that this is a standard way to enumerate eigenvalues of closed Riemannian manifolds, which is different from the one we used for the Neumann problem on Euclidean domains.

a way that $\text{Area}(\mathbb{S}^2, g) = 4\pi$. We need to show that $\lambda_1(g) \leq 2$. We claim that there exists a conformal map $\varphi : (\mathbb{S}^2, g) \to (\mathbb{S}^2, g_0)$ such that the pull-back metric $\varphi^* g_0 = \alpha(x)g$ with $\alpha(x) > 0$. Indeed, by the uniformisation theorem, $\mathbb{S}^2$ admits a unique complex structure up to a diffeomorphism; see [**Tay11b**, Proposition 9.8]. At the same time, there is a one-to-one correspondence between complex structures and conformal classes on a Riemannian surface (see, for instance, [**Bob11**, Theorem 4]). Therefore, up to a diffeomorphism, there is a unique conformal class on $\mathbb{S}^2$, and the claim follows.

Set $y_i = x_i \circ \varphi$ for $i = 1, 2, 3$. Recall that by (3.1.14), the Dirichlet energy is conformally invariant in two dimensions, and hence

$$(5.3.12) \qquad \int_{\mathbb{S}^2} |\nabla y_i|_g^2 \, dV_g = \int_{\mathbb{S}^2} |\nabla x_i|^2 \, dV = 2 \int_{\mathbb{S}^2} x_i^2 \, dV = \frac{8\pi}{3},$$

where dV_g and dV are the area forms corresponding to the metrics g and g_0, respectively. Note that the last equality follows from the symmetry considerations.

For each $p \in \mathbb{S}^2$, we have $y_1^2(p) + y_2^2(p) + y_3^2(p) = 1$. Thus,

$$(5.3.13) \qquad \sum_{i=1}^{3} \frac{1}{R[y_i]} = \sum_{i=1}^{3} \frac{\int_{\mathbb{S}^2} y_i^2 \, dV_g}{\int_{\mathbb{S}^2} |\nabla y_i|_g^2 \, dV_g} = \frac{3}{8\pi} \int_{\mathbb{S}^2} \sum_{i=1}^{3} y_i^2 \, dV_g = \frac{3}{8\pi} \cdot 4\pi = \frac{3}{2}.$$

Therefore, for at least one of $i = 1, 2, 3$, we have $\frac{1}{R[y_i]} \geq \frac{1}{2}$, and hence $R[y_i] \leq 2$. If we were able to take this particular y_i as a test function for λ_1, that would have been the end of the proof. However, a priori $\int_{\mathbb{S}^2} y_i \, dV_g \neq 0$. At the same time, we still have the freedom to choose the conformal map φ. Our goal is to do it in such a way that

$$(5.3.14) \qquad \int_{\mathbb{S}^2} y_i \, dV_g = 0, \qquad i = 1, 2, 3.$$

In other words, the map φ must keep the center of mass at the origin; cf. Lemma 5.3.4.

In order to construct such a map φ, we use the group of conformal automorphisms of the sphere. Let $\mathbb{B}^3 \subset \mathbb{R}^3$ be the open unit ball. Given $\xi \in \mathbb{B}^3$, define a transformation $K_\xi : \overline{\mathbb{B}^3} \to \overline{\mathbb{B}^3}$ by the formula

$$(5.3.15) \qquad K_\xi(x) = \frac{(1 - |\xi|^2)x + (1 + 2\langle \xi, x \rangle + |x|^2)\xi}{1 + 2\langle \xi, x \rangle + |\xi|^2 |x|^2}.$$

■ **Exercise 5.3.9.** Show that K_ξ is a conformal map from $\mathbb{S}^2$ to itself.

Let us now define a map $H = (H_1, H_2, H_3) : \mathbb{B}^3 \to \mathbb{B}^3$ as

$$H_i(\xi) = \frac{1}{4\pi} \int\limits_{\mathbb{S}^2} x_i \circ K_\xi \circ \varphi \, dV_g, \qquad i = 1, 2, 3.$$

Lemma 5.3.10 (Hersch's lemma). There exists $\xi \in \mathbb{B}^3$ such that $H(\xi) = 0$.

Proof of Hersch's lemma. Let $\xi \in \partial\mathbb{B}^3 = \mathbb{S}^2$. Then it follows from (5.3.15) that $K_\xi(\mathbb{S}^2 \setminus \{-\xi\}) = \xi$. Therefore,

$$H_i(\xi) = \frac{1}{4\pi} \int\limits_{\mathbb{S}^2} x_i \circ K_\xi \circ \varphi \, dV_g = \frac{1}{4\pi} \int\limits_{\mathbb{S}^2} x_i(\xi) \, dV_g = \xi_i.$$

In other words, the map $H : \mathbb{B}^3 \to \mathbb{B}^3$ can be continuously extended to $\partial\mathbb{B}^3 = \mathbb{S}^2$, and it is the identity on the boundary. Suppose there is no $\xi \in \mathbb{B}^3$ such that $H(\xi) = 0$. Then the map $\frac{H(\xi)}{|H(\xi)|} : \overline{\mathbb{B}^3} \to \mathbb{S}^2$ is a retraction, i.e., a continuous map which is identity on $\mathbb{S}^2$. We get a contradiction with the no-retraction theorem, or, equivalently, with Brouwer's fixed point theorem (cf. the proof of Lemma 5.3.4). Indeed, if such a map exists, we can compose it with a central symmetry with respect to the origin and get a continuous map of $\overline{\mathbb{B}^3}$ into itself without fixed points, which is impossible. □

Replacing now φ in $y_i = x_i \circ \varphi$ by $K_\xi \circ \varphi$, where ξ is given by Hersch's lemma, yields (5.3.14). It remains to show that the equality in (5.3.11) is attained if and only if the metric g is round. Without loss of generality, assume that $\mathrm{Area}(\mathbb{S}^2, g) = 4\pi$. Suppose also that φ keeps the center of mass at the origin (if not, we replace it by $K_\xi \circ \varphi$ as above). Then the functions $y_i = x_i \circ \varphi_i$ are orthogonal to constants with respect to the measure dV_g. The equality $\lambda_1(g) = 2$ together with the variational principle implies that $R[y_i] = 2$, $i = 1, 2, 3$, and that y_i are the first nontrivial eigenfunctions of the Laplacian $-\Delta_g$ with the eigenvalue 2. At the same time, consider the pull-back of the standard round metric $\varphi^* g_0 = \alpha(x)g$. Since x_i are the first nontrivial eigenfunctions of $-\Delta_{g_0}$ with the eigenvalue 2, the functions y_i are the first nontrivial eigenfunctions of $-\Delta_{\alpha(x)g}$ with the same eigenvalue. Therefore,

$$2y_i = -\Delta_{\alpha(x)g} y_i = -\frac{1}{\alpha(x)} \Delta_g y_i = \frac{2}{\alpha(x)} y_i, \qquad i = 1, 2, 3,$$

which implies $\alpha(x) \equiv 1$, and hence $g = \varphi^* g_0$ is a round metric. □

The proof of Hersch's theorem implies, in fact, a stronger statement.

Proposition 5.3.11. For any metric g on S^2,

$$\sum_{i=1}^{3} \frac{1}{\lambda_i(\mathbb{S}^2, g)} \geq \frac{3\,\mathrm{Area}(S^2, g)}{8\pi}.$$

The proof of this result uses the following generalisation of the variational principle.

■ **Exercise 5.3.12** (Variational principle for the sum of eigenvalue reciprocals). Show that

$$\sum_{i=1}^{k} \frac{1}{\lambda_i(M, g)} = \sup \sum_{i=1}^{k} \frac{1}{R[\varphi_i]},$$

where the supremum is taken over all $0 \neq \varphi_i \in H^1(M, g)$, such that $\int_M \varphi_i \, dV = 0$ for $i = 1, \ldots, k$ and $(\nabla\varphi_i, \nabla\varphi_j)_{L^2(M,g)} = 0$ for $i \neq j$. A proof of this statement can be found in [**Ban80**, formula (3.7)]; see also [**YanYau80**].

Proof of Proposition 5.3.11. In view of (5.3.13), it remains to check that y_i, $i = 1, 2, 3$, can be taken as test functions in the variational characterisation given in Exercise 5.3.12. Indeed, y_i are orthogonal to constants by Hersch's lemma. Moreover,

$$(\nabla y_i, \nabla y_j)_{L^2(\mathbb{S}^2, g)} = \int_{\mathbb{S}^2} \langle \nabla x_i, \nabla x_j \rangle \, dV = 2 \int_{\mathbb{S}^2} \langle x_i, x_j \rangle \, dV = 0$$

for $i \neq j$, where the first equality follows from the conformal equivalence of the Dirichlet energy via the relation $2 \langle \nabla y_i, \nabla y_j \rangle = |\nabla(y_i + y_j)|^2 - |\nabla y_i|^2 - |\nabla y_j|^2$.   □

5.3.3. Topological upper bounds for eigenvalues on surfaces. Hersch's theorem has been the starting point for the study of the isoperimetric inequalities for eigenvalues on surfaces. This is an active area of research, with a number of important recent advances. The goal of this subsection is to review some of the results in this subject.

Recall that each orientable surfaces is homeomorphic to a sphere with $\gamma \geq 0$ handles. The number of handles γ is called the *genus* of a surface. In particular, the sphere itself has genus zero. Let us start with an extension of Hersch's estimate to surfaces of higher genus.

Theorem 5.3.13 (The Yang–Yau bound [**YanYau80**]). Let M be an orientable surface of genus γ. Then, for any Riemannian metric g on M,

(5.3.16)
$$\sum_{i=1}^{3} \frac{1}{\lambda_i(M,g)} \geq \frac{3\,\mathrm{Area}(M,g)}{8\pi\left[\frac{\gamma+3}{2}\right]},$$

where $[\cdot]$ denotes the integer part. As a consequence,

(5.3.17)
$$\overline{\lambda}_1(M,g) \leq 8\pi\left[\frac{\gamma+3}{2}\right].$$

Proof. We follow the argument in [**YanYau80**]. Assume that there exists a conformal branched covering (or, equivalently, a nonconstant holomorphic map) $\psi : (M,g) \to (\mathbb{S}^2, g_0)$ of degree m (see [**Bob11**, §1.2] for definitions and background). Away from a finite number of branch points, ψ is a covering map with m sheets. Consider the push-forward metric

(5.3.18)
$$g_* = \sum_{j=1}^{m} \left(\psi_j^{-1}\right)^* g$$

on $\mathbb{S}^2$. Here ψ_j is a mapping from the jth sheet of the covering to $\mathbb{S}^2$ which is well-defined by ψ away from the branch points. The metric g_* is a smooth metric away from the branch points, and at those points it has conical singularities; see [**KarNPP19**, §6] for details. In fact, one can show that $g_* = \rho g_0$, where $0 \leq \rho \in L^p(\mathbb{S}^2, g_0)$ for some $p > 1$, and the Laplace eigenvalues for such metrics can be defined using the variational principle in the same manner as for the smooth metrics. However, for the purpose of the present argument, it suffices to verify that the area defined by g_* is finite, which can be done by a direct computation [**YanYau80**, p. 58].

It is also not hard to check that for any $u \in C^1(\mathbb{S}^2)$,

(5.3.19)
$$\int_{\mathbb{S}^2} u\,\mathrm{d}V_* = \int_{M} (u \circ \psi)\,\mathrm{d}V_g$$

and

(5.3.20)
$$\int_{\mathbb{S}^2} |\nabla u|^2 \,\mathrm{d}V = \int_{\mathbb{S}^2} |\nabla u|_{g_*}^2 \,\mathrm{d}V_* = \frac{1}{m}\int_{M} |\nabla(u \circ \psi)|_g^2 \,\mathrm{d}V_g,$$

where $\mathrm{d}V_g$ and $\mathrm{d}V_*$ are the area forms corresponding to the metrics g on M and g_* on $\mathbb{S}^2$, respectively. Indeed, (5.3.19) follows from the definition of the pull-back measure $\mathrm{d}V_*$, and (5.3.20) follows from (5.3.18) and conformal equivalence of the Dirichlet energy on each sheet of the covering.

Let us now proceed as in the proof of Hersch's theorem. As before, let x_i, $i = 1, 2, 3$, be the coordinate functions on the round sphere $\mathbb{S}^2$. By Hersch's lemma, choose a conformal map $\varphi : \mathbb{S}^2 \to \mathbb{S}^2$ such that the center of mass of the measure dV_* is at the origin. Then the functions $x_i \circ \varphi$, $i = 1, 2, 3$, are orthogonal to constants on $(\mathbb{S}^2, g_*)$, and hence by (5.3.19), the functions $v_i = x_i \circ \varphi \circ \psi$ are orthogonal to constants on (M, g). Therefore, setting $\lambda_i := \lambda_i(M, g)$ and arguing as in Proposition 5.3.11, we obtain

$$(5.3.21) \qquad \sum_{i=1}^{3} \frac{1}{\lambda_i} \geq \frac{\int_M v_i^2 \, dV_g}{\int_M |\nabla v_i|_g^2 \, dV_g}.$$

Note that by (5.3.20) and (5.3.12), the denominator in each term on the right-hand side is equal to $\frac{8\pi m}{3}$. Moreover, since $\sum_{i=1}^{3} v_i^2 = \sum_{i=1}^{3} x_i^2 = 1$ pointwise, it follows from (5.3.21) that

$$\sum_{i=1}^{3} \frac{1}{\lambda_i} \geq \frac{3 \operatorname{Area}(M, g)}{8\pi m}.$$

To complete the proof of (5.3.16) it remains to note that, as known from the theory of Riemann surfaces, one can choose $m = \left[\frac{\gamma+3}{2} \right]$ [**Gun72**, p. 186]; see also Remark 5.3.14. Inequality (5.3.17) immediately follows from (5.3.16) since λ_1 is the smallest of the three eigenvalues. $\square$

Remark 5.3.14. In the context of the Yang–Yau inequality, a possibility of choosing $m = \left[\frac{\gamma+3}{2} \right]$ was first observed in [**ElSIli84**]. Originally, the inequality was stated in [**YanYau80**] with $m = \gamma + 1$.

Substantial new ideas are needed to extend the Yang–Yau theorem to nonorientable surfaces. This has been done in [**Kar16**], improving upon the approach of [**LiYau82**].

Theorem 5.3.15 ([**Kar16**]). Let M be a nonorientable surface with an orientable double cover of genus γ. Then

$$(5.3.22) \qquad \overline{\lambda}_1(M, g) \leq 16\pi \left[\frac{\gamma + 3}{2} \right].$$

Estimates (5.3.17) and (5.3.22) imply that the quantity

$$(5.3.23) \qquad \Lambda_1(M) = \sup_g \overline{\lambda}_1(M, g)$$

is finite for any surface M. If there exists a metric attaining the supremum in (5.3.23) on a given surface M, we say that this metric is λ_1-*maximal*. The study of λ_1-maximal metrics on surfaces is a rapidly developing subject; see

[**KarNPP21**, §2] and references therein. It turns out that such metrics give rise to minimal isometric immersions of surfaces into spheres $\mathbb{S}^r$ by the first eigenfunctions, where $r+1$ is the multiplicity of the corresponding first eigenvalue. For the time being, λ_1-maximal metrics are explicitly known only for a few surfaces of low genus:

- $\Lambda_1(\mathbb{S}^2) = 8\pi$, attained on the standard round metric (Hersch's theorem).

- $\Lambda_1(\mathbb{RP}^2) = 12\pi$, attained on the standard round metric [**LiYau82**].

- $\Lambda_1(\mathbb{T}^2) = \frac{8\pi^2}{\sqrt{3}}$, attained on the flat equilateral torus. This was conjectured by M. Berger in [**Ber73**] and proved by N. Nadirashvili in [**Nad96**].

- $\Lambda_1(\mathbb{K}) = \overline{\lambda}_1(\mathbb{K}, g_{\max}) = 12\pi E\left(\frac{2\sqrt{2}}{3}\right)$, where $\mathbb{K}$ is the Klein bottle, $g_{\max}$ is a certain metric of revolution, and E is a complete elliptic integral of the second kind. The Klein bottle $(\mathbb{K}, g_{\max})$ is a bipolar surface for the Lawson $\tau_{3,1}$-torus and admits a minimal immersion by the first eigenfunctions into $\mathbb{S}^4$. Unlike the examples above, this metric does not have constant curvature. It was proved to be extremal for the first eigenvalue in [**JakNadPol06**] and conjectured to be maximal. It was proved in [**ElSGiaJaz06**] that there are no other extremal metrics on $\mathbb{K}$, and it was shown to be maximal in [**CiaKarMed19**].

  Interestingly enough, all the λ_1-maximal metrics above also maximise the multiplicity of the first eigenvalue on their respective surfaces. On the sphere and on the projective plane it was proved in Corollary 4.4.7, and on the torus as well as on the Klein bottle it follows from a refinement of (4.4.3) obtained in [**Nad87**].

- $\Lambda_1(\Sigma_2) = 16\pi$, where Σ_2 is the surface of genus two. The maximum is attained on a metric with conical singularities on the Bolza surface, induced from the round metric on the sphere using the standard branched double covering. This result was first stated in [**JakLNNP05**]; however, the last step of the proof there hinged upon a numerical calculation. A complete analytic proof was obtained in [**NaySho19**] using new ideas from algebraic geometry. Note that the Bolza surface is characterised among all surfaces of genus two as the one having the largest automorphism group.

Finding the explicit values of $\Lambda_1(M)$ and the corresponding maximising metrics is an open question for all other surfaces.

Remark 5.3.16. All the λ_1-maximising metrics above are unique up to isometries and dilations, except for the surface of genus two, on which there exists a continuous family of maximisers. Moreover, it was shown in [**KarNPS21**] that all these maximisers, once again with the exception of the genus two case, satisfy certain stability properties. We also note that all λ_1-maximal metrics are highly symmetric, and the multiplicity of the first eigenvalue in all the examples except for the surface of genus two is the largest possible (cf. Corollary 4.4.7 and [**Nad87**]). One can observe as well that the equality in Yang–Yau inequality (5.3.17) is attained for $\gamma = 0$ and $\gamma = 2$; as was shown in [**Kar19**], this is not the case for all other genera. Further improvements have been recently obtained in [**Ros22a, Ros22b**] and [**KarVin22**].

Let us now present a brief overview of related results for higher eigenvalues. It was conjectured in [**Yau82**] and proved by N. Korevaar in [**Kor93**] (see also [**GriNetYau04**]) that there exists a constant $C > 0$ such that for any $k \geq 1$,

$$(5.3.24) \qquad \overline{\lambda}_k(M, g) \leq Ck(\gamma + 1)$$

on any Riemannian surface (M, g). A substantial improvement of Korevaar's bound was obtained in [**Has11**]:

$$\overline{\lambda}_k(M) \leq C(k + \gamma).$$

As in the case of the first eigenvalue, these results lead to the question regarding the existence of λ_k-maximising metrics and the values of

$$\Lambda_k(M) := \sup_g \overline{\lambda}_k(M, g)$$

for various k and M. The latter question has been recently completely answered for the sphere and for the real projective plane. It was conjectured in [**Nad02**] and shown in [**KarNPP21**] that

$$(5.3.25) \qquad \Lambda_k(\mathbb{S}^2) = 8\pi k, \qquad k \geq 1$$

(see also [**Nad02, Pet14**] for $k = 2$ and [**NadSir17**] for $k = 3$), with the supremum attained in the limit by a sequence of metrics degenerating to a disjoint union of k identical round spheres; see Figure 5.4.

This is a manifestation of the "bubbling phenomenon" which arises for the maximisers of higher eigenvalues; see [**NadSir15, Pet18, KarNPP19, KarSte20**]. Similarly, it was conjectured in [**KarNPP21**] and proved in [**Kar21**] (see also [**NadPen18**] for $k = 2$) that

$$(5.3.26) \qquad \Lambda_k(\mathbb{RP}^2) = 4\pi(2k + 1), \qquad k \geq 1.$$

For $k \geq 2$ the supremum is attained in the limit by a sequence of metrics degenerating to a union of $k - 1$ identical round spheres and a standard projective plane touching each other, such that the ratio of the areas of the projective plane and the spheres is equal to $\Lambda_1(\mathbb{RP}^2) : \Lambda_1(\mathbb{S}^2) = 3 : 2$.

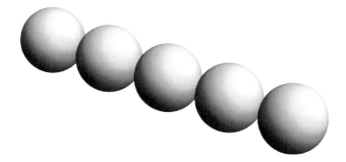

Figure 5.4. A disjoint union of five identical round spheres maximising $\lambda_5(\mathbb{S}^2, g)$.

Remark 5.3.17 (Korevaar's bound with an explicit constant). As was noted in [**KarNPP19**], using (5.3.25) and (5.3.26) one can make the constant C in Korevaar's bound (5.3.24) explicit. Indeed, a slight adaptation of the proof of Theorem 5.3.13 yields that (5.3.24) holds with $C = 8\pi \left\lceil \frac{\gamma+3}{2} \right\rceil$ for orientable surfaces and $C = 16\pi \left\lceil \frac{\gamma+3}{2} \right\rceil$ for nonorientable ones. In the latter case, γ is understood as the genus of the orientable double cover.

As was mentioned earlier, it was shown in [**ColDod94**] that $\Lambda_1(M) = +\infty$ for any Riemannian manifold M of dimension $d \geq 3$. Therefore, in higher dimensions, one needs to restrict the class of metrics over which the supremum is taken. For example, maximisation of the Laplace eigenvalues among metrics within a fixed conformal class is an interesting question in any dimension; see [**ColElS03**, **Kim22**, **KarSte22**, **Pet22**].

5.4. Universal inequalities

5.4.1. The Payne–Pólya–Weinberger inequality.
In 1956, L. E. Payne, G. Pólya, and H. F. Weinberger [**PayPólWei56**] proved the following.

Theorem 5.4.1 (The Payne–Pólya–Weinberger inequality). For any domain $\Omega \subset \mathbb{R}^d$, the eigenvalues of the Dirichlet Laplacian $\lambda_m = \lambda_m^D(\Omega)$ satisfy the gap estimates

$$(5.4.1) \qquad \lambda_{m+1} - \lambda_m \leq \frac{4}{dm} \sum_{j=1}^{m} \lambda_j$$

for each $m \in \mathbb{N}$.

The inequality (5.4.1) was improved to

$$(5.4.2) \qquad \sum_{j=1}^{m} \frac{\lambda_j}{\lambda_{m+1} - \lambda_j} \geq \frac{md}{4}$$

by G. N. Hile and M. H. Protter [**HilPro80**]. This is indeed stronger than
(5.4.1), which can be obtained from (5.4.2) by replacing all the λ_j in the
denominators in the left-hand side by λ_m.

Later, Hongcang Yang [**Yan91**] proved an even stronger inequality

$$(5.4.3) \qquad \sum_{j=1}^{m} (\lambda_{m+1} - \lambda_j)\left(\lambda_{m+1} - \left(1 + \frac{4}{d}\right), \lambda_j\right) \leq 0$$

which after some modifications implies the explicit estimate

$$(5.4.4) \qquad \lambda_{m+1} \leq \left(1 + \frac{4}{d}\right) \frac{1}{m} \sum_{j=1}^{m} \lambda_j.$$

These two inequalities are known as Yang's first and second inequalities,
respectively. We note that (5.4.3) still holds if we replace λ_{m+1} by an arbitrary $z \in (\lambda_m, \lambda_{m+1}]$ (see [**HarStu97**]) and that the sharpest explicit upper
bound on λ_{m+1} known so far is also derived from (5.4.3); see [**Ash99**, formula (3.33)].

The Payne–Pólya–Weinberger, Hile–Protter, and Yang inequalities are
commonly referred to as *universal estimates* for the eigenvalues of the Dirichlet Laplacian. These estimates are valid uniformly over all bounded domains
in $\mathbb{R}^d$ and depend only upon the dimension d. The derivation of all four results is similar and uses the variational principle with ingenious choices of
test functions, as well as the Cauchy–Schwarz inequality. We refer the reader
to the survey [**Ash99**] which provides the detailed proofs as well as the proof
of the implication

$$(5.4.3) \implies (5.4.4) \implies (5.4.2) \implies (5.4.1).$$

In 1997, E. M. Harrell and J. Stubbe [**HarStu97**] showed that all of
these results are consequences of a certain abstract operator identity and
that this identity has several other applications. This approach was further
simplified in [**LevPar02**], and we outline it in the next subsection. For
an alternative proof of Theorem 5.4.1 and other related results, see also
[**SchYau94**, §3.7] and [**Ura17**, Chapter 5].

5.4.2. Abstract commutator identities. We start with

> **Theorem 5.4.2** ([**LevPar02**, Theorem 2.2]). Let H and G be self-
> adjoint operators acting in a Hilbert space $\mathcal{H}$ with an inner product
> $(\cdot, \cdot) := (\cdot, \cdot)_{\mathcal{H}}$ and a norm $\| \cdot \| := \| \cdot \|_{\mathcal{H}}$. Assume that $G(\mathrm{Dom}(H)) \subseteq$
> $\mathrm{Dom}(H) \subseteq \mathrm{Dom}(G)$ and that H is semi-bounded from below. Let λ_j,
> $j \in \mathbb{N}$, be the eigenvalues of H (ordered nondecreasingly), and let u_j be

the corresponding orthonormal eigenvectors. Then for each fixed $j \in \mathbb{N}$

$$(5.4.5) \qquad \sum_{k=1}^{\infty} \frac{|([H,G]u_j, u_k)|^2}{\lambda_k - \lambda_j} = \sum_{k=1}^{\infty} (\lambda_k - \lambda_j) |(Gu_j, u_k)|^2$$

$$(5.4.6) \qquad\qquad\qquad = -\frac{1}{2}([[H,G],G]u_j, u_j).$$

Remark 5.4.3. Note that all the terms in the left-hand side of (5.4.5) with $\lambda_k = \lambda_j$ have vanishing denominators. However, as will be shown in the proof, these terms also have vanishing numerators and should be simply dropped from this and similar sums in the sequel.

Proof of Theorem 5.4.2. Obviously, we have

$$(5.4.7) \qquad [H,G]u_j = HGu_j - GHu_j = (H - \lambda_j)Gu_j.$$

Therefore,

$$(5.4.8) \qquad (G[H,G]u_j, u_j) = (G(H - \lambda_j)Gu_j, u_j).$$

Since G is selfadjoint, we have

$$(5.4.9) \quad \begin{aligned} (G(H - \lambda_j)Gu_j, u_j) &= ((H - \lambda_j)Gu_j, Gu_j) \\ &= \sum_{k=1}^{\infty} ((H - \lambda_j)Gu_j, u_k)(u_k, Gu_j) = \sum_{k=1}^{\infty} (\lambda_k - \lambda_j)|(Gu_j, u_k)|^2. \end{aligned}$$

We note that $[H,G]$ is skew-adjoint, since

$$[H,G]^* = (HG - GH)^* = GH - HG = -[H,G],$$

and therefore the left-hand side of (5.4.8) can be rewritten as

$$\begin{aligned} (G[H,G]u_j, u_j) &= -([[H,G],G]u_j, u_j) + ([H,G]Gu_j, u_j) \\ &= -([[H,G],G]u_j, u_j) - (u_j, G[H,G]u_j), \end{aligned}$$

so that

$$(G[H,G]u_j, u_j) = -\frac{1}{2}([[H,G],G]u_j, u_j)$$

(notice that $(G[H,G]u_j, u_j)$ is real; see (5.4.8) and (5.4.9)). This proves (5.4.6).

Since (5.4.7) implies

$$([H,G]u_j, u_k) = (\lambda_k - \lambda_j)(Gu_j, u_k),$$

this also proves (5.4.5). Obviously, $([H,G]u_j, u_k) = 0$ whenever $\lambda_k = \lambda_j$, and the notational convention of Remark 5.4.3 is therefore applicable. $\qquad \square$

We can now establish an abstract version of the Payne–Pólya–Weinberger inequality.

Theorem 5.4.4. Under the conditions of Theorem 5.4.2,

$$(5.4.10) \qquad -(\lambda_{m+1} - \lambda_m) \sum_{j=1}^{m} ([[H,G],G]u_j, u_j) \leq 2 \sum_{j=1}^{m} \|[H,G]u_j\|^2$$

for each $m \in \mathbb{N}$.

Proof. Let us sum up the equations (5.4.6) over $j = 1, \ldots, m$. Then we have

$$(5.4.11) \qquad \sum_{j=1}^{m} \sum_{k=1}^{\infty} \frac{|([H,G]u_j, u_k)|^2}{\lambda_k - \lambda_j} = -\frac{1}{2} \sum_{j=1}^{m} ([[H,G],G]u_j, u_j).$$

To estimate the left-hand side of (5.4.11) from above, we first note that since $[H,G]$ is skew-adjoint,

$$|([H,G]u_j, u_k)|^2 = |([H,G]u_k, u_j)|^2, \quad k, j \geq 1,$$

and all the terms with $k \leq m$ cancel out. Then we replace all the positive denominators by the smallest one $\lambda_{m+1} - \lambda_m$ and use Parseval's equality, giving

$$\sum_{j=1}^{m} \sum_{k=1}^{\infty} \frac{|([H,G]u_j, u_k)|^2}{\lambda_k - \lambda_j} = \sum_{j=1}^{m} \sum_{k=m+1}^{\infty} \frac{|([H,G]u_j, u_k)|^2}{\lambda_k - \lambda_j}$$

$$\leq \frac{1}{\lambda_{m+1} - \lambda_m} \sum_{j=1}^{m} \sum_{k=m+1}^{\infty} |([H,G]u_j, u_k)|^2$$

$$\leq \frac{1}{\lambda_{m+1} - \lambda_m} \sum_{j=1}^{m} \sum_{k=1}^{\infty} |([H,G]u_j, u_k)|^2$$

$$= \frac{1}{\lambda_{m+1} - \lambda_m} \sum_{j=1}^{m} \|[H,G]u_j\|^2.$$

Combining this with (5.4.11) proves (5.4.10). $\qquad \square$

An abstract version of Yang's inequality (5.4.3) is somewhat more complicated; for the proof of a slightly more general version see [**LevPar02**, Corollary 2.8].

Theorem 5.4.5. Under the condition of Theorem 5.4.2,

$$\sum_{j=1}^{m} (\lambda_{m+1} - \lambda_m) \|[H,G]u_j\|^2 \geq -\frac{1}{2} \sum_{j=1}^{m} (\lambda_{m+1} - \lambda_j)^2 ([[H,G],G]u_j, u_j)$$

for all $m \in \mathbb{N}$.

Although abstract inequalities in Theorems 5.4.4 and 5.4.5 are valid for any selfadjoint operators H and G such that the commutators involved make sense, in order to obtain meaningful bounds, a choice of G should be adjusted to a particular H, as illustrated below for the case $H = -\Delta_\Omega^{\mathrm{D}}$.

5.4.3. Applications to Dirichlet eigenvalues. Fix a bounded domain $\Omega \subset \mathbb{R}^d$, and let $H = -\Delta_\Omega^{\mathrm{D}}$ be the Dirichlel Laplacian on Ω with eigenvalues λ_m and orthonormalised eigenfunctions u_m. Let G be an operator of multiplication by the coordinate x_l, where l is between 1 and d. Obviously, the action of G preserves the domain $H_0^1(\Omega)$ of $-\Delta_\Omega^{\mathrm{D}}$.

An easy computation shows that in this case

$$[H,G]u = -\Delta(x_l u) + x_l \Delta u = -2\langle \nabla x_l, \nabla u\rangle = -2\frac{\partial u}{\partial x_l}$$

and

$$[[H,G],G]u = -2\frac{\partial(x_l u)}{\partial x_l} + 2x_l\frac{\partial u}{\partial x_l} = -2u;$$

therefore (5.4.5)–(5.4.6) simplify to

$$(5.4.12) \qquad 4\sum_{k=1}^{\infty}\frac{\left(\int_\Omega \frac{\partial u_j}{\partial x_l}u_k\,\mathrm{d}x\right)^2}{\lambda_k - \lambda_j} = \sum_{k=1}^{\infty}(\lambda_k - \lambda_j)\left(\int_\Omega x_l u_j u_k\,\mathrm{d}x\right)^2 = 1$$

for any fixed $j \in \mathbb{N}$. These relations have a long history — the second equation in (5.4.12), in the context of a Schrödinger operator acting in $\mathbb{R}^d$, is known as the *Thomas–Reiche–Kuhn sum rule* in the physics literature. It was derived by W. Heisenberg in 1925 [**Hei30**]. The name attached to the sum rule comes from the fact that W. Thomas, F. Reiche, and W. Kuhn had derived some semi-classical analogues of this formula in their study of the width of the lines of the atomic spectra [**Kuh25, ReiTho25**].

We are now in position to prove the original Payne–Pólya–Weinberger inequality (5.4.1).

Proof of Theorem 5.4.1. We use Theorem 5.4.4 with $H = -\Delta^{\mathrm{D}}$ and $G = x_l$, which gives

$$\lambda_{m+1} - \lambda_m \le \frac{4}{m}\sum_{j=1}^{m}\left\|\frac{\partial u_j}{\partial x_l}\right\|_{L^2(\Omega)}^2.$$

Summing up these inequalities over $l = 1, \ldots, d$ and using $\|\nabla u_j\|_{L^2(\Omega)}^2 = \lambda_j$ gives (5.4.1). $\qquad\square$

Using in the similar manner Theorem 5.4.5 produces (5.4.3).

We further demonstrate the use of commutator trace identity by deducing a bound on a sum of d consecutive eigenvalues, where d is the dimension. Fix $j \in \mathbb{N}$, $l \in \{1, \ldots, d\}$, and consider again the first equality (5.4.12) rewritten as

$$(5.4.13) \qquad \sum_{k=1}^{\infty} \frac{w_{lk}^2}{\lambda_k - \lambda_j} = \frac{1}{4},$$

where

$$w_{lk} := \int_{\Omega} \frac{\partial u_j}{\partial x_l} u_k \, \mathrm{d}x.$$

Consider the $d \times d$ matrix $W = (w_{lk})$, $l = 1, \ldots, d$, $k = j+1, \ldots, j+d$. We can rewrite it for brevity as

$$W = \int_{\Omega} (\nabla u_j) U \, \mathrm{d}x,$$

where ∇u_j is the gradient written as a column vector, U is the row vector $(u_{j+1}, \ldots, u_{j+d})$, and the integration is performed entry by entry. Let Q be the matrix of an orthogonal coordinate change $x \mapsto Qx$. Under this coordinate change, the gradient vector is transformed as $\nabla u_j \mapsto Q^t \nabla u_j$, and therefore the matrix W is transformed as $W \mapsto Q^t W$. On the other hand, we can always choose an orthogonal matrix Q_0 such that $W = Q_0 R$, where R is an upper-triangular matrix (QR-decomposition), and choosing the change of coordinates with $Q = Q_0$ thus makes W upper-triangular. We now fix this coordinate system, so that

$$w_{lk} = 0 \qquad \text{for} \quad l = 1, \ldots, d, \qquad k = j+1, \ldots, j+l-1.$$

We proceed to estimate the left-hand side of (5.4.13) by dropping all the negative terms (with $k < j$), replacing all the denominators in nonzero terms by the lowest possible one, extending summation to all k starting from one, and using Parseval's equality, arriving at

$$\frac{1}{\lambda_{j+l} - \lambda_j} \left\| \frac{\partial u_j}{\partial x_l} \right\|_{L^2(\Omega)}^2 \geq \sum_{k=1}^{\infty} \frac{w_{lk}^2}{\lambda_k - \lambda_j} = \frac{1}{4},$$

or

$$\lambda_{j+l} - \lambda_j \leq 4 \left\| \frac{\partial u_j}{\partial x_l} \right\|_{L^2(\Omega)}^2.$$

Summing up these inequalities over $l = 1, \ldots, d$, we obtain

Theorem 5.4.6 (**[LevPar02]**). The eigenvalues λ_j of the Dirichlet Laplacian on a bounded domain $\Omega \subset \mathbb{R}^d$ satisfy

$$(5.4.14) \qquad \sum_{l=1}^{d} \lambda_{j+l} \leq (4+d)\lambda_j$$

for all $j \in \mathbb{N}$. In particular, in the planar case $d = 2$,

$$\lambda_{j+1} + \lambda_{j+2} \leq 6\lambda_j.$$

One of the main drawbacks of the type of universal estimates we have considered is that by their very nature they are not supposed to be sharp. For example, the Payne–Pólya–Weinberger bound (5.4.1), the Hile–Protter bound (5.4.2), and Yang's bound (5.4.4), taken with $m = 1$, all yield, for the Dirichlet eigenvalues of bounded domains in $\mathbb{R}^d$,

$$(5.4.15) \qquad \frac{\lambda_2}{\lambda_1} \leq \frac{d+4}{d}.$$

At the same time, M. S. Ashbaugh and R. D. Benguria proved, using a more accurate approach involving symmetrisation, the optimal bound for the ratio of the first two Dirichlet eigenvalues, originally conjectured by Payne, Pólya, and Weinberger.

Theorem 5.4.7 (**[AshBen91]**). Let $\Omega \subset \mathbb{R}^d$ be a bounded domain. Then its first two Dirichlet eigenvalues $\lambda_m = \lambda_m^{\mathrm{D}}(\Omega)$, $m = 1, 2$, satisfy

$$(5.4.16) \qquad \frac{\lambda_2}{\lambda_1} \leq \frac{\lambda_2(\mathbb{B}^d)}{\lambda_1(\mathbb{B}^d)} = \frac{j_{\frac{d}{2},1}^2}{j_{\frac{d}{2}-1,1}^2},$$

where $j_{p,1}$ is the first positive zero of the Bessel function $J_p(x)$. The equality in (5.4.16) is attained if and only if Ω is a ball.

The bound (5.4.16) is stronger than (5.4.15); for example, in dimension $d = 2$ the constants in the right-hand sides of these bounds are 2.539 (approximately) and 3, respectively. Nonoptimality of universal estimates is even more noticeable for higher eigenvalues; see for example [**LevYag03**].

Remark 5.4.8 (Fundamental gap). Inequality (5.4.16) means in a way that the first and the second Dirichlet eigenvalues of a Euclidean domain cannot be too far apart. Can they be arbitrarily close to each other? Without further restrictions, the answer is positive; indeed, take a domain which is a union of two identical balls joined by a thin short passage. However, under the additional *convexity* assumption this question can be made interesting if instead of the ratio we consider the difference $\lambda_2^{\mathrm{D}} - \lambda_1^{\mathrm{D}}$. This quantity is called the *fundamental gap*. It was shown in [**AndClu11**] that

$$(5.4.17) \qquad \lambda_2^{\mathrm{D}}(\Omega) - \lambda_1^{\mathrm{D}}(\Omega) \geq \frac{3\pi^2}{\mathrm{diam}(\Omega)^2},$$

where diam(Ω) denotes the diameter of Ω. This inequality has been previously known as the fundamental gap conjecture, which originated in [**vdB83**, **Yau86**, **AshBen89**]. The equality in (5.4.17) is attained in the limit as a thin rectangular box degenerates into an interval. There is also a Neumann analogue of (5.4.17) called the *Payne–Weinberger inequality*:

$$(5.4.18) \qquad \lambda_2^{\mathrm{N}}(\Omega) \geq \frac{\pi^2}{\mathrm{diam}(\Omega)^2},$$

with equality once again achieved in the limit as a thin rectangular box degenerates into an interval. Since $\lambda_1^{\mathrm{N}}(\Omega) = 0$, it can be viewed as a bound on the Neumann fundamental gap. Inequality (5.4.18) was proved in [**PayWei60**]; see also [**Beb03**] for a slight correction in dimensions $d \geq 3$.

5.4.4. Spectral prescription. What about universal bounds for the eigenvalues of the Neumann Laplacian $-\Delta_\Omega^{\mathrm{N}}$, $\Omega \subset \mathbb{R}^d$? One technical difficulty in applying commutator trace identities in this situation is making sure that the commutators are well-defined; necessarily, a choice of G such that $G\left(\mathrm{Dom}\left(-\Delta_\Omega^{\mathrm{N}}\right)\right) \subseteq \mathrm{Dom}\left(-\Delta_\Omega^{\mathrm{N}}\right)$ is more complicated than in the Dirichlet case. The resulting bounds are not, strictly speaking, universal, but depend on some geometric properties of either Ω of M; see, e.g., [**HarMic95**] and some further improvements in [**LevPar02**] by analogy with [**ChuGriYau96**].

There is however a fundamental obstacle for the existence of universal eigenvalue bounds in the Neumann case. Consider the following general question of *spectral prescription*: given a finite monotone sequence of positive (or nonnegative) real numbers, can it coincide with the beginning of the sequence of eigenvalues of either the Dirichlet or Neumann Laplacian in a domain $\Omega \subset \mathbb{R}^d$? Obviously, in the Dirichlet case, the universal bounds (5.4.4) and (5.4.14) for the eigenvalues should hold: if a finite positive sequence $\{\lambda_1, \ldots, \lambda_K\}$ does not satisfy either of these conditions, it cannot form the lower part of the spectrum of a Dirichlet Laplacian for a domain in $\mathbb{R}^d$.

Rather surprisingly, in the Neumann case for higher dimensions there are no significant obstructions to spectral prescription as demonstrated by the following result of Y. Colin de Verdière.

Theorem 5.4.9 ([**CdV87**, Theorem 1.4]). *Let $0 = \eta_1 < \eta_2 \leq \cdots \leq \eta_K$ be a finite monotone increasing sequence of real numbers. Then for any $d \geq 3$ there exists a domain $\Omega \subset \mathbb{R}^d$ with piecewise C^1 boundary such* that $\eta_j = \lambda_j^{\mathrm{N}}(\Omega)$ for $j = 1, \ldots, K$. The same is true for $d = 2$ if and only if $K \leq 4$. If, moreover, the sequence $\{\eta_j\}_{j=1}^K$ is strictly increasing, then such a domain exists for any $d \geq 2$ and any K.

A similar result holds in the Riemannian case.

Theorem 5.4.10 ([**CdV87**, Theorems 1.2 and 1.3]). Let M be a closed manifold of dimension $d \geq 3$, and let $0 = \mu_0 < \mu_1 \leq \mu_2 \leq \cdots \leq \mu_K$ be a finite monotone increasing sequence of real numbers. Then there exists a Riemannian metric g on M such that $\mu_j = \lambda_j(M, g)$ for $j = 1, \ldots, K$. If, moreover, the sequence $\{\mu_j\}_{j=1}^K$ is strictly increasing, this is also true in dimension two.

Note that in dimension two the condition that the sequence is strictly increasing cannot be completely removed in either the Riemannian or the Neumann case due to the multiplicity bound (4.4.3) and its Neumann analogue.

Chapter 6

Heat equation, spectral invariants, and isospectrality

Subbaramiah
**Minakshi-
sundaram**
(1913–1968)

In this chapter, we construct the heat kernel on a Riemannian manifold and study its asymptotics at small times. As an application, we prove Weyl's law for eigenvalues of the Laplace–Beltrami operator on a closed manifold. We also discuss spectral invariants arising from the heat asymptotics and the related question "Can one hear the shape of a drum?", leading to the notion of isospectrality. We present Milnor's example of isospectral sixteen-dimensional tori as well as a more general Sunada construction of isospectral manifolds. The transplantation of eigenfunctions and related examples of isospectral planar domains with Dirichlet, Neumann, and mixed boundary conditions are also presented. We conclude the chapter with a brief overview of results and open problems concerning spectral rigidity.

Mark
Kac
(1914–1984)

6.1. Heat equation and spectral invariants

6.1.1. Heat kernel on a Riemannian manifold. Let (M, g) be a closed Riemannian manifold. Consider the initial value problem for the heat equation,

$$(6.1.1) \qquad \begin{cases} \frac{\partial u}{\partial t}(t, y) = \Delta_y u(t, y), & t \in \mathbb{R}_+ = (0, +\infty), y \in M, \\ u(0, y) = \varphi(y), & y \in M. \end{cases}$$

Recall that the physical meaning of the heat equation is as follows: given initial temperature distribution $\varphi(y)$, find the temperature $u(t, y)$ at the point y at the time t. Equation (6.1.1) is also often referred to as the *diffusion equation*; in this case $u(t, y)$ is understood as the density of the diffusing substance.

To simplify notation, throughout this section when integrating over the Riemannian measure dV_g with respect to some variable z we denote the measure simply by dz.

Definition 6.1.1. A *fundamental solution* of the heat equation (or the *heat kernel*) is a function $e(t, x, y)$ for $t \in \mathbb{R}_+$, $(x, y) \in M \times M$, which is continuous in all three variables, C^1 in t, C^2 in y, and satisfies (6.1.1) for all $(t, x, y) \in \mathbb{R}_+ \times M \times M$ with the initial temperature distribution $\varphi(y) = \delta_x(y)$. The initial condition is understood in a weak sense:

$$(6.1.2) \qquad \lim_{t \to 0^+} \int_M e(t, x, y) f(y) \, dy = f(x)$$

for any $f \in C(M)$. Here δ_x denotes the Dirac δ-function supported at the point $x \in M$.

The following important result holds.

Theorem 6.1.2 (Existence and uniqueness of a heat kernel). Let (M, g) be a closed Riemannian manifold. There exists a unique heat kernel $e(t, x, y)$ on $\mathbb{R}_+ \times M \times M$ which is a C^∞ function. Moreover,

$$(6.1.3) \qquad e(t, x, y) = \sum_{j=0}^{\infty} e^{-\lambda_j t} u_j(x) u_j(y),$$

where $\{u_j\}_{j=0}^{\infty}$ is an orthonormal basis of eigenfunctions of the Laplace–Beltrami operator $-\Delta_M$ corresponding to the eigenvalues λ_j and the series in the right-hand side converges pointwise in $\mathbb{R}_+ \times M \times M$.

We follow the exposition in [**BerGauMaz71**] and [**Ros97**]. Let us first assume that a heat kernel exists and use the method of [**Gaf58**] to prove that it is unique and is given by (6.1.3).

Proposition 6.1.3. Let M be a closed Riemannian manifold. Suppose that a heat kernel $e(t, x, y)$ exists. Then it is unique, and the series (6.1.3) converges pointwise in $\mathbb{R}_+ \times M \times M$.

Proof. For any fixed $t > 0$ and $x \in M$, we can write, by expanding in an orthonormal basis of eigenfunctions u_j in $L^2(M)$,

$$e(t, x, y) = \sum_{j=0}^{\infty} e_j(t, x) u_j(y)$$

as a function of y. The coefficients of this expansion are given by

(6.1.4)
$$e_j(t, x) = \int_M e(t, x, y) u_j(y) \, \mathrm{d}y.$$

Therefore, differentiating with respect to t, we get

$$\frac{\mathrm{d}}{\mathrm{d}t} e_j(t, x) = \int_M (\Delta_y e(t, x, y)) \, u_j(y) \, \mathrm{d}y$$

$$= \int_M e(t, x, y) \, (\Delta_y u_j(y)) \, \mathrm{d}y = -\lambda_j e_j(t, x),$$

where we first used the fact that $e(t, x, y)$ solves the heat equation and then integrated by parts. Hence, we get an ordinary differential equation for $e_j(t, x)$ which yields

$$e_j(t, x) = c_j(x) \mathrm{e}^{-\lambda_j t},$$

with the coefficients $c_j(x)$ still to be determined. From the expression (6.1.4) and property (6.1.2) we get that $c_j(x) = u_j(x)$. Hence,

(6.1.5)
$$e(t, x, y) = \sum_{j=0}^{\infty} \mathrm{e}^{-\lambda_j t} u_j(x) u_j(y)$$

in $L^2(M)$ (in the variable y for given t, x). The convergence of the series in $L^2(M)$ implies that for any fixed t, x there exists a subsequence $j_m \to \infty$ such that

(6.1.6)
$$\sum_{j=0}^{j_m} \mathrm{e}^{-\lambda_j t} u_j(x) u_j(y) \to e(t, x, y)$$

for almost every y. At the same time, by Parseval's theorem,

(6.1.7)
$$\left(e\left(\frac{t}{2}, x, z\right), e\left(\frac{t}{2}, y, z\right) \right)_{L^2(M)} = \sum_{j=0}^{\infty} \mathrm{e}^{-\frac{\lambda_j t}{2}} u_j(x) \mathrm{e}^{-\frac{\lambda_j t}{2}} u_j(y)$$

$$= \sum_{j=0}^{\infty} \mathrm{e}^{-\lambda_j t} u_j(x) u_j(y)$$

for any $x, y \in M$. In particular, the right-hand side of (6.1.7) converges pointwise. Since, by definition, the heat kernel is continuous in all three variables, the left-hand side of (6.1.7) is a continuous function in t, x, y.

Therefore, the right-hand side defines a continuous function in $\mathbb{R}_+ \times M \times M$. Combining this with the almost everywhere convergence of the series (6.1.6), we obtain that the right-hand side of (6.1.5) converges pointwise everywhere (since two continuous functions which are equal almost everywhere are equal). In particular, this implies that the heat kernel is unique provided it exists. $\square$

Definition 6.1.4. The *heat trace* of a closed Riemannian manifold (M, g) is defined by

$$e_M(t) := \sum_{j=0}^{\infty} e^{-\lambda_j t} = \operatorname{Tr} e^{t\Delta_M}.$$

Corollary 6.1.5. The heat trace $e_M(t)$ is a convergent series for $t > 0$, and its sum equals $\int_M e(t, x, x)\,dx$.

Proof. Setting $x = y$ in the heat kernel expression, we get

$$e(t, x, x) = \sum_{j=0}^{\infty} e^{-\lambda_j t} u_j(x)^2.$$

Since all terms are nonnegative, we can integrate the series in the right-hand side term by term and obtain

$$\int_M e(t, x, x)\,dx = \sum_{j=0}^{\infty} e^{-\lambda_j t} \int_M u_j(x)^2\,dx = \sum_{j=0}^{\infty} e^{-\lambda_j t}$$

given that all the eigenfunctions have been chosen to have the unit L^2 norm. $\square$

Let us now describe the main ideas of the proof of the existence of the heat kernel.

Existence of the heat kernel: Sketch of the proof. First, recall that on $\mathbb{R}^d$,

$$e(t, x, y) = (4\pi t)^{-\frac{d}{2}} e^{-\frac{r^2}{4t}},$$

where $r = |x - y|$. Note that the Euclidean heat kernel is small unless both r and t are small. We expect a similar property to hold on an arbitrary Riemannian manifold. Moreover, any Riemannian metric is locally close to a Euclidean metric. Hence, we may attempt to construct an approximate heat kernel for x close to y and t small, by using an appropriate perturbation of the Euclidean heat kernel, and then modify it slightly to obtain a global solution.

Let us express the Riemannian metric g in Riemannian normal coordinates centred at x and set $\theta(y) = \sqrt{\det g(y)}$. We look for approximations of the heat kernel as $t \to 0^+$ of the form

$$(6.1.8) \quad S_k(t, x, y) = (4\pi t)^{-\frac{d}{2}} e^{-\frac{r^2}{4t}} \left(v_0(x, y) + v_1(x, y)t + \cdots + v_k(x, y)t^k \right),$$

where $k \in \mathbb{N}_0$, $r = \mathrm{dist}(x, y) < \varepsilon$ is now the Riemannian distance, $\varepsilon > 0$ is small enough, and the functions $v_j(x, y)$ depend on the local geometry of the manifold. We choose $\varepsilon < \rho_{\mathrm{inj}}(M)$, where $\rho_{\mathrm{inj}}(M)$ denotes the injectivity radius, so that $B_{x,\varepsilon}$ is a geodesic ball for any $x \in M$. Let us define $v_j(x, y)$ recursively as follows; see [**BerGauMaz71**, §III.E.III]. Set $v_0(x, y) = \theta^{-\frac{1}{2}}(y)$ and

$$v_j(x, y) = \theta^{-\frac{1}{2}}(y) r^{-j} \int_0^r \theta^{\frac{1}{2}}(\gamma(s)) \Delta_y v_{j-1}(\gamma(s), y) s^{j-1} \, \mathrm{d}s, \qquad j \in \mathbb{N},$$

where $\gamma(s)$ is a unit speed minimal geodesic emanating from x to y. Then for k large enough, S_k is "almost" a solution of the heat equation as $t \to 0^+$ in the following sense:

$$(6.1.9) \qquad L_y S_k(x, y, t) = (4\pi)^{-\frac{d}{2}} t^{k-\frac{d}{2}} e^{\frac{-r^2}{4t}} \Delta_y v_k(x, y) = O\left(t^{k-\frac{d}{2}}\right),$$

where $L_y = \frac{\partial}{\partial t} - \Delta_y$ is the heat operator.

Let $H_k = \eta S_k$, where η is a smooth cut-off function with $\eta \equiv 1$ near the diagonal $x = y$, and $\eta \equiv 0$ when $\mathrm{dist}(x, y) \geq \varepsilon$. One can show that

   (i) the functions H_k are smooth for $x, y \in M$ and $t > 0$,

   (ii) $\lim\limits_{t \to 0+} H_k(t, x, y) = \delta_x(y)$ for all $y \in M$ (as in (6.1.2) with e replaced by H_k).

The properties (i) and (ii) hold for any $k \geq 0$. Moreover,

   (iii) for any $k > \frac{d}{2}$, $L_y H_k$ can be extended to a continuous function in $\mathbb{R}_{\geq 0} \times M \times M$.

Note that $t = 0$ is included; this is the most nontrivial point of the statement (iii) which can be deduced using (6.1.9).

Remark 6.1.6. A function satisfying the conditions (i)–(iii) is called a *parametrix* for the heat equation. In fact, one can show that $L_y H_k \in C^l(\mathbb{R}_{\geq 0} \times M \times M)$ for $k > \frac{d}{2} + l$, $l \geq 0$.

Let us now modify a parametrix to a fundamental solution. Recall the notion of a convolution of two continuous functions $F, H \in C(\mathbb{R}_{\geq 0} \times M \times M)$:

$$(F * H)(t, x, y) := \int_0^t \int_M F(s, x, z) H(t - s, z, y) \, \mathrm{d}z \, \mathrm{d}s.$$

We will also denote the iterated convolutions by $F^{*j} = F * \cdots * F$, where F is repeated $j \geq 1$ times.

■ **Exercise 6.1.7.** Let $F \in C(\mathbb{R}_{\geq 0} \times M \times M)$. Show that for any $k > \frac{d}{2} + 2$, $F * H_k \in C^2(\mathbb{R}_+ \times M \times M)$ and $L_y(F * H_k) = F + F * (L_y H_k)$. For a solution, see [**BerGauMaz71**, Lemme E.III.7]. Compare this exercise with Duhamel's principle [**Eva10**, §2.3.1.c], [**Cha84**, §VI.1].

Fix some $k > \frac{d}{2} + 2$, and set $F = F_k = \sum_{j=1}^{\infty} (-1)^{j+1} (L_y H_k)^{*j}$. One can show that the series defining F_k converges and $F_k \in C^2(\mathbb{R}_{\geq 0} \times M \times M)$. We claim that the function $P_k(t, x, y) := H_k - F_k * H_k$ is the fundamental solution of the heat equation. Indeed, by Exercise 6.1.7, $P_k(t, x, y) \in C^2(\mathbb{R}_+ \times M \times M)$ and

$$
\begin{aligned}
L_y P_k &= L_y (H_k - F_k * H_k) = L_y(H_k) - L_y(F_k * H_k) \\
&= L_y H_k - F_k - F_k * (L_y H_k) \\
&= L_y H_k - \sum_{j=1}^{\infty} (-1)^{j+1} (L_y H_k)^{*j} - \sum_{j=1}^{\infty} (-1)^{j+1} (L_y H_k)^{*(j+1)} = 0.
\end{aligned}
$$

It remains to check that

$$
(6.1.10) \qquad\qquad\qquad P_k(t, x, y) \to \delta_x(y)
$$

as $t \to 0^+$. Indeed, $\lim_{t \to 0^+} H_k(t, x, y) = \delta_x(y)$. At the same time, one can show that there exists $C > 0$ such that

$$
(6.1.11) \qquad\qquad\qquad F_k(t, x, y) \leq C t^{k - \frac{d}{2}}
$$

for all $x, y \in M$ and $0 \leq t < 1$; see [**BerGauMaz71**, Lemme E.III.6]. A direct computation then implies that

$$
F_k * H_k \to 0 \qquad \text{as } t \to 0^+
$$

for any $k > \frac{d}{2}$ (where convergence is understood in the sense of measures), which proves (6.1.10). Since the uniqueness of the heat kernel has already been established, we have $P_k(t, x, y) = e(t, x, y)$ (note that this implies that the definition of $P_k(t, x, y)$ does not depend on the choice of $k > \frac{d}{2} + 2$). This completes the proof of Theorem 6.1.2. □

6.1.2. Heat kernel asymptotics. From the viewpoint of spectral geometry, of particular interest is the behaviour of the heat kernel on the diagonal $x = y$ as $t \to 0^+$.

Theorem 6.1.8 (Minakshisundaram–Pleijel asymptotic expansion [**MinPle49**]). Let (M, g) be a closed Riemannian manifold, $\dim M = d$. The following asymptotic expansion of the heat kernel holds for $t \to 0^+$:

$$e(t, x, x) = (4\pi t)^{-\frac{d}{2}} \left(\sum_{j=0}^{k} a_j(x) t^j + O\left(t^{k+1}\right) \right),$$

for all $k > 0$. The heat kernel coefficients $a_j(x)$ are called the *local heat invariants* and are calculated in terms of the local geometry of M near x.

Proof. We have $e(t, x, y) = H_k - F_k * H_k$ for all $k > \frac{d}{2} + 2$. Since on the diagonal $y = x$ one has $H_k(t, x, x) = S_k(t, x, x)$, with $S_k(t, x, y)$ given by (6.1.8), we obtain

$$(4\pi t)^{\frac{d}{2}} H_{k+1}(t, x, x) = \sum_{j=0}^{k+1} v_j(x, x) t^j.$$

Set

$$a_j(x) := v_j(x, x);$$

then

$$(4\pi t)^{\frac{d}{2}} e(t, x, x) = \sum_{j=0}^{k} a_j(x) t^j + a_{k+1}(x) t^{k+1} - (4\pi t)^{\frac{d}{2}} \left(F_{k+1} * H_{k+1} \right)(t, x, x).$$

In view of (6.1.11), we get, for $0 < t < 1$,

$$|(F_{k+1} * H_{k+1})(t, x, x)| = \left| \int_0^t \int_M F_{k+1}(s, x, z) H_{k+1}(t - s, z, x) \, dz \, ds \right|$$

$$\leq C_1 t^{k+1-\frac{d}{2}} \int_0^t \int_M |H_{k+1}(t - s, z, x)| \, dz \, ds$$

$$= C_1 t^{k+1-\frac{d}{2}} \int_0^t \int_M |H_{k+1}(s, z, x)| \, dz \, ds,$$

where throughout this proof C_j denote some positive constants which may depend on k. Note that $H_{k+1}(s, z, x)$ is nonzero only near the diagonal $z = x$, so we can assume that $\mathrm{dist}(z, x) < \rho$ with $\rho \in (\varepsilon, \rho_{\mathrm{inj}}(M))$, where ε is defined after (6.1.8). Then $|H_{k+1}(s, z, x)|$ is bounded by $C s^{-d/2} \mathrm{e}^{-\mathrm{dist}(z,x)^2/(4s)}$, with

C independent of z and x. We therefore get

$$|(F_{k+1} * H_{k+1})(t, x, x)| \leq C_2 t^{k+1-\frac{d}{2}} \int_0^t \int_{B_{x,\rho}} s^{-\frac{d}{2}} e^{-\frac{\text{dist}(z,x)^2}{4s}} \, dz \, ds$$

$$\leq C_3 t^{k+1-\frac{d}{2}} \int_0^1 \int_{B(0,\rho) \subset \mathbb{R}^d} s^{-\frac{d}{2}} e^{-\frac{|y|^2}{4s}} \, dy \, ds$$

$$\leq C_3 t^{k+1-\frac{d}{2}} \int_0^1 \int_{\mathbb{R}^d} e^{-\frac{|w|^2}{4}} \, dw \, ds$$

$$= C_4 t^{k+1-\frac{d}{2}},$$

where we changed the variables as $w = y/\sqrt{s}$. This completes the proof of the theorem. $\qquad\square$

Recall now that $e(t, x, x) = \sum_{j=0}^{\infty} e^{-\lambda_j t} u_j(x)^2$. Therefore, as $t \to 0^+$,

(6.1.12) $$\sum_{j=0}^{\infty} e^{-\lambda_j t} = (4\pi t)^{-\frac{d}{2}} \sum_{j=0}^{\infty} a_j t^j,$$

where $a_j := a_j(M) = \int_M a_j(x) \, dx$. The coefficients a_j are called the *heat invariants* of the Riemannian manifold M.

The heat trace asymptotics is an important tool in the study of the *inverse spectral problem*, which is concerned with the recovery of the geometric properties of the manifold M from the spectrum of the corresponding Laplace–Beltrami operator. Following Mark Kac, this problem is often described by the celebrated question "Can one hear the shape of a drum?" [**Kac66**]. We say that a property of M is a *spectral invariant* (or that it can be "heard") if it is completely determined by the Laplace spectrum. For example, the left-hand side in (6.1.12) is determined by the Laplace eigenvalues of M. This immediately implies that the dimension d and the heat invariants a_j are spectral invariants. Using explicit calculations in Riemannian normal coordinates one obtains (see [**Ros97**, §3.3])

$$a_0(x) = 1, \qquad a_1(x) = \frac{1}{6}\tau(x),$$

where $\tau(x)$ is the scalar curvature. Hence, $a_0 = \text{Vol}(M)$, and therefore the volume of a Riemannian manifold is a spectral invariant. Similarly, the total scalar curvature $\int_M \tau(x) \, dx$ is determined by the spectrum. Moreover, if M

is two dimensional, its Euler characteristic is given by

$$\chi(M) = \frac{1}{4\pi} \int_M \tau(x) \, \mathrm{d}x = \frac{3}{2\pi} a_1(M).$$

Therefore, the Euler characteristic of a surface is a spectral invariant; in particular, one can hear the number of handles of an orientable surface!

There is a vast literature on the computation of heat invariants (see, for instance, [**Gil04**], [**Pol00**] and references therein), and there exist various ways to express them. Geometrically, the most natural way is to present the local heat invariants in terms of curvatures and their derivatives. However, the complexity of this task rapidly increases for higher heat invariants, and the geometric information becomes difficult to extract. Still, heat invariants are quite useful in the study of spectral rigidity; see §6.2.6 for further details.

6.1.3. Weyl's law on a Riemannian manifold. Let us now use the heat trace expansion (6.1.12) to prove Weyl's law for the eigenvalue counting function on closed manifolds. We have already stated this result with a sharp remainder estimate; see Theorem 3.3.4. As was mentioned in Remark 3.3.5, its proof uses techniques that are beyond the scope of this book. Below we present a proof of Weyl's law based on the heat trace expansion, albeit with a weaker remainder estimate.

Theorem 6.1.9 (Weyl's law for manifolds). *Let M be a closed Riemannian manifold, $\dim M = d$. The counting function $\mathcal{N}_M(\lambda)$ of Laplace–Beltrami eigenvalues on M satisfies the asymptotics*

(6.1.13) $$\mathcal{N}_M(\lambda) = C_d \operatorname{Vol}(M) \lambda^{\frac{d}{2}} + o\left(\lambda^{\frac{d}{2}}\right).$$

As before, the numerical coefficient is $C_d = \dfrac{1}{(4\pi)^{\frac{d}{2}} \Gamma(\frac{d}{2}+1)} = \dfrac{\omega_d}{(2\pi)^d}$, where ω_d is the volume of the unit ball in $\mathbb{R}^d$.

The proof of Theorem 6.1.9 will use the following well-known result; see, for example, [**Fel71**, §XIII.5].

Theorem 6.1.10 (Hardy–Littlewood–Karamata Tauberian theorem). *Let $N(\lambda)$ be a monotone increasing function such that*

$$\int_0^\infty e^{-\lambda t} \, \mathrm{d}N(\lambda) = ct^{-\alpha} + o\left(t^{-\alpha}\right) \qquad \text{as } t \to 0^+.$$

Then

$$N(\lambda) = \frac{c}{\Gamma(\alpha+1)} \lambda^\alpha + o\left(\lambda^\alpha\right) \qquad \text{as } \lambda \to \infty.$$

Proof of Theorem 6.1.9. Since $a_0 = \text{Vol}(M)$, it follows from the heat trace expansion that

$$(6.1.14) \qquad \int_0^\infty e^{-\lambda t}\, dN(\lambda) = \sum_{j=0}^\infty e^{-\lambda_j t} = \frac{1}{(4\pi t)^{\frac{d}{2}}}(\text{Vol}(M) + O(t)).$$

Taking $\alpha = \frac{d}{2}$ and applying the Hardy–Littlewood–Karamata theorem to the right-hand side of (6.1.14) completes the proof of Theorem 6.1.9. □

Remark 6.1.11. The heat trace expansion (6.1.12) can be extended to manifolds with Dirichlet or Neumann boundary conditions; see [**Gil04**]. For manifolds with boundary, the expansion has twice as many terms:

$$\sum_{j=1}^\infty e^{-\lambda_j t} \underset{t \to 0^+}{\sim} (4\pi t)^{-\frac{d}{2}} \sum_{k=0}^\infty a_{\frac{k}{2}} t^{\frac{k}{2}}.$$

As before, $a_0 = \text{Vol}(M)$, but the terms inside the sum corresponding to $k = m + \frac{1}{2}$ with integer m arise from the boundary contributions. In particular,

$$(6.1.15) \qquad a_{\frac{1}{2}} = \pm\frac{\sqrt{\pi}}{2} \text{Vol}_{d-1}(\partial M),$$

where the plus sign is taken for the Neumann boundary condition and the minus sign for the Dirichlet boundary condition. It follows that the volume of the boundary is a spectral invariant.

■ **Exercise 6.1.12.** Assume that the conjectured two-term asymptotic formula (3.3.5) in Weyl's law holds. Use Theorem 6.1.10 to show that formula (6.1.15) agrees with the second term in (3.3.5) .

Remark 6.1.13. The main term of the heat trace asymptotics (and, hence, of Weyl's asymptotics (6.1.13) for the eigenvalue counting function) is not affected by the boundary condition. This can be explained using Kac's principle of "not feeling the boundary". It is best illustrated using the model of diffusion: for small times, the particles in the interior do not feel the boundary, and the diffusion process is not influenced by the boundary conditions. We refer to [**Kac51**] for further details.

Example 6.1.14 (Heat trace asymptotics for planar domains). Let Ω be a smooth planar domain with r boundary components. Then the Dirichlet heat trace of Ω satisfies

$$\sum_{j=1}^\infty e^{-\lambda_j^{\text{D}}(\Omega)t} = \frac{\text{Area}(\Omega)}{4\pi t} - \frac{L(\partial\Omega)}{8\sqrt{\pi t}} + \frac{(2-r)}{6} + o(1).$$

For the Neumann boundary condition, the second term should be taken with a plus sign:

$$\sum_{j=1}^{\infty} e^{-\lambda_j^{\mathrm{N}}(\Omega)t} = \frac{\mathrm{Area}(\Omega)}{4\pi t} + \frac{L(\partial\Omega)}{8\sqrt{\pi t}} + \frac{(2-r)}{6} + o(1).$$

In the presence of corners, the third term becomes more complicated and depends on the angles at the corner points; see [**vdBSri88**], [**NurRowShe19**] and references therein.

6.2. Isospectral manifolds and domains

6.2.1. Isospectrality. We start with

Definition 6.2.1 (Isospectral manifolds). We say that two closed Riemannian manifolds (M, g) and (N, h) are *isospectral* if $\mathrm{Spec}\left(-\Delta_{(M,g)}\right) = \mathrm{Spec}\left(-\Delta_{(N,h)}\right)$, understood as the equality of multisets with account of multiplicities.

Similarly, one can define *isospectrality* for manifolds with boundary and for Euclidean domains; in this case, boundary conditions have to be specified. One of the central questions in spectral geometry is to understand the possible mechanisms of isospectrality: how do we construct manifolds or domains that are isospectral and not isometric? A counterpoint to this question is *spectral rigidity*: which manifolds or domains are uniquely defined by their spectrum, or at least do not admit isospectral deformations? We focus on these problems in the present section.

It turns out that the heat trace is an important tool in the study of isospectrality. The following simple observation is useful.

■ Exercise 6.2.2. Let (M, g) and (N, h) be two compact Riemannian manifolds; if their boundaries are nonempty, we assume that the same selfadjoint boundary condition is specified on each boundary. Suppose that the corresponding heat traces coincide for all times: $e_M(t) = e_N(t)$, $t > 0$. Then (M, g) and (N, h) are isospectral.

Below we present two elegant constructions of isospectral and not isometric Riemannian manifolds, relying on the heat trace. The first one is due to J. Milnor [**Mil64**] and the second one was discovered by T. Sunada [**Sun85**]. In fact, Sunada's construction has lead to a whole variety of examples of isospectral manifolds and domains. Somewhat surprisingly, Milnor's and Sunada's examples are based on methods coming from different areas of mathematics which are seemingly distant from spectral geometry: the theory of modular forms and group theory.

6.2.2. Milnor's example. In this subsection we follow the exposition of **[BerGauMaz71**, §III.B.III]. The argument is based on the Poisson summation formula for lattices. First, let us recall the usual *Poisson summation formula*: given a Schwartz (see §B.2) function $f \in \mathcal{S}(\mathbb{R}^d)$, we have

$$(6.2.1) \qquad \sum_{k \in \mathbb{Z}^d} f(k) = \sum_{m \in \mathbb{Z}^d} \hat{f}(m).$$

Here

$$\hat{f}(y) := \int_{\mathbb{R}^d} f(x) e^{-2\pi i \langle x, y \rangle} \, \mathrm{d}x = (2\pi)^{d/2} \left(\mathcal{F}f\right)(2\pi y)$$

is the rescaled Fourier transform of f; cf. (2.1.3).

Baron Siméon Denis **Poisson**

(1781–1840)

■ **Exercise 6.2.3.** Prove the Poisson summation formula (6.2.1) for $d = 1$. *Hint:* Compute the Fourier coefficients of the 1-periodic function $F(x) := \sum_{k \in \mathbb{Z}} f(x+k)$ and evaluate the resulting Fourier series at $x = 0$.

The Poisson summation formula can be generalised to an arbitrary *lattice* Γ in $\mathbb{R}^d$ (that is, a discrete additive subgroup of $\mathbb{R}^d$ such that $\mathbb{R}^d/\Gamma$ is compact). If Γ is a lattice, let Γ^* be the dual lattice; i.e., Γ^* consists of all elements $x \in \mathbb{R}^n$ such that the scalar product $\langle x, y \rangle \in \mathbb{Z}$ for all $y \in \Gamma$. The Poisson summation formula for lattices states that

$$\sum_{k \in \Gamma} f(k) = \frac{1}{\mathrm{Vol}(\Gamma)} \sum_{m \in \Gamma^*} \hat{f}(m),$$

where the volume of a lattice is understood as the volume of $\mathbb{R}^d/\Gamma$. Take $f(x) = e^{-a|x|^2} \in \mathcal{S}(\mathbb{R}^d)$, where $a > 0$. Then,

$$\hat{f}(y) = \left(\frac{\pi}{a}\right)^{\frac{d}{2}} e^{-\frac{\pi^2 |y|^2}{a}}.$$

Plugging $f(x)$ with $a = \frac{1}{4t}$ into the Poisson summation formula and switching the variables x and y, we obtain

$$(6.2.2) \qquad \sum_{x \in \Gamma^*} e^{-4\pi^2 t |x|^2} = \frac{\mathrm{Vol}(\Gamma)}{(4\pi t)^{\frac{d}{2}}} \sum_{y \in \Gamma} e^{-\frac{|y|^2}{4t}}.$$

Note that the left-hand side of (6.2.2) is precisely the heat trace of the flat torus $\mathbb{R}^d/\Gamma$, because its eigenvalues are given by $4\pi^2|x|^2$, $x \in \Gamma^*$; cf. Exercise 1.2.10. The right-hand side can be interpreted as follows: the values of $|y|$ with $y \in \Gamma$ are the lengths of the closed geodesics in $\mathbb{R}^d/\Gamma$, and in the sum we take one closed geodesic in each free homotopy class.

Remark 6.2.4. The Poisson formula is a manifestation of a link between the spectral (quantum) and dynamical (classical) quantities, which can be explained via Bohr's correspondence principle in quantum mechanics. This important connection has already been mentioned in §3.3.2, and we will revisit it in §6.2.6. There exist various generalisations of the Poisson formula, such as the Selberg trace formula, the Balian–Bloch trace formula, the wave-trace formula, etc. For a generalisation based on the heat trace we refer to [**CdV73**].

Consider now the following special class of lattices in $\mathbb{R}^d$ with $d = 8k$, $k \in \mathbb{N}$. Let Γ_2 be the lattice in $\mathbb{R}^d$ consisting of $(x_1, \ldots, x_d) \in \mathbb{Z}^d$ such that $\sum_{j=1}^d x_j$ is even. It is a sublattice (i.e., a subgroup) of $\mathbb{Z}^d$ of index two. Let $\Gamma(d)$ be the lattice in $\mathbb{R}^d$ generated by Γ_2 and the vector $w_d = \left(\frac{1}{2}, \ldots, \frac{1}{2}\right)$. Since $2w_d \in \Gamma_2$ (recall that $d = 8k$ is even), it is easy to check that Γ_2 is a sublattice of index two in $\Gamma(d)$. Hence $\mathrm{Vol}(\Gamma(d)) = \frac{1}{2}\mathrm{Vol}(\Gamma_2) = \mathrm{Vol}(\mathbb{Z}^d) = 1$.

■ **Exercise 6.2.5.** Let $\Gamma = \Gamma(d)$ for $d = 8k$, $k \in \mathbb{N}$. Show the following:

(i) For all $x \in \Gamma$, $|x|^2$ is even.

(ii) $\Gamma^* = \Gamma$.

Consider two sixteen-dimensional lattices $\Gamma(16)$ and $\Gamma(8,8) := \Gamma(8) \oplus \Gamma(8)$.

■ **Exercise 6.2.6.** Show that the lattice $\Gamma(8)$ is generated by the elements of norm $\sqrt{2}$, while $\Gamma(16)$ is not.

Theorem 6.2.7 ([**Mil64**]). The two flat sixteen-dimensional tori, $M_1 = \mathbb{R}^{16}/\Gamma(16)$ and $M_2 = \mathbb{R}^{16}/\Gamma(8,8)$, are isospectral but nonisometric.

Proof. It immediately follows from Exercise 6.2.6 that the tori M_1 and M_2 are not isometric. Let us show that M_1 and M_2 are isospectral by comparing their heat traces. Given an arbitrary lattice $\Gamma \subset \mathbb{R}^d$, consider its theta-function

$$\theta_\Gamma(t) := e_{\mathbb{R}^d/\Gamma}\left(\frac{t}{4\pi}\right) = \sum_{x \in \Gamma^*} e^{-\pi|x|^2 t}.$$

Let $\Gamma \subset \mathbb{Z}^{16}$ be a lattice satisfying the properties (i) and (ii) in Exercise 6.2.5; clearly, this is true for both $\Gamma(16)$ and $\Gamma(8,8)$. Property (ii) implies, in particular, that Γ is unimodular; i.e., $\mathrm{Vol}(\Gamma) = 1$. Therefore, the Poisson summation formula yields

$$\theta_\Gamma(t) = \sum_{x \in \Gamma} e^{-\pi|x|^2 t} = t^{-8} \sum_{y \in \Gamma} e^{-\pi\frac{|y|^2}{t}}.$$

Hence, $\theta_\Gamma(t) = t^{-8}\theta_\Gamma\left(t^{-1}\right)$. One can show, using the Weierstrass theorem, that $\theta_\Gamma(t)$ extends to a holomorphic function on the complex half-plane $\mathrm{Re}\, z > 0$ and that

$$\theta_\Gamma(z) - z^{-8}\theta_\Gamma(z^{-1}) = 0.$$

Indeed, this equality holds for any real positive z, and since holomorphic functions have isolated zeros, it must hold for all $\mathrm{Re}\, z > 0$. Set

$$\widetilde{\theta}_\Gamma(z) := \theta_\Gamma(-\mathrm{i}z).$$

The function $\widetilde{\theta}_\Gamma$ is holomorphic in the upper half-plane $\mathrm{Im}\, z > 0$ and satisfies

(6.2.3) $$\widetilde{\theta}_\Gamma(z) = z^{-8}\widetilde{\theta}_\Gamma\left(-z^{-1}\right).$$

Moreover,

(6.2.4) $$\widetilde{\theta}_\Gamma(z+1) = \sum_{x \in \Gamma} e^{\mathrm{i}\pi|x|^2 z} e^{\mathrm{i}\pi|x|^2} = \widetilde{\theta}_\Gamma(z),$$

since $|x|^2$ is even by the first assertion of Exercise 6.2.5.

■ **Exercise 6.2.8.** Using (6.2.3) and (6.2.4), show that

$$\widetilde{\theta}_\Gamma\left(\frac{az+b}{cz+d}\right) = (cz+d)^8\widetilde{\theta}_\Gamma(z)$$

for any $a, b, c, d \in \mathbb{Z}$ such that $ad - bc = 1$; i.e., the matrix

$$\begin{pmatrix} a & b \\ c & d \end{pmatrix} \in \mathrm{SL}_2(\mathbb{Z}).$$

Note also that if $z = u + \mathrm{i}v$ and $v \to \infty$, then all the terms in the sum (6.2.4) vanish in the limit except for $|x| = 0$, and hence

(6.2.5) $$\widetilde{\theta}_\Gamma(u + \mathrm{i}v) \to 1 \quad \text{as } v \to \infty.$$

This condition, together with the result of Exercise 6.2.8, implies that $\widetilde{\theta}_\Gamma(z)$ is a *modular form* of weight 8. However, it is known that such a form is unique up to multiplication; see [**Ser73**, §VII.3.2, Theorem 4]. Therefore, condition (6.2.5) determines $\widetilde{\theta}_{\Gamma}(z)$ uniquely, and hence $\theta_{\Gamma}(z)$ does not depend on the choice of Γ. In particular, the heat traces for $\Gamma = \Gamma(16)$ and $\Gamma = \Gamma(8) \oplus \Gamma(8)$ coincide. Therefore, it follows from Exercise 6.2.2 that the corresponding sixteen-dimensional tori M_1 and M_2 are isospectral. This completes the proof of the theorem.   □

■ **Exercise 6.2.9.** Show that any two isospectral two-dimensional flat tori are isometric.

In fact, the minimal dimension in which there exist isospectral but not isometric flat tori is equal to four; see [**Sch90, ConSlo92, Sch97**].

6.2.3. Sunada's construction. In this subsection we follow the exposition of R. Brooks [**Bro88**], [**Bro98**]. Let M and N be two closed smooth manifolds. Recall that $p : M \to N$ is a *covering map* if it is a surjective and continuous map such that every point in N has an open neighbourhood whose pre-image is a disjoint union of open sets, and the restriction of p to each of them is a homeomorphism. A covering (or *deck*) transformation corresponding to a smooth covering map p is a diffeomorphism ψ such that $p \circ \psi = p$:

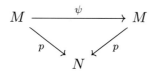

In other words, a deck transformation permutes the elements of the fiber $p^{-1}(x)$, $x \in N$. The set of all covering transformations is called a *covering group*. If, in addition, the manifolds M and N are Riemannian and p is a local isometry, we say that p is a *Riemannian covering map*. If ω is a Riemannian metric on N, then $\widetilde{\omega} = p^*\omega$ is a Riemannian metric on M which is invariant under the deck transformations, and $p : (M, \widetilde{\omega}) \to (N, \omega)$ is a Riemannian covering.

Example 6.2.10. If $p : \mathbb{S}^d \to \mathbb{RP}^d$ is the standard double cover, its deck transformation group is $\mathbb{Z}_2$.

Let $\pi_1(N, b)$ be the fundamental group of N with the base point $b \in N$, and let $\widetilde{b} \in M$ be such that $p(\widetilde{b}) = b$. A covering map $p : M \to N$ is called *normal* if $p_*\left(\pi_1\left(M, \widetilde{b}\right)\right)$ is a normal subgroup of $\pi_1(N, b)$. It is easy to verify that this definition does not depend on the choice of the base points. One can show that a covering map p is normal if and only if its group of deck transformations G acts transitively on the fibers; i.e., for any $x \in N$

and any $\widetilde{x}_1, \widetilde{x}_2$ such that $p(\widetilde{x}_i) = x$, $i = 1, 2$, there exists $g \in G$ such that $g\widetilde{x}_1 = \widetilde{x}_2$; see [**Hat01**, Proposition 1.39].

Theorem 6.2.11. Let $p : M \to N$ be a normal Riemannian covering with a finite covering group G. Then the heat kernels on M and N are related by

$$(6.2.6) \qquad e_N(t, x, y) = \sum_{g \in G} e_M(t, \widetilde{x}, g\widetilde{y}),$$

where $p(\widetilde{x}) = x$ and $p(\widetilde{y}) = y$.

Note that since p is a normal covering, the right-hand side of (6.2.6) does not depend on the particular choice of the pre-images $\widetilde{x}$ and $\widetilde{y}$.

■ **Exercise 6.2.12.** Prove Theorem 6.2.11. *Hint:* Use a direct computation to show that the right-hand side of (6.2.6) satisfies the heat equation and the initial condition.

Therefore, the heat trace on the Riemannian manifold N can be represented as

$$e_N(t) = \int_N e_N(t, x, x)\,\mathrm{d}x = \frac{1}{\operatorname{card} G} \sum_{g \in G} \int_M e_M(t, \widetilde{x}, g\widetilde{x})\,\mathrm{d}\widetilde{x},$$

where $\operatorname{card} G$ is the cardinality of the group G. The last equality follows by replacing the integration over M by the integration over $(\operatorname{card} G)$ copies of N.

Let h be an isometry of M. Then $e_M(t, h\widetilde{x}, h\widetilde{y}) = e_M(t, \widetilde{x}, \widetilde{y})$ and

$$\int_M e_M(t, \widetilde{x}, hgh^{-1}\widetilde{x})\,\mathrm{d}\widetilde{x} = \int_M e_M(t, h^{-1}\widetilde{x}, gh^{-1}\widetilde{x})\,\mathrm{d}\widetilde{x} = \int_M e_M(t, \widetilde{x}, g\widetilde{x})\,\mathrm{d}\widetilde{x}.$$

Therefore, one can rewrite the formula for the heat trace as

$$(6.2.7) \qquad e_N(t) = \sum_{[g] \subset G} \frac{\operatorname{card}[g]}{\operatorname{card} G} \int_M e_M(t, \widetilde{x}, g\widetilde{x})\,\mathrm{d}\widetilde{x},$$

where $[g]$ denotes the conjugacy class of the element $g \in G$.

Definition 6.2.13 (Sunada triple). Let G be a finite group and let H_1, H_2 be two subgroups of G. We say that (G, H_1, H_2) is a *Sunada triple* if for any $g \in G$,

$$\operatorname{card}\{[g] \cap H_1\} = \operatorname{card}\{[g] \cap H_2\}.$$

Definition 6.2.13 implies that if (G, H_1, H_2) is a Sunada triple, then $\operatorname{card} H_1 = \operatorname{card} H_2$.

In group theory, the subgroups satisfying Definition 6.2.13 have been first considered by F. Gassmann, and thus Sunada triples are sometimes referred to as Gassmann triples.

■ **Exercise 6.2.14** (Gassmann's example [**Gas26**]). Let $G = \mathrm{Sym}(6)$, a symmetric group acting on six elements $\{a, b, c, d, e, f\}$, and let

$$H_1 = \{1, (ab)(cd), (ac)(bd), (ad)(bc)\}$$

and

$$H_2 = \{1, (ab)(cd), (ab)(ef), (cd)(ef)\}$$

be two subgroups of G. Show that (G, H_1, H_2) is a Sunada triple and the subgroups H_1 and H_2 are not conjugate in G (i.e., there is no $g \in G$ such that $gH_1g^{-1} = H_2$).

We can now describe the Sunada construction of isospectral manifolds. Consider the following diagram of coverings where p is normal (and hence p_1 and p_2 are normal as well):

(6.2.8)

For example, we may assume that N is a four-dimensional manifold with the fundamental group G (it is known that any finite group can be realised as the fundamental group of a four-manifold) and M is its universal cover.

Theorem 6.2.15 ([**Sun85**]). Suppose that (G, H_1, H_2) is a Sunada triple, and let manifolds M, N, N_1, N_2 be as in diagram (6.2.8). Take any Riemannian metric on N and lift it to the coverings N_1 and N_2. Then the Riemannian manifolds N_1 and N_2 are isospectral.

Proof. In view of formula (6.2.7) for the heat trace, we have for $i = 1, 2$,

$$e_{N_i} = \sum_{[g] \subset H_i} \frac{\mathrm{card}([g])}{\mathrm{card}\, H_i} \int_M e_M(t, \tilde{x}, g\tilde{x})\, d\tilde{x}$$

$$= \sum_{[g] \subset G} \frac{\mathrm{card}([g] \cap H_i)}{\mathrm{card}\, H_i} \int_M e_M(t, \tilde{x}, g\tilde{x})\, d\tilde{x},$$

where the metric on M is the lift of the metric on N. Since (G, H_1, H_2) is a Sunada triple, the right-hand side is independent of i. Therefore, $e_{N_1}(t) = e_{N_2}(t)$ for all $t > 0$, and by Exercise 6.2.2 it follows that N_1 and N_2 are isospectral. $\square$

It remains to show that there exist Sunada triples leading to nonisometric manifolds N_1 and N_2. Suppose that H_1 and H_2 are not conjugate in G (cf. Exercise 6.2.14) and M is the universal cover of N. If the metric on N (which we are free to choose) is bumpy enough so that M has no isometries that are not in G, then N_1 and N_2 are not isometric. Indeed, in that case any isometry between N_1 and N_2 lifts to an isometry of M which conjugates H_1 and H_2 and hence does not belong to the deck transformation group G. Moreover, there exist examples of Sunada triples such that H_1 and H_2 are not isomorphic (see [**Sun85**], [**Ros97**] for details). In this case, N_1 and N_2 have nonisomorphic fundamental groups and are thus nonhomeomorphic and hence nonisometric.

While isospectral and nonisometric manifolds have been known prior to Sunada's work (like Milnor's example described in the previous subsection), Sunada's construction provided the first "machine" to produce an abundance of such examples. Moreover, an adaptation of Sunada's method to planar domains has lead to a breakthrough paper [**GorWebWol92**] by C. Gordon, D. Webb, and S. Wolpert, who have produced the first examples of isospectral nonisometric planar domains with either Dirichlet or Neumann boundary conditions; see Figure 6.1. We discuss some related examples in the next subsection and show that the algebraic techniques of Sunada can be in fact replaced by a rather elementary idea called the *transplantation of eigenfunctions*, originating in [**Bér92**].

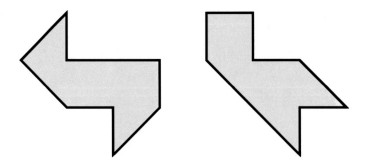

Figure 6.1. Isospectral domains of C. Gordon, D. Webb, and S. Wolpert, each constructed of seven isosceles right-angled triangles.

▣ **Numerical Exercise 6.2.16.** Compute the eigenvalues of the domains in Figure 6.1 to check their isospectrality, with either Dirichlet or Neumann boundary conditions. Check if isospectrality still holds for Robin conditions.

6.2.4. Transplantation of eigenfunctions and mixed Dirichlet–Neumann isospectrality. Let us first apply the transplantation technique to a simplified problem: find isospectral nonisometric domains with *mixed* Dirichlet and Neumann boundary conditions. The possibility to impose mixed conditions, as shown in [**LevParPol06**], provides more freedom and leads to simpler examples, while capturing the main idea of the method.

> **Theorem 6.2.17** ([**LevParPol06**]). The following two boundary value problems, see Figure 6.2, are isospectral:
>
> (i) A unit square Ω, with the Dirichlet condition imposed on three sides and the Neumann condition on the remaining side.
>
> (ii) A right isosceles triangle $\widetilde{\Omega}$ with the Dirichlet condition imposed on the hypotenuse of length 2 and on one of the sides, and the Neumann condition on the other side.

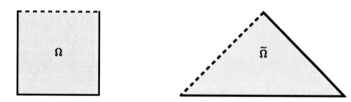

Figure 6.2. A pair of isospectral domains with mixed boundary conditions. The solid lines denote the Dirichlet boundaries, and the dashed lines denote the Neumann boundaries.

Proof. It is convenient to position Ω and $\widetilde{\Omega}$ as shown in Figure 6.3. Let $K = \Omega \cap \widetilde{\Omega}$ be the triangle shown with the vertical side denoted a, the horizontal side b, and the hypotenuse c, so that

$$\Omega = K \cup c \cup \tau_c K, \qquad \widetilde{\Omega} = K \cup a \cup \tau_a K,$$

where τ_a and τ_c denote the mirror symmetries with respect to a and c.

Let u be some eigenfunction of the corresponding mixed problem on Ω. We represent u as a pair of functions $(u_1, u_2) : K \times K \to \mathbb{R}$ as follows:

$$u(x) = \begin{cases} u_1(x), & x \in K, \\ u_2(\tau_c x), & x \in \tau_c K. \end{cases}$$

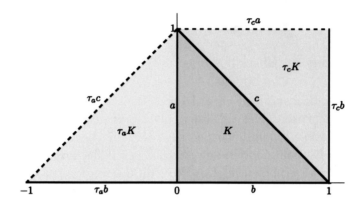

Figure 6.3. The construction for the proof of Theorem 6.2.17.

Note that since u is smooth inside Ω we have the matching conditions

(6.2.9) $$u_1\big|_c = u_2\big|_c, \qquad \partial_n u_1\big|_c = -\partial_n u_2\big|_c.$$

The minus sign appears in the second condition since reflections change the direction of the normal (and therefore the sign of the normal derivative) to the opposite one. We also have the boundary conditions

(6.2.10) $$u_1\big|_a = \partial_n u_2\big|_a = 0, \qquad u_1\big|_b = u_2\big|_b = 0.$$

Let us now transplant u to $\widetilde{\Omega}$. Introduce a new pair of functions (v_1, v_2) : $K \times K \to \mathbb{R}$ by setting

$$v_1(x) = u_2(x) - u_1(x), \qquad v_2(x) = u_1(x) + u_2(x).$$

Consider the function v on $\widetilde{\Omega}$ defined by

(6.2.11) $$v(x) = \begin{cases} v_1(x), & x \in K, \\ v_2(\tau_a x), & x \in \tau_a K, \end{cases}$$

and let us show that v is an eigenfunction of the corresponding mixed problem on $\widetilde{\Omega}$. It is easy to check that v satisfies the correct boundary conditions on $\partial\widetilde{\Omega}$. Indeed, we have

$$v\big|_b = v_1\big|_b = (u_2 - u_1)\big|_b = 0, \qquad v\big|_{\tau_a b} = v_2\big|_b = (u_1 + u_2)\big|_b = 0,$$
$$v\big|_c = v_1\big|_c = (u_2 - u_1)\big|_c = 0, \quad \partial_n v\big|_{\tau_a c} = \partial_n v_2\big|_c = \partial_n(u_1 + u_2)\big|_c = 0,$$

by (6.2.10) and (6.2.9).

Obviously, v satisfies the eigenvalue equation on $\widetilde{\Omega} \setminus a$ but we need to verify that it is true on the whole domain $\widetilde{\Omega}$. A nontrivial point here is that u extends smoothly across the line of reflection a; cf. Remark 3.2.22. Recall that the function u_1 satisfies the Dirichlet condition on a, and u_2 the Neumann condition. Therefore, by the reflection principle of Proposition 3.2.20, u_1 reflects antisymmetrically about a, and u_2 reflects symmetrically. As a result, $v_1 = u_2 - u_1$ becomes $u_2 + u_1$ after the reflection, thus matching v_2, and the definition (6.2.11) therefore indeed produces a smooth eigenfunction on $\widetilde{\Omega}$.

In order to complete the proof it remains to note that the operations used to construct v out of u are invertible and linear, and hence there is a one-to-one correspondence between linearly independent eigenfunctions of the two problems, which therefore have the same eigenvalue. Thus, the domains Ω_1 and Ω_2 with the boundary conditions specified above are isospectral.   □

■ **Exercise 6.2.18.** Prove Theorem 6.2.17 by an explicit computation of the spectra for both problems using separation of variables.

Remark 6.2.19. Alternative approaches to proving Theorem 6.2.17 and its generalisations can be found in [**LevParPol06**] and [**BanParBSh09**].

■ **Exercise 6.2.20.** Show, in each case, that the following Zaremba problems are isospectral:

(a) Two domains shown in the top row of Figure 6.4, one simply connected and another not simply connected.

(b) Two Zaremba problems on half-disk, shown in the second row of Figure 6.4, obtained from each other by swapping Dirichlet and Neumann boundary conditions. The central arc where the boundary conditions change is a quarter-circle. This result, first stated in [**JakLNP06**], plays a role in studying the first eigenvalue of the Laplace–Beltrami operator on the Bolza surface mentioned in §5.3.3.

(c) Four Zaremba problems shown in the last two rows of Figure 6.4.

Remark 6.2.21. One can show using the heat trace asymptotics that isospectral planar domains with mixed boundary conditions must have the same area (corresponding to the coefficient a_0 in the heat trace expansion) and the same difference between the lengths of the Dirichlet and Neumann parts of the boundary (this quantity corresponds to the heat trace coefficient $a_{\frac{1}{2}}$ of the mixed problem; see [**NurRowShe19**]). One can observe that this is indeed the case for all isospectral domains shown in Figures 6.2 and 6.4. At the same time, it was shown in [**vdBDryKap14**, Example 6] that isospectral problems in Figure 6.2 can be distinguished by their heat contents (5.1.2) corresponding to the unit initial temperature distributions.

Figure 6.4. Two pairs and a quadruple of mixed isospectral problems from Exercise 6.2.20. In each case, the solid lines denote the Dirichlet boundaries, and the dashed lines denote the Neumann boundaries.

6.2.5. Isospectral drums. Let us now apply the transplantation method to the case of pure Dirichlet (or Neumann) boundary conditions. We start with the following simple example, where the isospectral regions are disconnected.

Example 6.2.22 ([**Cha95**]). With either Neumann or Dirichlet boundary conditions, the disjoint union of a square of side 1 and an isosceles right triangle of side 2 is isospectral to the disjoint union of a 1×2 rectangle and an isosceles right triangle of side $\sqrt{2}$; see Figure 6.5. This is essentially a variation of the construction presented in Theorem 6.2.17.

As we have mentioned previously, the first examples of planar isospectral connected domains were constructed in [**GorWebWol92**]; see Figure 6.1. A bit later, a whole zoo of isospectral pairs was produced using a similar approach in [**BusCDS94**]. In fact, one can find an underlying Sunada triple behind each of those pairs. At the same time, in this case isospectrality can also be verified directly using the elementary transplantation method.

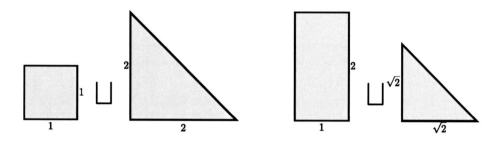

Figure 6.5. Two disjoint isospectral regions from Example 6.2.22.

The simplest example of isospectral domains constructed in [**BusCDS94**] is presented in Figure 6.6. These domains are called "warped propellers", and we will denote them by Ω and $\widetilde{\Omega}$.

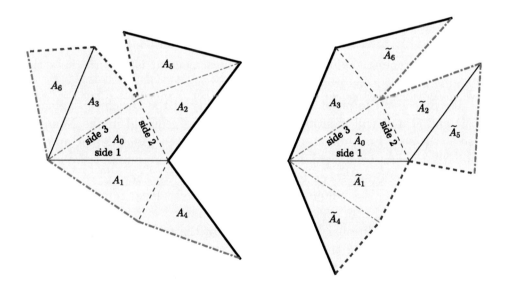

Figure 6.6. Two isospectral warped propellers Ω and $\widetilde{\Omega}$.

Each of the warped propellers is a union[5] of seven identical copies of the same given scalene triangle[6], arranged in a particular manner; we will denote these copies by A_j and $\widetilde{A}_j$, with $j = 0, \ldots, 6$. To construct Ω, we start with the given triangle A_0, enumerating its sides from one to three.

---

[5]Strictly speaking, the interior of the closure of the union.
[6]Some restrictions on the angles of this given triangle are required in order to avoid self-intersecting propellers.

We then construct[7] A_j, $j = 1, 2, 3$, as

$$A_j = \tau_{0,j} A_0,$$

where $\tau_{m,n}$ denotes the reflection with respect to the straight line containing the nth side of A_m. We do the same for $\widetilde{A}_j$, $j = 1, 2, 3$, starting from $\widetilde{A}_0 = A_0$, so at this stage the propellers are identical. We preserve the enumeration of sides under reflections.[8]

We now construct the remaining triangles, numbered 4 to 6, in two different ways. For Ω, we set

$$A_4 = \tau_{1,2} A_1, \quad A_5 = \tau_{2,3} A_2, \quad A_6 = \tau_{3,1} A_3,$$

whereas for $\widetilde{\Omega}$ we reflect as

$$\widetilde{A}_4 = \tau_{1,3} \widetilde{A}_1, \quad \widetilde{A}_5 = \tau_{2,1} \widetilde{A}_2, \quad \widetilde{A}_6 = \tau_{3,2} \widetilde{A}_3.$$

Theorem 6.2.23 ([**BusCDS94**]). *The domains Ω and $\widetilde{\Omega}$ are nonisometric and isospectral for both Dirichlet and Neumann boundary conditions.*

Proof. Since the triangles are chosen to be scalene, it is easy to check that Ω and $\widetilde{\Omega}$ are not isometric.

We first give the proof of the isospectrality of the Dirichlet Laplacians on Ω and $\widetilde{\Omega}$ and will mention the modifications required in the Neumann case at the end. Let u be an eigenfunction of the Dirichlet Laplacian on Ω. Similarly to what we have done in the proof of Theorem 6.2.17, we identify u with a collection of seven functions $u_j : A_0 \to \mathbb{R}$, where $u|_{A_j} = u_j \circ \kappa_j$, and $\kappa_j : A_j \to A_0$ is a unique (since triangles are scalene) isometry between triangles, $j = 0, \ldots, 6$, $\kappa_0 = \mathrm{Id}$. The functions u_j satisfy some boundary and matching conditions. Firstly, if a side of the triangle A_j is part of the external boundary $\partial\Omega$, then on that side we have $u_j = 0$. Secondly, if two triangles A_j and A_k are reflections of each other across a common side, then on that side

$$u_j = u_k \quad \text{and} \quad \partial_n u_j = -\partial_n u_k.$$

We now describe the transplantation of the eigenfunction u from Ω to an eigenfunction v on $\widetilde{\Omega}$. We once more identify v with a collection of seven functions $v_j : \widetilde{A}_0 \to \mathbb{R}$, where

(6.2.12) $$v|_{\widetilde{A}_j} = v_j \circ \widetilde{\kappa}_j,$$

[7]Our enumeration of triangles and other notation differ sometimes from those in [**BusCDS94**].
[8]To help distinguishing the sides, the different sides of the original triangles and their reflections are marked in different line styles in both Figures 6.6 and 6.7.

and $\widetilde{\kappa}_j : \widetilde{A}_j \to \widetilde{A}_0$ is a unique isometry between triangles, $j = 0, \ldots, 6$, $\widetilde{\kappa}_0 = \text{Id}$. We assume for simplicity that the propellers are positioned in such a way that $\widetilde{A}_0 = A_0$.

We start by assigning

(6.2.13) $$v_0 = u_1 + u_2 + u_3.$$

We now have to "propagate" this eigenfunction across the boundary of the triangles in the following way. We start by reflecting (6.2.13) across the joint side 1 of $\widetilde{A}_0$ and $\widetilde{A}_1$. We note that on Ω, u_1 smoothly matches u_0 across side 1 and u_3 smoothly matches u_6 across the common side 1 of triangles A_3 and A_6. Finally, side 1 of the triangle A_2 is a part of the exterior boundary of $\partial\Omega$; thus by the reflection principle of Proposition 3.2.20, u_2 reflects antisymmetrically across side 1 and becomes $-u_2$. We therefore assign

(6.2.14) $$v_1 = u_0 - u_2 + u_6;$$

see Figure 6.7.

We now reflect (6.2.13) across the joint side 2 of $\widetilde{A}_0$ and $\widetilde{A}_2$. In the same manner, u_1 smoothly reflects to u_4 across the joint side 2 of triangles A_1 and A_4, u_2 smoothly reflects to u_0 across the joint side 2 of triangles A_2 and A_0, and since side 2 is an exterior side of triangle A_3, u_3 smoothly reflects to $-u_3$ across this side. We therefore assign

(6.2.15) $$v_2 = u_4 + u_0 - u_3.$$

Continuing in the same manner, we further obtain

(6.2.16) $$\begin{aligned} v_3 &= -u_1 + u_5 + u_0, & v_4 &= u_3 - u_5 - u_6, \\ v_5 &= -u_4 + u_1 - u_6, & v_6 &= -u_4 - u_5 + u_2; \end{aligned}$$

see Figure 6.7.

By construction, the resulting function v defined by (6.2.12)–(6.2.16) satisfies the eigenvalue equation and is smooth in $\widetilde{\Omega}$. It remains to verify that it satisfies the Dirichlet boundary condition on $\partial\widetilde{\Omega}$. This is done triangle by triangle. Let us look, for example, at the triangle $\widetilde{A}_4$ which has two exterior sides: side 1 and side 2. Recalling the definition of v_4 in (6.2.16), we observe that on side 1 we have $u_3 = u_6$ since they match across this side; we also have on this side $u_5 = 0$, since side 1 is an exterior side of the triangle A_5. Thus $v_4 = u_3 - u_5 - u_6 = 0$ on side 1. Similarly, on side 2 we have $u_3 = u_5 = u_6 = 0$ since it is an exterior side for all three triangles A_3, A_5, and A_6, and therefore $v_4 = 0$ on side 2. The remaining triangles and their exterior sides are checked similarly.

Thus, our transplantation $u \mapsto v$ indeed generates an eigenfunction v of the Dirichlet Laplacian on $\widetilde{\Omega}$ corresponding to the same eigenvalue as

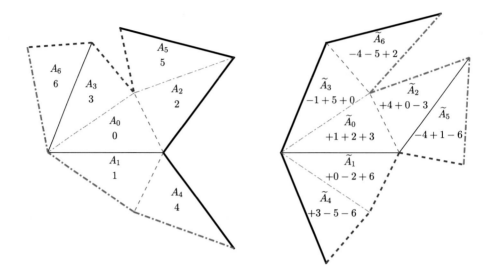

Figure 6.7. A transplantation of a Dirichlet eigenfunction u on a warped propeller Ω to a Dirichlet eigenfunction v on $\widetilde{\Omega}$. The number j inside the triangle A_j is a shorthand for writing $u|_{A_j} = u_j \circ \kappa_j$. The expression of the form $\pm l \pm m \pm n$ inside the triangle $\widetilde{A}_j$ is a shorthand for writing $v|_{\widetilde{A}_j} = \pm u_l \circ \widetilde{\kappa}_l \pm u_m \circ \widetilde{\kappa}_m \pm u_n \circ \widetilde{\kappa}_n$. For a transplantation of a Neumann eigenfunction, replace all minuses by pluses.

u. Moreover, as in the proof of Theorem 6.2.17, the operator $u \mapsto v$ is linear and invertible, and hence we obtain that Ω and $\widetilde{\Omega}$ are indeed Dirichlet isospectral.

For the Neumann boundary conditions the argument is the same, but instead of reflecting the functions u_j antisymmetrically across the sides of the triangle A_j which lie on the boundary of Ω, we apply the symmetric reflection. As a result, all minuses in formulas (6.2.14)–(6.2.16) and in Figure 6.7 should be replaced by pluses. Verifying that the resulting function v satisfies the Neumann conditions on $\partial\widetilde{\Omega}$ is straightforward. $\qquad\square$

The starting transplantation (6.2.13) used in the proof of Theorem 6.2.23 above is not unique.

■ **Exercise 6.2.24.** Give another proof of this theorem by choosing a different starting transplantation defined by

$$(6.2.17) \qquad\qquad v_0 = u_0 + u_4 + u_5 + u_6.$$

Show that any nontrivial linear combination of the transplantations defined by (6.2.13) and (6.2.17) is also a transplantation.

Note that the transplantation method uses in an essential way the fact that the boundary conditions on each part of the boundary are either Dirichlet or Neumann. In particular, it does not work for the Robin eigenvalue problem; cf. Numerical Exercise 6.2.16.

Open Problem 6.2.25. Do there exist Robin isospectral planar domains with a nonzero Robin parameter?

A similar question is also open for the Steklov problem that will be considered in Chapter 7. Some higher-dimensional examples of Robin and Steklov isospectral manifolds can be found in [**GorHerWeb21**].

6.2.6. Spectral rigidity. In this subsection we discuss some results in the opposite direction to isospectrality. Namely, we would like to understand which manifolds and domains are uniquely determined (in an appropriate sense) by their spectra. This is an active area of research, and there is still very little known on this subject. For example, while we have seen in the previous subsection that there exist isospectral nonisometric planar domains, all known examples of isospectral pairs are nonsmooth and nonconvex.

Open Problem 6.2.26.

(i) Do there exist smooth Dirichlet (or Neumann) isospectral nonisometric planar domains? (In the Dirichlet case, this is precisely the question posed in [**Kac66**].)

(ii) Do there exist Dirichlet (or Neumann) isospectral nonisometric convex planar domains?

Let us discuss some results in the negative direction.

Theorem 6.2.27. Let $\Omega \subset \mathbb{R}^d$ be a bounded Lipschitz domain, and suppose that the Dirichlet (respectively, Neumann) spectrum of Ω coincides with the Dirichlet (respectively, Neumann) spectrum of a ball $B \subset \mathbb{R}^d$. Then Ω coincides with B up to a rigid motion.

Proof. Consider the Dirichlet case first. It follows from Weyl's law that $\mathrm{Vol}(\Omega) = \mathrm{Vol}(B)$. Putting this together with the equality $\lambda_1(\Omega) = \lambda_1(B)$ and recalling that the equality in Faber–Krahn's theorem is attained among Lipschitz domains only for a ball (see Remark 5.1.14), it follows that Ω is a ball of the same volume as $B = \Omega^*$.

The argument in the Neumann case is identical, with the equality in the Faber–Krahn inequality replaced by the equality in the Szegő–Weinberger inequality; see Theorem 5.3.2.   □

Remark 6.2.28. As follows from the Ashbaugh–Benguria theorem (Theorem 5.4.7), the ratio between the first two Dirichlet eigenvalues attains its maximum if and only if the domain is a ball. Therefore, in the Dirichlet case, the ball is uniquely determined by only two lowest eigenvalues. An analogue of this result in the Neumann case is false in dimensions $n \geq 3$; in two dimensions, it is not known whether a disk is uniquely determined by any finite part of its Neumann spectrum.

Remark 6.2.29. One can alternatively prove Theorem 6.2.27 using the heat trace asymptotics (see [**Bro93**] for the two-term heat trace expansion on Lipschitz domains) and the classical isoperimetric inequality. Indeed, the heat trace coefficients a_0 and $a_{\frac{1}{2}}$ determine the volumes of Ω and of $\partial\Omega$, and the equality in the isoperimetric inequality is attained among Lipschitz domains if and only if the domain is a ball.

Beyond Theorem 6.2.27, rather little is known about domains which are spectrally determined in full generality. Some important advances have been achieved in the class of real analytic domains satisfying certain symmetry assumptions; see, for example, [**Zel09**]. To illustrate how difficult the questions on spectral rigidity are, let us note that it is unknown whether any ellipse is spectrally determined among all smooth planar domains. Recently, this has been shown in [**HezZel22**] for ellipses of small eccentricity (i.e., that are close to a disk) using a highly sophisticated machinery coming from billiard dynamics developed in [**KalSor18**], [**AviDSiKal16**]. As we have mentioned earlier in Remark 6.2.4, there is a deep connection between the Laplace spectrum and the dynamics of the geodesic (or billiard) flow. In particular, the Laplace spectrum contains a lot of information about the *length spectrum*, i.e., the set of lengths of closed trajectories, which in some cases allows control of the geometry.

Consider also a "local" version of Open Problem 6.2.26 which was formulated by P. Sarnak.

Conjecture 6.2.30 ([**Sar90**]). There exist no nonisometric isospectral continuous deformations of smooth planar domains.

In other words, the claim is that all isospectral pairs are "isolated". For domains close to a disk, some progress on this conjecture and its dynamical counterpart, for which isospectrality is understood in the sense of the length spectrum, has been obtained in [**DeSKalWei17**]. The best known general result in this direction is the compactness in C^∞ topology of the set of Dirichlet isospectral planar domains [**OsgPhiSar88**]. In the same paper, a similar compactness result was also obtained for Riemannian metrics on closed surfaces. Interestingly enough, the proof in [**OsgPhiSar88**] uses a certain property of the heat trace coefficients. Another related result in the Riemannian setting states that closed negatively curved manifolds do not admit nonisometric isospectral deformations. It was proved in [**GuiKaz80**] in dimension two, and in [**CroSha98**] in arbitrary dimensions.

Let us conclude this chapter by the following interesting result obtained by S. Tanno [**Tan73**]. It uses the explicit expressions for the heat trace coefficients a_1, a_2, and a_3 of a closed Riemannian manifold.

Theorem 6.2.31 ([**Tan73**]). Let M be a closed Riemannian manifold of dimension $d \leq 6$ with $\text{Spec}\left(-\Delta_{(M,g)}\right) = \text{Spec}\left(-\Delta_{(\mathbb{S}^d, g_0)}\right)$, where g_0 is a round metric. Then (M, g) is isometric to $(\mathbb{S}^d, g_0)$.

In dimension $d \geq 7$, the geometric information contained in the first three heat invariants becomes insufficient to prove the result of Theorem 6.2.31.

Open Problem 6.2.32. Is a round sphere uniquely determined by its Laplace–Beltrami spectrum among all compact closed Riemannian manifolds in any dimension?

Chapter 7

The Steklov problem and the Dirichlet-to-Neumann map

Vladimir
Andreevich
Steklov
(or **Stekloff**)
(1864–1926)

In this chapter, we focus on the spectral geometry of the Steklov eigenvalue problem and the Dirichlet-to-Neumann map. We state the variational principle for the Steklov spectrum and prove the Hersch–Payne–Schiffer inequalities for Steklov eigenvalues on simply connected planar domains. We also use the Hörmander–Pohozhaev identity to investigate the link between the Dirichlet-to-Neumann map and the boundary Laplacian. As an application, we derive the spectral asymptotics for the Steklov problem on smooth Riemannian manifolds with boundary. We also discuss the asymptotics of Steklov eigenvalues on planar domains with corners, as well as the spectrum of the Dirichlet-to-Neumann map for the Helmholtz equation.

Lars Valter
Hörmander
(1931–2012)

7.1. The Steklov eigenvalue problem

7.1.1. Definition and variational principle. Let Ω be a bounded domain in a complete Riemannian manifold of dimension $d \geq 2$. This includes bounded Euclidean domains and compact Riemannian manifolds with boundary. We denote the boundary of Ω by $M = \partial\Omega$ and assume that M is at least Lipschitz. The *Steklov eigenvalue problem* on Ω is stated as follows:

(7.1.1)
$$\begin{cases} \Delta U = 0 & \text{in } \Omega, \\ \partial_n U = \sigma U & \text{on } M. \end{cases}$$

Note that, unlike the Dirichlet and Neumann problems, the spectral parameter σ for the Steklov problem is in the boundary condition. Sometimes, a more general Steklov-type boundary condition is considered:

(7.1.2)
$$\partial_n U = \sigma \rho U, \qquad \text{on } M,$$

where $L^\infty(M) \ni \rho \geq 0$ is a nonzero weight function.

The Steklov problem was introduced by Vladimir Steklov at the turn of the twentieth century; see [**KuzKKNPPS14**] for a historical overview. It arises in various contexts, in particular, in inverse problems, hydrodynamics, and differential geometry. Some of these applications will be discussed later on. We can alternatively interpret the Steklov eigenvalue problem as a spectral problem for the *Dirichlet-to-Neumann map* $\mathcal{D}_0$ defined in the following way. Let $u \in H^{1/2}(M)$, and let us consider the nonhomogeneous Dirichlet problem

(7.1.3)
$$\begin{cases} \Delta U = 0 & \text{in } \Omega, \\ U = u & \text{on } M. \end{cases}$$

This problem has a unique (weak) solution $U \in H^1(\Omega)$; see, e.g., [**McL00**, Theorem 4.10]. We will call this solution the *harmonic extension* of u into Ω and denote it by

$$U = \mathcal{E}_0 u.$$

The Dirichlet-to-Neumann map for the Laplacian,

$$\mathcal{D}_0 : H^{1/2}(M) \to H^{-1/2}(M),$$

is defined as a linear operator $\mathcal{D}_0 : u \mapsto (\partial_n U)|_M = (\partial_n(\mathcal{E}_0 u))|_M$, which maps the boundary Dirichlet datum of a harmonic function U into its Neumann datum. Here, we define the normal derivative $\partial_n U \in H^{-1/2}(M)$ by the relation

$$\int_M (\partial_n U)\, v\, \mathrm{d}s = \int_\Omega \langle \nabla U, \nabla V \rangle \, \mathrm{d}x$$

for every $V \in H^1(\Omega)$ such that $\Delta V \in L^2(\Omega)$, where $v := V|_M \in H^{1/2}(M)$; see [**ChWGLS12**, p. 280].

Note that the operator $\mathcal{D}_0$ is nonlocal and thus is not differential. If the boundary M is smooth, then $\mathcal{D}_0$ is an elliptic selfadjoint pseudodifferential operator of order one. Its principal symbol is given by $|\xi|$, which is the square root of the principal symbol of the boundary Laplacian $-\Delta_M$. The close link between $\mathcal{D}_0$ and $\sqrt{-\Delta_M}$ will be particularly important for spectral asymptotics; see also Remark 7.1.5.

Remark 7.1.1. It is customary to call the function $U \neq 0$ in (7.1.1) an *eigenfunction of the Steklov problem* corresponding to an eigenvalue σ. At the same time, an eigenfunction of the corresponding Dirichlet-to-Neumann map $\mathcal{D}_0$ (which acts on the functions defined on the boundary) is $U|_M$.

Let

$$(7.1.4) \qquad \mathcal{H}_0(\Omega) := \{U \in H^1(\Omega) : \Delta U = 0\} = \left\{\mathcal{E}_0 u : u \in H^{1/2}(M)\right\}$$

be the subspace of harmonic functions in $H^1(\Omega)$. If $U \in \mathcal{H}_0(\Omega)$ satisfies (7.1.1), i.e., it is a Steklov eigenfunction, then by Green's formula we get

$$(\nabla U, \nabla V)_{L^2(\Omega)} = (-\Delta U, V)_{L^2(\Omega)} + (\partial_n U, V)_{L^2(M)} = \sigma (U, V)_{L^2(M)}$$

for any $V \in H^1(\Omega)$, since $\Delta U = 0$. The weak spectral problem

$$(7.1.5) \qquad (\nabla U, \nabla V)_{L^2(\Omega)} = \sigma (U, V)_{L^2(M)} \qquad \text{for all } V \in H^1(\Omega)$$

is a weak version of the Steklov problem (7.1.1). Any weak eigenfunction $U \in H^1(\Omega)$ of (7.1.5) automatically belongs to $\mathcal{H}_0(\Omega)$ and is therefore harmonic; see, e.g., [**AreMaz12**].

Using a similar approach to that in §2.1, one can show that the spectrum of the Steklov problem (or of the Dirichlet-to-Neumann map $\mathcal{D}_0$) is discrete provided that the composition of the trace map and the embedding $H^1(\Omega) \to H^{1/2}(M) \hookrightarrow L^2(M)$ is compact. This condition will be assumed throughout this chapter. It is true, for instance, if Ω has Lipschitz boundary M, in which case the trace map is continuous and the embedding is compact (see, for example, [**AreMaz12**]).

Moreover, taking in (7.1.5) $V = U$, we immediately deduce that the eigenvalues of the Steklov problem are nonnegative. We denote the Steklov eigenvalues by

$$0 = \sigma_1 = \sigma_1(\Omega) < \sigma_2 = \sigma_2(\Omega) \leq \cdots \nearrow +\infty,$$

where the eigenfunction corresponding to $\sigma_1 = 0$ is constant, as for the Neumann boundary conditions. The eigenfunctions $u_j = U_j|_M$ of the Dirichlet-to-Neumann map (which coincide with the boundary traces of the Steklov eigenfunctions U_j) form an orthogonal basis in $L^2(M)$.

■ **Exercise 7.1.2.** Let $\Omega \subset \mathbb{R}^d$ be a bounded domain, and let Ω_a be its homothety with a coefficient $a > 0$. Show that $\sigma_k(\Omega_a) = \frac{1}{a}\sigma_k(\Omega)$; cf. Lemma 2.1.30.

Example 7.1.3 (The Steklov eigenvalues of the unit disk). The Steklov eigenvalues of the unit disk $\mathbb{D}$ are given by

$$\sigma_1(\mathbb{D}) = 0, \quad \sigma_{2k}(\mathbb{D}) = \sigma_{2k+1}(\mathbb{D}) = k, \qquad k \in \mathbb{N}.$$

The eigenfunction corresponding to $\sigma_1 = 0$ is a constant function, and the eigenspace corresponding to $\sigma_{2k} = \sigma_{2k+1} = k$ is spanned by the functions $r^k \sin k\theta$ and $r^k \cos k\theta$, written in polar coordinates (r, θ). Indeed, recall that

$$\Delta = \frac{\partial^2}{\partial r^2} + \frac{1}{r}\frac{\partial}{\partial r} + \frac{1}{r^2}\frac{\partial^2}{\partial \theta^2},$$

and it is easy to see that all these functions are harmonic. This can alternatively be seen from the fact that these functions are just the real and imaginary parts of holomorphic functions z^k, where $z = re^{i\theta} = x + iy$. We note that the eigenspace corresponding to $\sigma_2 = \sigma_3 = 1$ is spanned by the Cartesian coordinate functions x and y; cf. Exercise 1.2.19 for a basis of the first eigenspace of the Laplace–Beltrami operator on the round sphere.

Moreover, since the normal derivative on the boundary coincides with the partial derivative with respect to r,

$$\left(\frac{\partial}{\partial r}\left(r^k \sin k\theta\right)\right)\Big|_{r=1} = k\left(r^k \sin k\theta\right)\Big|_{r=1},$$

$$\left(\frac{\partial}{\partial r}\left(r^k \cos k\theta\right)\right)\Big|_{r=1} = k\left(r^k \cos k\theta\right)\Big|_{r=1}.$$

There are no other eigenvalues as the boundary traces of the Steklov eigenfunctions

$$\{1, \sin\theta, \cos\theta, \ldots, \sin k\theta, \cos k\theta, \ldots\}$$

form a basis in $L^2(M) = L^2(\mathbb{S}^1)$.

Remark 7.1.4 (Steklov–Robin duality). It is easily seen that σ is a Steklov eigenvalue if and only if 0 is an eigenvalue of the Robin Laplacian $-\Delta^{\mathrm{R},-\sigma}$; moreover, the dimensions of the corresponding eigenspaces coincide. See also Remark 3.1.19.

Throughout this chapter, let

$$0 = \nu_1(M) \le \nu_2(M) \le \cdots$$

denote the eigenvalues of the Laplace–Beltrami operator $-\Delta_M$ on the boundary $M = \partial\Omega$, assuming that this boundary is sufficiently smooth.[9]

---

[9] This enumeration of eigenvalues differs from the standard one used in the rest of the book; cf. footnote on page 165. In terms of our usual notation, $\nu_k(M) = \lambda_{k-1}(M)$, $k \in \mathbb{N}$.

Remark 7.1.5. Note that $\sigma_k^2(\mathbb{D}) = \nu_k(\mathbb{S}^1)$, $k \in \mathbb{N}$. Moreover, if U_k are the Steklov eigenfunctions on $\mathbb{D}$, then $u_k = U_k|_{\mathbb{S}^1}$ are the Laplace–Beltrami eigenfunctions on $\mathbb{S}^1$.

Let us mention as well that the Steklov eigenfunctions U_k behave as r^k for $r < 1$; i.e., they decay rapidly in the interior. This decay is a general feature of Steklov eigenfunctions; see [**HisLut01**, Theorem 1.1].

■ **Exercise 7.1.6.** Calculate the Steklov eigenvalues and eigenfunctions of the unit ball $\mathbb{B}^d$ in $\mathbb{R}^d$, and compare the results with the Laplace–Beltrami eigenvalues and eigenfunctions of the round sphere $\mathbb{S}^{d-1}$.

■ **Exercise 7.1.7** (Steklov problem on a generalised cylinder [**ColElSGir11**]). Let Σ be a closed Riemannian manifold. Let $0 = \nu_1 < \nu_2 \leq \cdots$ be its Laplace–Beltrami spectrum, and let $\{u_k\}$ be the corresponding orthonormal basis of eigenfunctions satisfying $-\Delta_\Sigma u_k = \nu_k u_k$. Given any $l > 0$, consider a cylinder $\Omega = (-l, l) \times \Sigma \subset \mathbb{R} \times \Sigma$. Show that the Steklov spectrum of Ω is given by

$$0, \frac{1}{l}, \sqrt{\nu_k}\tanh(\sqrt{\nu_k}l), \sqrt{\nu_k}\coth(\sqrt{\nu_k}l), \qquad k \geq 2,$$

and the corresponding eigenfunctions are

$$1, t, \cosh(\sqrt{\nu_k}t)u_k(x), \sinh(\sqrt{\nu_k}t)u_k(x), \qquad t \in (-l, l), x \in \Sigma.$$

Compare also with Exercise 4.3.17.

For the remainder of this subsection we assume for simplicity that $\Omega \subset \mathbb{R}^d$ is a Euclidean domain. The extension of the variational principles to the Riemannian case is essentially verbatim.

Let $u \in H^{1/2}(M) = \mathrm{Dom}(\mathcal{D}_0)$. The quadratic form of the Dirichlet-to-Neumann map is given by

$$(7.1.6) \qquad (\mathcal{D}_0 u, u)_{L^2(M)} = (\partial_n U, u)_{L^2(M)} = \|\nabla U\|_{L^2(\Omega)}^2,$$

where $U = \mathcal{E}_0 u \in \mathcal{H}_0(\Omega)$. Therefore, the Rayleigh quotient for the Dirichlet-to-Neumann map is given by

$$(7.1.7) \qquad R^S[u] := \frac{\|\nabla \mathcal{E}_0 u\|_{L^2(\Omega)}^2}{\|u\|_{L^2(M)}}, \qquad u \in H^{1/2}(M) \setminus \{0\}.$$

Using (7.1.7) and arguing in the same way as in §3.1, we obtain the following variational characterisation of the Steklov eigenvalues:
(7.1.8)

$$\sigma_k = \min_{\substack{\widetilde{\mathcal{L}} \subset H^{1/2}(M) \\ \dim \widetilde{\mathcal{L}} = k}} \max_{u \in \widetilde{\mathcal{L}} \setminus \{0\}} \frac{\|\nabla \mathcal{E}_0 u\|_{L^2(\Omega)}^2}{\|u\|_{L^2(M)}^2} = \min_{\substack{\mathcal{L} \subset \mathcal{H}_0(\Omega) \\ \dim \mathcal{L} = k}} \max_{\substack{U \in \mathcal{L} \\ U \neq 0}} \frac{\|\nabla U\|_{L^2(\Omega)}^2}{\||U|_M\|_{L^2(M)}^2}, \qquad k \in \mathbb{N}.$$

Note that in the first min-max of (7.1.8) the minimum is taken over subspaces $\tilde{\mathcal{L}}$ of $H^{1/2}(M)$, and in the second one over subspaces $\mathcal{L}$ of the space $\mathcal{H}_0(\Omega)$ of harmonic functions. We can in fact replace $\mathcal{H}_0(\Omega)$ there by the usual Sobolev space $H^1(\Omega)$ but to show this we need

Proposition 7.1.8. Let $\Omega \subset \mathbb{R}^d$ be a bounded open set. Then

$$(7.1.9) \qquad H^1(\Omega) = \mathcal{H}_0(\Omega) \oplus H_0^1(\Omega).$$

The direct sum in (7.1.9) is not orthogonal; however

$$(7.1.10) \qquad (\nabla U, \nabla V)_{L^2(\Omega)} = 0 \qquad \text{for any } U \in \mathcal{H}_0(\Omega), V \in H_0^1(\Omega).$$

Proof. Let $W \in H^1(\Omega)$. Set $u = W|_M$, and let $U = \mathcal{E}_0 u \in \mathcal{H}_0(\Omega)$ be the unique solution of (7.1.3). Then $V = W - U$ belongs to $H_0^1(\Omega)$ since $V|_M = 0$. As $\mathcal{H}_0(\Omega) \cap H_0^1(\Omega) = \{0\}$, (7.1.9) follows.

To prove (7.1.10), we integrate by parts:

$$(\nabla U, \nabla V)_{L^2(\Omega)} = (-\Delta U, V)_{L^2(\Omega)} + (\partial_n U, V)_{L^2(M)} = 0,$$

since $\Delta U = 0$ in Ω and $V|_M = 0$. $\qquad\square$

We can now prove

Theorem 7.1.9 (The variational principle for the Steklov problem). Let Ω be a bounded open set in $\mathbb{R}^d$ with a Lipschitz boundary $M = \partial\Omega$, and let $\sigma_k(\Omega)$ be the eigenvalues of the Steklov problem on Ω. Then[10]

$$(7.1.11) \qquad \sigma_k(\Omega) = \min_{\substack{\mathcal{L} \subset H^1(\Omega) \\ \dim \mathcal{L} = k}} \max_{\substack{W \in \mathcal{L} \\ W \neq 0}} \frac{\|\nabla W\|_{L^2(\Omega)}^2}{\|W|_M\|_{L^2(M)}^2}, \qquad k \in \mathbb{N}.$$

In particular,

$$\sigma_2(\Omega) = \min_{\substack{0 \neq W \in H^1(\Omega) \\ \int_M W|_M \, ds = 0}} \frac{\|\nabla W\|_{L^2(\Omega)}^2}{\|W|_M\|_{L^2(M)}^2}.$$

Proof. Using Proposition 7.1.8, we represent any $W \in H^1(\Omega)$ as $W = U + V$, where $U \in \mathcal{H}_0(\Omega)$, $V \in H_0^1(\Omega)$. We note that $\|(U+V)|_M\|_{L^2(M)}^2 = \|U|_M\|_{L^2(M)}^2$. Moreover, by (7.1.10) and the variational principle for the principal Dirichlet eigenvalue on Ω we have

$$\|\nabla(U+V)\|_{L^2(\Omega)}^2 = \|\nabla U\|_{L^2(\Omega)}^2 + \|\nabla V\|_{L^2(\Omega)}^2$$
$$\geq \|\nabla U\|_{L^2(\Omega)}^2 + \lambda_1^D(\Omega)\|V\|_{L^2(\Omega)}^2.$$

---

[10]In what follows, we use the convention $\frac{Q}{0} = +\infty$ for $Q > 0$.

The minimisation procedure now requires taking $V = 0$, and thus (7.1.11) is equivalent to (7.1.8). $\qquad\square$

7.1.2. The sloshing problem. Steklov eigenvalues of a square. Similarly to Zaremba problems considered in §3.1.3, we will also be looking at the *mixed Steklov–Neumann–Dirichlet* spectral problems, with the spectral parameter in the boundary conditions, stated as follows. Let $\Omega \subset \mathbb{R}^d$ be a bounded simply connected domain with a Lipschitz boundary $M = \partial\Omega$. We decompose M into a disjoint union $M = \mathcal{S} \sqcup \mathcal{W}_{\mathrm{D}} \sqcup \mathcal{W}_{\mathrm{N}}$, where either of $\mathcal{W}_{\mathrm{D,N}}$ may be empty, and consider the spectral problem

$$(7.1.12) \qquad \begin{cases} \Delta U = 0 & \text{in } \Omega, \\ \partial_n U = \sigma U & \text{on } \mathcal{S}, \\ \partial_n U = 0 & \text{on } \mathcal{W}_{\mathrm{N}}, \\ U = 0 & \text{on } \mathcal{W}_{\mathrm{D}}. \end{cases}$$

Sir Horace **Lamb**

(1849–1934)

The problem (7.1.12) goes back to the important special case considered by H. Lamb and A. G. Greenhill already in the 19th century [**Lam93**, Chapter 9], [**Gre86**]; see also [**FoxKut83**, **LevPPS22a**]. Let $\Omega \subset \mathbb{R}^2$ be in the lower half-plane, let $\mathcal{S}$ be an interval of the real line, called the *sloshing surface*, and let $\mathcal{W}_{\mathrm{D}} = \emptyset$; see Figure 7.1. Then (7.1.12) models small gravitational oscillations of an ideal fluid in an infinite canal with the cross-section Ω and the *walls* $\mathcal{W} = \mathcal{W}_{\mathrm{N}}$ and is called the *sloshing problem*. The square roots of *sloshing eigenvalues* σ are proportional to the frequencies of the fluid oscillations, and the sloshing eigenfunctions U represent the fluid velocity potential.

We may equivalently consider the mixed problem (7.1.12) as an example of a Steklov problem with a variable weight boundary condition (7.1.2),

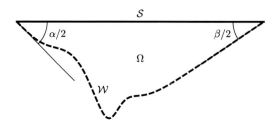

Figure 7.1. Geometry of a sloshing problem. The angles between the sloshing surface $\mathcal{S}$ and the walls $\mathcal{W}$ (which will be convenient to denote by $\frac{\alpha}{2}$ and $\frac{\beta}{2}$) play a significant role in the asymptotics of the sloshing eigenvalues; see §7.3.5.

where we formally take

$$\rho = \begin{cases} 1 & \text{on } \mathcal{S}, \\ 0 & \text{on } \mathcal{W}_{\mathrm{N}}, \\ +\infty & \text{on } \mathcal{W}_{\mathrm{D}}. \end{cases}$$

The weak statement of the mixed problem (7.1.12) is to find $\sigma \in \mathbb{R}$ and $U \in H^1_{0,\mathcal{W}_{\mathrm{D}}}(\Omega) \setminus \{0\}$ such that

$$(\nabla U, \nabla V)_{L^2(\Omega)} = \sigma \, (U, V)_{L^2(\mathcal{S})} \qquad \text{for all } V \in H^1_{0,\mathcal{W}_{\mathrm{D}}}(\Omega).$$

Similarly to the Steklov problem, the spectrum of (7.1.12) is discrete and nonnegative, and the eigenfunctions can be chosen so that their traces on $\mathcal{S}$ form an orthogonal basis in $L^2(\mathcal{S})$.

■ **Exercise 7.1.10.** Let Ω be a rectangle $(0,1) \times (-h,0)$, $h > 0$, and let $\mathcal{S} = (0,1)$. Find the eigenvalues and eigenfunctions of (7.1.12) assuming either the Neumann or the Dirichlet boundary conditions on the rest of the boundary.

We will now use the properties of some mixed Steklov–Neumann–Dirichlet problems to find the Steklov spectrum of a square, following [**GirPol17**]. Let $\Omega = (-1,1)^2 \subset \mathbb{R}^2$ be a square of side 2. Looking for the eigenfunctions of the Steklov problem on Ω using separation of variables, we easily obtain the following eigenfunctions and the equations for the separation parameter κ, which is assumed to be positive; the eigenvalues are then easily expressed in terms of the positive solutions of the corresponding equations; see Table 7.1 and Figure 7.2.

It remains to prove that there are no other eigenvalues. To do so, it is sufficient to demonstrate that the traces of the eigenfunctions U^0, U^1, U^j_κ, $j = 2, \ldots, 9$, form a basis in $L^2(\partial\Omega)$. We observe that the Steklov problem on

Table 7.1. The Steklov eigenfunctions and eigenvalues of the square $(-1,1)^2$ obtained by the separation of variables.

Eigenfunction	Equation for κ	Eigenvalue σ	Multiplicity
$U^0 := 1$		0	1
$U^1 := xy$		1	1
$U^2_\kappa := \cos(\kappa x)\cosh(\kappa y)$ $U^3_\kappa := \cosh(\kappa x)\cos(\kappa y)$	$\tan\kappa + \tanh\kappa = 0$	$\kappa\tanh\kappa$	2
$U^4_\kappa := \sin(\kappa x)\cosh(\kappa y)$ $U^5_\kappa := \cosh(\kappa x)\sin(\kappa y)$	$\tan\kappa - \coth\kappa = 0$	$\kappa\tanh\kappa$	2
$U^6_\kappa := \cos(\kappa x)\sinh(\kappa y)$ $U^7_\kappa := \sinh(\kappa x)\cos(\kappa y)$	$\tan\kappa + \coth\kappa = 0$	$\kappa\coth\kappa$	2
$U^8_\kappa := \sin(\kappa x)\sinh(\kappa y)$ $U^9_\kappa := \sinh(\kappa x)\sin(\kappa y)$	$\tan\kappa - \tanh\kappa = 0$	$\kappa\coth\kappa$	2

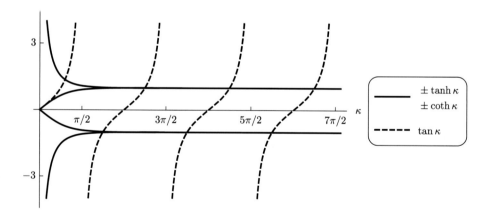

Figure 7.2. Equations for the Steklov eigenvalues of the square: the κ-coordinate of each intersection of a dashed curve with one of the solid curves for $\kappa > 0$ corresponds to a double Steklov eigenvalue of the square.

the square is symmetric with respect to the two diagonals $\{(x,y) : x = \pm y\}$. Reasoning as in the proof of the symmetry decomposition (3.2.7) for the Dirichlet Laplacian, we obtain that the Steklov problem on the square decomposes into the union of four mixed Neumann–Steklov, Dirichlet–Steklov, or Neumann–Dirichlet–Steklov problems on an isosceles right-angled triangle of side $\sqrt{2}$, with the Steklov condition on the hypothenuse; see Figure 7.3.

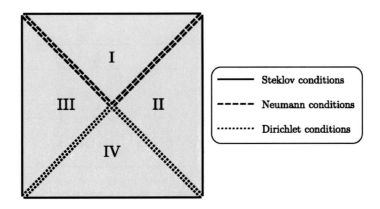

Figure 7.3. Decomposition of the Steklov problem on the square into four mixed problems on isosceles right-angled triangles.

We can now identify the eigenfunctions of the Steklov problem given in the first column of Table 7.1, after transformations of the basis, with each of the mixed problems from Figure 7.3; see Table 7.2.

Table 7.2. The correspondence between the Steklov eigenfunctions from Table 7.1 and the mixed problems from Figure 7.3.

The Steklov eigenfunction	The corresponding mixed problem
U^0	Mixed problem I
U^1	Mixed problem I
$U_\kappa^2 + U_\kappa^3$ $U_\kappa^2 - U_\kappa^3$	Mixed problem I Mixed problem IV
$U_\kappa^4 + U_\kappa^5$ $U_\kappa^4 - U_\kappa^5$	Mixed problem II Mixed problem III
$U_\kappa^6 + U_\kappa^7$ $U_\kappa^6 - U_\kappa^7$	Mixed problem II Mixed problem III
$U_\kappa^8 + U_\kappa^9$ $U_\kappa^8 - U_\kappa^9$	Mixed problem I Mixed problem IV

Consider now the mixed problem I, which is in fact a sloshing problem. To ensure that we have encountered all of its eigenvalues it is enough to demonstrate that the traces on $\mathcal{S} = (-1,1) \times \{1\}$ of the corresponding Steklov eigenfunctions

(7.1.13)
$$U^0, U^1, U_\kappa^2 + U_\kappa^3, U_\kappa^8 + U_\kappa^9,$$

selected from Table 7.2, form an orthogonal basis in $L^2(\mathcal{S})$. To do so, we use the following result which was already known to Lamb (see also [**FoxKut83**], [**LevPPS22a**]): the traces on $\mathcal{S}$ of the eigenfunctions of the sloshing problem I coincide with the eigenfunctions of the one-dimensional *vibrating free beam problem*

(7.1.14)
$$\begin{cases} f^{(\mathrm{iv})}(x) = \kappa^4 f(x), & x \in (-1,1), \\ f''(\pm 1) = f'''(\pm 1) = 0, \end{cases}$$

where κ^4 plays the role of the spectral parameter. It is now easy to verify that the traces of (7.1.13), that is, the functions

$$1, x,$$
$$\cosh(1)\cos(\kappa x) + \cos(1)\cosh(\kappa x) \quad \text{with } \tan\kappa + \tanh\kappa = 0,$$
$$\sinh(1)\sin(\kappa x) + \sin(1)\sinh(\kappa x) \quad \text{with } \tan\kappa - \tanh\kappa = 0,$$

are indeed the only eigenfunctions of (7.1.14). As (7.1.14) is a selfadjoint fourth-order Sturm–Liouville problem, its eigenfunctions form a basis in $L^2(\mathcal{S})$ as required.

The mixed problems II–IV can be treated in a similar manner. They are again linked to the boundary value spectral problems for the fourth derivative on the sloshing surface, the only difference being the boundary conditions; at the ends adjoining the Dirichlet walls we need to impose the Dirichlet conditions $f = f' = 0$ rather than the free ones as in (7.1.14). Combining all the results together we confirm that Table 7.1 gives the full list of eigenvalues and eigenfunctions of the Steklov problem on $(-1,1)^2$.

■ **Exercise 7.1.11.** Using Table 7.1 and Figure 7.2, show that asymptotically the Steklov eigenvalues of the square $(-1,1)^2$ satisfy

$$\sigma_{4m-k} = \left(m - \frac{1}{2}\right)\frac{\pi}{2} + O\left(m^{-\infty}\right), \qquad k = 0,1,2,3, \qquad \text{as } m \to +\infty.$$

Remark 7.1.12. A calculation of Steklov eigenvalues of rectangles and higher-dimensional boxes and an alternative proof of completeness of the set of eigenfunctions which uses the Steklov–Robin duality mentioned in Remark 7.1.4 can be found in [**GirLPS19**].

7.1.3. Isoperimetric inequalities for the Steklov eigenvalues. As was indicated in §7.1.1, the Steklov eigenvalue problem shares some common features with the Neumann problem. Recall that, in two dimensions, the Neumann problem models vibrations of a homogeneous free membrane. Similarly, the Steklov problem (7.1.1) can be viewed as a model for a vibrating free membrane with all the weight uniformly distributed along the boundary (or with a density ρ, if a more general boundary condition (7.1.2) is considered). It is therefore natural to look for an analogue of the Szegő–Weinberger inequality (Theorem 5.3.2) in the Steklov case.

The following result was obtained by R. Weinstock in [**Wei54**], using a modification of Szegő's approach.

Theorem 7.1.13 (Weinstock's inequality [**Wei54**]). *Let $\Omega \subset \mathbb{R}^2$ be a bounded simply connected domain with a Lipschitz boundary of length $L(\partial\Omega) := \mathrm{Vol}_1(\partial\Omega)$. Then*

$$\sigma_2(\Omega)L(\partial\Omega) \leq 2\pi,$$

with the equality attained if and only if Ω is a disk.

■ **Exercise 7.1.14.** Prove Theorem 7.1.13 by adapting the proof of Hersch's theorem (Theorem 5.3.8). Note that the first nontrivial Steklov eigenfunctions of the disk are the coordinate functions (see Example 7.1.3), similarly to the first nontrivial Laplace eigenfunctions on the round sphere. For a solution, see [**GirPol10a**], as well as [**FreLau20**, §7] for details on the equality case for general Lipschitz boundaries.

While Weinstock's theorem is a direct analogue of Szegő's result, there are significant differences between the isoperimetric inequalities for the Steklov and the Neumann eigenvalues. In particular, one can observe that Weinstock's inequality does not admit a generalisation to nonsimply connected planar domains.

Example 7.1.15. Using separation of variables, one can investigate the first Steklov eigenvalues and eigenfunctions of an annulus $A_\varepsilon := \mathbb{D} \setminus B_\varepsilon^2$; see [**GirPol17**, Example 4.2.5]. In particular, if $\varepsilon > 0$ is small enough, then $\sigma_2(A_\varepsilon)L(\partial A_\varepsilon) > 2\pi$.

Remark 7.1.16 (Nonsimply connected planar domains). The previous example indicates that an appropriate perforation of a domain increases the first nontrivial Steklov eigenvalue. This is indeed the case; as was shown in [**GirKarLag21**], there exists a sequence of planar domains Ω_k, with the number of boundary components of Ω_k tending to infinity as $k \to \infty$, and such that $\sigma_2(\Omega_k)L(\partial\Omega_k) \to 8\pi$. Moreover, this is the maximal possible value of the limit, since, as was shown in [**Kok14**],

$$(7.1.15) \qquad\qquad \sigma_2(\Omega)L(\partial\Omega) \leq 8\pi$$

on any surface with boundary. The proof of (7.1.15) uses Hersch's estimate, which explains why the constant on the right-hand side of (7.1.15) is precisely the same as in (5.3.11).

Remark 7.1.17 (Higher dimensions). Weinstock's theorem does not admit a straightforward generalisation to higher dimensions. However, among convex domains of given boundary volume, the ball maximises the first nonzero Steklov eigenvalue [**BucFNT21**]. One can also use a different normalisation: fix the volume of the domain itself, rather than of its boundary. F. Brock has shown in [**Bro01**] that the ball maximises σ_2 among all Euclidean domains of given volume. Note that for simply connected planar domains this result is an easy consequence of Theorem 7.1.13 and the classical isoperimetric inequality. It is also interesting to note that Brock's inequality is stable similarly to the Szegő–Weinberger inequality (see [**BraDeP17**, §7.5]), whereas Weinstock's inequality is extremely unstable [**BucNah21**].

For the remainder of this subsection let us focus on simply connected planar domains. Surprisingly enough, in this case one can obtain sharp isoperimetric inequalities for *all* Steklov eigenvalues.

Theorem 7.1.18 (The Hersch–Payne–Schiffer inequality [**HerPaySch75**]). Let $\Omega \subset \mathbb{R}^2$ be a bounded simply connected domain with a Lipschitz boundary. Then for any $p, q \geq 1$,

$$(7.1.16) \qquad \sigma_{p+1}\sigma_{q+1}L(\partial\Omega)^2 \leq \begin{cases} \pi^2(p+q-1)^2 & \text{if } p+q \text{ is odd,} \\ \pi^2(p+q)^2 & \text{if } p+q \text{ is even.} \end{cases}$$

In particular, with $p = q = k$,

$$(7.1.17) \qquad\qquad \sigma_{k+1}(\Omega)L(\partial\Omega) \leq 2\pi k$$

for all $k \in \mathbb{N}$.

Remark 7.1.19. Note that (7.1.17) is precisely Weinstock's inequality for $k = 1$. Moreover, it was shown in [**GirPol10b**] that this inequality is sharp for any k. The equality is attained in the limit by a union of k identical disks touching each other (cf. (5.3.25) and the corresponding construction of maximisers for the Laplace eigenvalues on the sphere).

Before proceeding to the proof of Theorem 7.1.18, we give a brief reminder of some facts from complex analysis.

Definition 7.1.20 (Harmonic conjugate). Given a harmonic function $U : \Omega \to \mathbb{R}$ defined in a simply connected planar domain Ω, its *harmonic conjugate* $V : \Omega \to \mathbb{R}$ is a harmonic function such that $U + iV$ is holomorphic in Ω.

■ **Exercise 7.1.21.** Show that for any harmonic function on a bounded simply connected planar domain, its harmonic conjugate exists and is uniquely defined up to an additive constant.

■ **Exercise 7.1.22.** Let Ω be a bounded simply connected planar domain with Lipschitz boundary. Let $u \in H^1(\Omega)$ be a harmonic function and let v be its harmonic conjugate. It easily follows from the Cauchy–Riemann equations that $v \in H^1(\Omega)$. Show that

$$(7.1.18) \qquad \partial_n u = -\partial_\tau v$$

on $\partial\Omega$, where the normal derivative $\partial_n u$ and the tangential derivative of $\partial_\tau v$ are understood as elements of the Sobolev space $H^{-1/2}(\partial\Omega)$. *Hint:* Use the Cauchy–Riemann equations and Lemma 2.1.12. For a complete solution, see [**BarBouLeb16**, §6.2.1].

Proof of Theorem 7.1.18. Since Ω is simply connected, by the Riemann mapping theorem there exists a conformal diffeomorphism $\psi : \Omega \to \mathbb{D}$. Moreover, by Carathéodory's theorem, ψ extends continuously to the boundary; see [**Pom92**, Chapter 2]. Let $\mathrm{d}s$ be the measure on $\partial\Omega$ and let $\mathrm{d}\mu = \psi_*\mathrm{d}s$ be the push-forward measure on $\mathbb{S}^1 = \partial\mathbb{D}$.

Let us introduce the "mass parameter"

$$m(\theta) = \int\limits_0^\theta \mathrm{d}\mu,$$

where θ is the coordinate on $\mathbb{S}^1$. Then $\mathrm{d}\mu = m'(\theta)\mathrm{d}\theta$, and

$$m(2\pi) = \int\limits_{\mathbb{S}^1} \mathrm{d}\mu = L(\partial\Omega).$$

Let $h : \mathbb{R} \to \mathbb{R}$ be a smooth periodic function (to be chosen later) with period $L := L(\partial\Omega)$. Let $u : \mathbb{S}^1 \to \mathbb{R}$ be defined by

$$u(\theta) = h(m(\theta)).$$

The function u admits a unique harmonic extension to the disk, which we denote by $U = \mathcal{E}_0 u$.

Choosing an appropriate additive constant, we can choose a harmonic conjugate V of U such that $\int_{\mathbb{S}^1} V \, d\mu = 0$. Let $A, B : \Omega \to \mathbb{R}$ be defined as $A = U \circ \psi$ and $B = V \circ \psi$. By conformal equivalence of the Dirichlet energy,

$$(7.1.19) \qquad \int_\Omega |\nabla A|^2 \, dx = \int_{\mathbb{D}} |\nabla U|^2 \, dx$$

and

$$(7.1.20) \qquad \int_\Omega |\nabla B|^2 \, dx = \int_{\mathbb{D}} |\nabla V|^2 \, dx.$$

Let $x = (x_1, x_2)$. By the Cauchy–Riemann equations, we have

$$\frac{\partial U}{\partial x_1} = \frac{\partial V}{\partial x_2}, \qquad \frac{\partial U}{\partial x_2} = -\frac{\partial V}{\partial x_1},$$

and hence $|\nabla U|^2 = |\nabla V|^2$. Therefore, denoting

$$v(\theta) := V|_{\mathbb{S}^1},$$

we get

$$(7.1.21) \qquad \int_{\mathbb{D}} |\nabla U|^2 \, dx = \int_{\mathbb{D}} |\nabla V|^2 \, dx = \int_{\mathbb{S}^1} v \frac{\partial V}{\partial r} \, d\theta,$$

where the last equality follows from Green's formula since V is harmonic. Putting together (7.1.19), (7.1.20), and (7.1.21) yields

$$(7.1.22) \qquad \left(\int_\Omega |\nabla A|^2 \, dx \right) \left(\int_\Omega |\nabla B|^2 \, dx \right) = \left(\int_{\mathbb{S}^1} v \frac{\partial V}{\partial r} \, d\theta \right)^2.$$

Recall that $U = u$ on $\mathbb{S}^1$. It follows from (7.1.18) that

$$\frac{\partial V}{\partial r} \Big|_{\mathbb{S}^1} = -u'(\theta)$$

as elements of $H^{-1/2}(\mathbb{S}^1)$. Plugging the last equation into (7.1.22) and taking into account that $u'(\theta) = h'(m(\theta))m'(\theta)$, we get

$$\left(\int_\Omega |\nabla A|^2 \, dx \right) \left(\int_\Omega |\nabla B|^2 \, dx \right) = \left(\int_{\mathbb{S}^1} v(\theta) h'(m(\theta)) m'(\theta) \, d\theta \right)^2$$

$$\leq \left(\int_{\mathbb{S}^1} v(\theta)^2 \, d\mu \right) \left(\int_{\mathbb{S}^1} h'(m(\theta))^2 \, d\mu \right),$$

where we have used the Cauchy–Schwarz inequality and the fact that $\mathrm{d}\mu = m'(\theta)\,\mathrm{d}\theta$. At the same time, by the definition of the push-forward measure $\mathrm{d}\mu$,

$$\int_{\partial\Omega} a^2 \,\mathrm{d}s = \int_{\mathbb{S}^1} u^2 \,\mathrm{d}\mu, \qquad \int_{\partial\Omega} b^2 \,\mathrm{d}s = \int_{\mathbb{S}^1} v^2 \,\mathrm{d}\mu,$$

where we set

$$a := A|_{\partial\Omega}, \qquad b := B|_{\partial\Omega}.$$

Therefore, it follows that the product of the Steklov Rayleigh quotients on Ω of A and B can be estimated as

$$R^{\mathrm{S}}[A]R^{\mathrm{S}}[B] \leq \frac{\left(\int_{\mathbb{S}^1} v^2 \,\mathrm{d}\mu\right)\left(\int_{\mathbb{S}^1} h'(m(\theta))^2 \,\mathrm{d}\mu\right)}{\left(\int_{\mathbb{S}^1} v^2 \,\mathrm{d}\mu\right)\left(\int_{\mathbb{S}^1} h(m(\theta))^2 \,\mathrm{d}\mu\right)} = R_L[h],$$

where

$$R_L[h] = \frac{\int_0^L h'(\eta)^2 \,\mathrm{d}\eta}{\int_0^L h(\eta)^2 \,\mathrm{d}\eta}$$

is the usual Rayleigh quotient of h with respect to the Laplacian on the circle of length L (to simplify notation below, we have introduced a new variable $\eta := m(\theta)$). In other words, we have reduced the problem to the boundary. Note that the term $\int_{\mathbb{S}^1} v^2 \,\mathrm{d}\mu$ cancels out, which is the key feature of the method.

Let Φ_j denote the eigenfunctions of the Steklov problem on Ω corresponding to eigenvalues σ_j, $j \in \mathbb{N}$, and chosen in such a way that their boundary traces $\varphi_j := \Phi_j|_{\partial\Omega}$ form an orthogonal basis in $L^2(\partial\Omega)$. We will now specify the choice of the function h. The main idea is to use the resulting functions A and B as the test functions for σ_{p+1} and σ_{q+1}, respectively. Therefore, a should be orthogonal, in $L^2(\partial\Omega)$, to φ_j with $j = 1, \ldots, p$, and b should be orthogonal to φ_j with $j = 1, \ldots, q$.

Let $h_k : \mathbb{R} \to \mathbb{R}$, $k \in \mathbb{N}$, be the Laplace–Beltrami eigenfunctions on the circle of length L, extended by periodicity,

$$h_k(\eta) = \begin{cases} \cos\left(\frac{2\pi n\eta}{L}\right), & \text{if } k = 2n+1, \\ \sin\left(\frac{2\pi n\eta}{L}\right), & \text{if } k = 2n, \end{cases}$$

where $n \in \mathbb{N}_0$ (we ignore the function $h_0 = 0$). Clearly, $R_L[h_{2n}] = R_L[h_{2n+1}] = \left(\frac{2\pi n}{L}\right)^2$. Set $N = p + q$, and consider

$$u = \sum_{k=2}^{N} c_k u_k,$$

where $c_k \in \mathbb{R}$ and the functions $u_k : \mathbb{S}^1 \to \mathbb{R}$ are defined by $u_k(\theta) = h_k(m(\theta))$, $k \in \mathbb{N}$. The functions u_k are $\mathrm{d}\mu$-orthogonal and hence linearly independent. Therefore, their harmonic extensions $U_k = \mathcal{E}_0 u_k$ onto the unit disk are also linearly independent, and so are the harmonic conjugates V_k of U_k. Moreover, since $u_1 = 1$, the functions u_k are $d\mu$-orthogonal to constants for $k \geq 2$, and hence $\int_{\partial\Omega} a_k \, ds = 0$ for all $k \geq 2$, where $a_k = u_k \circ \psi$. At the same time, by our normalisation of harmonic conjugates, $\int_{\partial\Omega} b_k \, ds = 0$, where $b_k = (V_k|_{\mathbb{S}^1}) \circ \psi$.

Set now

$$U = \sum_{k=2}^{N} c_k U_k, \quad V = \sum_{k=2}^{N} c_k V_k, \quad h = \sum_{k=2}^{N} c_k h_k.$$

We have $p - 1 + q - 1 = N - 2$ orthogonality conditions on $a = (U \circ \psi)|_{\partial\Omega}$ and $b = (V \circ \psi)|_{\partial\Omega}$ left to be satisfied, and there are $N - 1$ coefficients to choose. Therefore, there exists a nontrivial choice of the coefficients c_k for $k = 1, \ldots, N$ such that all the orthogonality conditions are fulfilled, and hence

$$\sigma_{p+1}\sigma_{q+1} \leq R_L^{\mathrm{S}}[A]R_L^{\mathrm{S}}[B] \leq R_L[h]$$

$$\leq R_L[h_N] = \left(\frac{\pi}{L}\right)^2 \begin{cases} (p+q-1)^2 & \text{if } p+q \text{ is odd,} \\ (p+q)^2 & \text{if } p+q \text{ is even.} \end{cases}$$

This completes the proof of the theorem. $\qquad\square$

Remark 7.1.23. It has been already mentioned in Remark 7.1.19 that the inequalities (7.1.16) are sharp for $p = q = k$. It immediately follows that they are also sharp for $p = k$, $q = k + 1$. In particular, $\sigma_2 \sigma_3 L^2 \leq 4\pi^2$.

Remark 7.1.24. There exist various generalisations of the Hersch–Payne–Schiffer inequalities. In particular, for the Steklov problem with a weight $\rho \geq 0$ in the boundary condition (7.1.2), the inequalities (7.1.16) hold provided the perimeter is replaced by the "mass"

$$L_\rho(\partial\Omega) = \int\limits_{\partial\Omega} \rho(s) \, ds.$$

One can also extend the inequalities (7.1.16) to arbitrary surfaces with boundary; see [**GirPol12**]. In particular, it was shown in [**Kar18**] that

$$\sigma_k L(\partial\Sigma) \leq 2\pi(k + \gamma + \ell - 1),$$

where γ is the genus of the surface Σ and ℓ is the number of its boundary components.

7.2. The Dirichlet-to-Neumann map and the boundary Laplacian

7.2.1. Weyl's law for Steklov eigenvalues.
The goal of this section is to prove Weyl's law for the Steklov eigenvalues of a bounded domain Ω, or,

equivalently, for the eigenvalues of the Dirichlet-to-Neumann map $\mathcal{D}_0$ acting on its boundary $M = \partial\Omega$. Let

$$\mathcal{N}^{\mathrm{S}}(\sigma) = \mathcal{N}^{\mathrm{S}}_\Omega(\sigma) = \mathcal{N}^{\mathcal{D}_0}(\sigma) := \#\{j : \sigma_j(\Omega) \le \sigma\}$$

be the counting function of the Steklov eigenvalues.

Theorem 7.2.1 (Weyl's law for Steklov eigenvalues). Let Ω be a bounded domain in a complete Riemannian manifold, and assume that $\partial\Omega = M$ is smooth. Then the following asymptotic relation holds:

(7.2.1) $\mathcal{N}^{\mathrm{S}}(\sigma) = C_{d-1} \operatorname{Vol}(M)\sigma^{d-1} + o\left(\sigma^{d-1}\right)$ as $\sigma \to +\infty$.

Here, as before, $C_{d-1} = \frac{\omega_{d-1}}{(2\pi)^{d-1}}$ denotes the Weyl constant, and ω_{d-1} is the volume of a unit ball in $\mathbb{R}^{d-1}$.

The standard approach to establishing Theorem 7.2.1 uses the theory of pseudodifferential operators, which is beyond the scope of this book. The key observation is that the principal symbol of the operator $\mathcal{D}_0$ is precisely the square root of the principal symbol of the boundary Laplacian $-\Delta_M$ on M. This implies that $\mathcal{D}_0$ and $\sqrt{-\Delta_M}$ have similar eigenvalue asymptotics. Here we take a different route which is based on rather elementary tools and at the same time provides a more geometric way to understand the link between the Dirichlet-to-Neumann operator and the boundary Laplacian. Our exposition mostly follows [**GirKLP22**] and is based on the so-called Pohozhaev identity [**Poh65**] and its generalisations, which in turn is an application of the method of multipliers going back to F. Rellich (see [**ChWGLS12**, p. 205] for a discussion) and to an old unpublished work of L. Hörmander [**Hör18**] that was originally written in the 1950s (see also [**Hör54**] where an identity similar to Pohozhaev's has been obtained).

For simplicity, we will prove Theorem 7.2.1 in the Euclidean setting, and we will outline the necessary modifications for the Riemannian case and some relaxations of the conditions of the theorem at the end; see Remark 7.2.11. We also note that for Euclidean domains, Weyl's law was first obtained by L. Sandgren in [**San55**] using a different approach under the assumption that the boundary is C^2 regular. Using heavier machinery, the result can also be proved for Euclidean domains with piecewise C^1 boundaries [**Agr06**].

Remark 7.2.2. The validity of Weyl's law for the Steklov problem in a Lipschitz domain $\Omega \subset \mathbb{R}^d$ has been a longstanding open problem attributed to M. S. Agranovich. In the two-dimensional case, it was proved in [**KarLagPol23**] using the theory of conformal mappings. While this book was in the final preparation stage, G. Rozenblum [**Roz23**] established Weyl's law for domains with Lipschitz boundary in any dimension.

7.2.2. The Hörmander–Pohozhaev identities. We start with a reminder of some notions and identities from vector calculus. Let $\Omega \subset \mathbb{R}^d$ be an open set, and let $a : \Omega \to \mathbb{R}$ and $\mathbf{A}, \mathbf{B} : \Omega \to \mathbb{R}^d$ be a scalar function and vector fields on Ω, respectively, which we assume to be sufficiently smooth.[11] We denote by

$$\operatorname{Jac}_{\mathbf{A}} := \left(\frac{\partial A_i}{\partial x_j} \right)_{i,j=1,\dots,d}$$

the *Jacobian* of $\mathbf{A}$ and by

$$\operatorname{Hes}_a := \operatorname{Jac}_{\nabla a} = \left(\frac{\partial^2 a}{\partial x_i \partial x_j} \right)_{i,j=1,\dots,d}$$

the *Hessian* of a. Additionally, for any linear operator (that is, a matrix) C acting in $\mathbb{R}^d$, we will denote by C^* its adjoint (that is, a transposed matrix) and by

$$\mathsf{C}[\mathbf{A}, \mathbf{B}] := \langle \mathsf{C}\mathbf{A}, \mathbf{B} \rangle$$

its quadratic form.

■ **Exercise 7.2.3.** Prove the following identities:

(7.2.2) $\operatorname{div}(a\mathbf{A}) = \langle \nabla a, \mathbf{A} \rangle + a \operatorname{div} \mathbf{A},$

(7.2.3) $\nabla(\langle \mathbf{A}, \mathbf{B} \rangle) = \operatorname{Jac}_{\mathbf{A}}^* \mathbf{B} + \operatorname{Jac}_{\mathbf{B}}^* \mathbf{A},$

(7.2.4) $\nabla(|\nabla a|^2) = 2\operatorname{Hes}_a \nabla a.$

Let us now state a useful Pohozhaev-type identity which has various applications; see [**ColGirHas18**, Lemma 20].

Theorem 7.2.4 (Generalised Pohozhaev identity for the Laplacian). Let $\Omega \subset \mathbb{R}^d$ be a bounded domain with smooth boundary $M = \partial\Omega$. Let $\mathbf{F}$ be a smooth vector field on $\overline{\Omega}$, let $u \in H^1(M)$, and let $U = \mathcal{E}_0 u$ be the unique harmonic extension of u onto Ω. Then

(7.2.5)
$$\int_M \langle \mathbf{F}, \nabla U \rangle \, \partial_n U \, ds - \frac{1}{2} \int_M |\nabla U|^2 \langle \mathbf{F}, \mathbf{n} \rangle \, ds$$
$$+ \frac{1}{2} \int_\Omega |\nabla U|^2 \operatorname{div} F \, d\mathbf{x} - \int_\Omega \operatorname{Jac}_{\mathbf{F}}[\nabla U, \nabla U] \, d\mathbf{x} = 0.$$

Proof. Since $\Delta U = \operatorname{div} \nabla U = 0$ in Ω, using (7.2.2) and (7.2.3), we obtain

$$\operatorname{div}(\langle \mathbf{F}, \nabla U \rangle \nabla U) = \langle \nabla \langle \mathbf{F}, \nabla U \rangle, \nabla U \rangle = \operatorname{Jac}_{\mathbf{F}}[\nabla U, \nabla U] + \operatorname{Hes}_U[\mathbf{F}, \nabla U]$$

---

[11] Throughout this chapter, we distinguish the vector fields by bold font, in particular the exterior normal vector on the boundary of Ω will be denoted $\mathbf{n}$.

(note that the Hessian of U is well-defined since U is harmonic). At the same time, using (7.2.2) once more together with (7.2.4),

$$\frac{1}{2} \operatorname{div} \left(|\nabla U|^2 \mathbf{F} \right) = \operatorname{Hes}_U [\mathbf{F}, \nabla U] + \frac{1}{2} |\nabla U|^2 \operatorname{div} F.$$

Subtracting the second equality from the first one, we get

$$\operatorname{div} \left(\langle \mathbf{F}, \nabla U \rangle \nabla U - \frac{1}{2} |\nabla U|^2 \mathbf{F} \right) = \operatorname{Jac}_{\mathbf{F}} [\nabla U, \nabla U] - \frac{1}{2} |\nabla U|^2 \operatorname{div} \mathbf{F}.$$

Finally, we integrate this identity over Ω and use the divergence theorem, noting that $(\nabla U)|_M \in L^2(M)$ since we have assumed $u = U|_M \in H^1(M)$ (see [**ChWGLS12**, Theorem A.5]). $\qquad\square$

We now make a choice of a vector field $\mathbf{F}$ in Theorem 7.2.4, leading to the following result, which was originally obtained by L. Hörmander in the 1950s.

> **Theorem 7.2.5** (Hörmander's identity [**Hör18**]). Let $\Omega \subset \mathbb{R}^d$ be a bounded domain with a smooth boundary $M = \partial\Omega$. Let $\mathbf{F}$ be a smooth vector field on $\overline{\Omega}$ which on the boundary of Ω coincides with the exterior unit normal, $\mathbf{F}|_M = \mathbf{n}$. Let $u \in H^1(M)$, and let $U = \mathcal{E}_0 u$ be the unique harmonic extension of u onto Ω. Then
> $$(\mathcal{D}_0 u, \mathcal{D}_0 u)_{L^2(M)} - (-\Delta_M u, u)_{L^2(M)}$$
> (7.2.6)
> $$= \int\limits_\Omega \left(2 \operatorname{Jac}_{\mathbf{F}} [\nabla U, \nabla U] - |\nabla U|^2 \operatorname{div} \mathbf{F} \right) \, \mathrm{dx}.$$

Proof. Using $\mathbf{F}|_M = \mathbf{n}$ and the definition of the Dirichlet-to-Neumann map $\mathcal{D}_0$, we substitute into (7.2.5) the relations

$$\langle \mathbf{F}|_M, \mathbf{n} \rangle = 1, \qquad\qquad \langle \mathbf{F}, \nabla U \rangle |_M = \partial_n U,$$
$$|\nabla U|^2 \big|_M = |\nabla_M u|^2 + (\partial_n U)^2, \qquad\qquad \mathcal{D}_0 u = \partial_n U,$$

and (7.2.6) then follows immediately taking into account the expression for the quadratic form of the Laplace–Beltrami operator on M, $(-\Delta_M u, u)_{L^2(M)} = (\nabla_M u, \nabla_M u)_{L^2(M)}$. $\qquad\square$

7.2.3. The Steklov spectrum and the spectrum of the boundary Laplacian. Theorem 7.2.5 almost immediately implies

> **Corollary 7.2.6.** Let $\Omega \subset \mathbb{R}^d$ be a bounded domain with a smooth boundary $M = \partial\Omega$. Then there exists a constant $C > 0$ such that for any $u \in H^1(M)$,
> $$(7.2.7) \quad \left| (\mathcal{D}_0 u, \mathcal{D}_0 u)_{L^2(M)} - (-\Delta_M u, u)_{L^2(M)} \right| \leq C \, (\mathcal{D}_0 u, u)_{L^2(M)}.$$

Proof. We note that the integrand in the right-hand side of (7.2.6) is a quadratic form in ∇U with bounded coefficients, since the vector field $\mathbf{F}$ is smooth. Hence, there exists a constant $C > 0$ such that

$$\left| \int_{\Omega} \left(2 \mathrm{Jac}_{\mathbf{F}} \left[\nabla U, \nabla U \right] - |\nabla U|^2 \operatorname{div} \mathbf{F} \right) \mathrm{d}\mathbf{x} \right| \leq C \left\| \nabla U \right\|^2_{L^2(\Omega)}$$

$$= C \left(\mathcal{D}_0 u, u \right)_{L^2(M)}. \qquad \square$$

Remark 7.2.7. In fact, the constant C appearing in the right-hand side of (7.2.7) may be chosen to depend only on the geometry of Ω in a small neighbourhood of M. To see this, we may choose $\mathbf{F}(x) = \nabla \left(d_M(\mathbf{x}) \chi(\mathbf{x}) \right)$, where $d_M(\mathbf{x})$ is a distance from $\mathbf{x}$ to the boundary and $\chi(\mathbf{x})$ is a smooth cut-off function equal to one near M and zero outside a small neighbourhood of M. Then $\mathbf{F}(x)$ satisfies the assumptions of Theorem 7.2.5; see [**ProStu19**, §5.3]. For explicit expressions on C in terms of geometric characteristics of Ω and M see [**ProStu19**], [**Xio18**], [**ColGirHas18**].

Corollary 7.2.6 already links, in a way, the Dirichlet-to-Neumann map and the Laplace-Beltrami operator on M. We will now use it to compare the eigenvalues of this operator, using the following abstract result, essentially due to L. Hörmander.

Proposition 7.2.8 ([**GirKLP22**, Proposition 3.3], generalising [**Hör18**]). Let $\mathcal{H}$ be a Hilbert space with an inner product $(\cdot, \cdot)_{\mathcal{H}}$. Let $\mathcal{A}, \mathcal{B}$ be two nonnegative selfadjoint operators in $\mathcal{H}$ with discrete spectra $\mathrm{Spec}(\mathcal{A}) = \{\alpha_1 \leq \alpha_2 \leq \cdots\}$ and $\mathrm{Spec}(\mathcal{B}) = \{\beta_1 \leq \beta_2 \leq \cdots\}$ and the corresponding orthonormal bases of eigenfunctions $\{a_k\}, \{b_k\}$. Assume additionally that $a_k \in \mathrm{Dom}(\mathcal{B})$ and $b_k \in \mathrm{Dom}(\mathcal{A}^2)$, $k \in \mathbb{N}$, where the domains are understood in the sense of quadratic forms. Suppose that for some $C > 0$,

$$(7.2.8) \qquad \left| (\mathcal{A}u, \mathcal{A}u)_{\mathcal{H}} - (\mathcal{B}u, u)_{\mathcal{H}} \right| \leq C \left(\mathcal{A}u, u \right)_{\mathcal{H}}$$

$$\text{for all } u \in D := \mathrm{Dom}(\mathcal{B}) \cap \mathrm{Dom}(\mathcal{A}^2).$$

Then

$$(7.2.9) \qquad \left| \alpha_k^2 - \beta_k \right| \leq C \alpha_k,$$

and consequently

$$(7.2.10) \qquad \left| \alpha_k - \sqrt{\beta_k} \right| \leq C$$

for all $k \in \mathbb{N}$, with the same constant C as in (7.2.8).

Proof. We note that (7.2.8) is equivalent to

$$
(7.2.11) \qquad
\begin{cases}
(\mathcal{B}u, u)_{\mathcal{H}} \leq (\mathcal{A}u, \mathcal{A}u)_{\mathcal{H}} + C\,(\mathcal{A}u, u)_{\mathcal{H}}, \\
(\mathcal{A}u, \mathcal{A}u)_{\mathcal{H}} - C\,(\mathcal{A}u, u)_{\mathcal{H}} \leq (\mathcal{B}u, u)_{\mathcal{H}},
\end{cases}
$$

and (7.2.9) is equivalent to

$$
(7.2.12) \qquad
\begin{cases}
\beta_k \leq \alpha_k^2 + C\alpha_k, \\
\beta_k \geq \alpha_k^2 - C\alpha_k.
\end{cases}
$$

From the variational principle for the eigenvalues of $\mathcal{B}$ and the first inequality in (7.2.11) we have

$$
(7.2.13) \qquad
\begin{aligned}
\beta_k &\leq \sup_{0 \neq u \in V_k \subset \mathrm{Dom}(\mathcal{B})} \frac{(\mathcal{B}u, u)_{\mathcal{H}}}{(u, u)_{\mathcal{H}}} \\
&\leq \sup_{0 \neq u \in V_k \subset \mathrm{Dom}(\mathcal{B})} \frac{(\mathcal{A}u, \mathcal{A}u)_{\mathcal{H}} + C\,(\mathcal{A}u, u)_{\mathcal{H}}}{(u, u)_{\mathcal{H}}}
\end{aligned}
$$

for any subspace V_k with $\dim V_k = k$. Take $V_k = \mathrm{Span}\{a_1, \ldots, a_k\}$. As for any $u = c_1 a_1 + \cdots + c_k a_k \in V_k$ with $|c_1|^2 + \cdots + |c_k|^2 = 1$ we have, due to orthogonality,

$$
\frac{(\mathcal{A}u, \mathcal{A}u)_{\mathcal{H}} + C\,(\mathcal{A}u, u)_{\mathcal{H}}}{(u, u)_{\mathcal{H}}} = \sum_{j=1}^{k} |c_j|^2 (\alpha_j^2 + C\alpha_j) \leq \alpha_k^2 + C\alpha_k,
$$

the first inequality (7.2.12) follows immediately from (7.2.13).

We now prove the second inequality (7.2.12). Let $K_0 := \max\{k \in \mathbb{N} : \alpha_k \leq C\}$. We note that for $k \leq K_0$ the second inequality (7.2.12) is automatically satisfied since in this case $\beta_k \geq 0 \geq \alpha_k\,(\alpha_k - C)$, so we need to consider only $k > K_0$. We rewrite the second inequality (7.2.11) as

$$
\left(\widetilde{\mathcal{A}}u, \widetilde{\mathcal{A}}u\right)_{\mathcal{H}} \leq (\mathcal{B}u, u)_{\mathcal{H}} + \frac{C^2}{4}\,(u, u)_{\mathcal{H}},
$$

where $\widetilde{\mathcal{A}} := \mathcal{A} - \frac{C}{2}$. Let $\widetilde{\alpha}_k^2$ denote the eigenvalues of $\widetilde{\mathcal{A}}^{\,2}$ enumerated in nondecreasing order. We note that $\widetilde{\alpha}_k^2 = \left(\alpha_k - \frac{C}{2}\right)^2$ for $k > K_0$ (this may not be the case for $k \leq K_0$ but as mentioned above we can ignore these values of k). Writing down the variational principle for $\widetilde{\alpha}_k^2$ similarly to (7.2.13) and choosing a test subspace $V_k = \mathrm{Span}\{b_1, \ldots, b_k\}$ leads in a similar manner to

$$
\widetilde{\alpha}_k^2 = \left(\alpha_k - \frac{C}{2}\right)^2 \leq \beta_k + \frac{C^2}{4},
$$

which gives the second inequality (7.2.12) after a simplification.

Finally, we note that (7.2.9) implies, for $\alpha_k \beta_k \neq 0$,

$$\left| \alpha_k - \sqrt{\beta_k} \right| \leq C \frac{\alpha_k}{\alpha_k + \sqrt{\beta_k}} \leq C,$$

yielding (7.2.10). Note that $\alpha_k = 0$ implies $\beta_k = 0$ by (7.2.12). □

Using Proposition 7.2.8, we are now able to obtain a uniform bound comparing the Steklov eigenvalues with the ones of the Laplace–Beltrami operator on the boundary.

Theorem 7.2.9. Let $\Omega \subset \mathbb{R}^d$ be a bounded domain with a smooth boundary $M = \partial\Omega$, and let σ_k, ν_k, $k \in \mathbb{N}$, be the Steklov eigenvalues of Ω and the eigenvalues of the Laplace–Beltrami operator on M, respectively. Then

(7.2.14) $$\left| \sigma_k - \sqrt{\nu_k} \right| \leq C$$

holds for all $k \in \mathbb{N}$ with the same constant C as in (7.2.7).

Proof. We apply Proposition 7.2.8 with $\mathcal{A} = \mathcal{D}_0$, $\mathcal{B} = -\Delta_M$, and therefore $\alpha_k = \sigma_k$, and $\beta_k = \nu_k$, taking into account Corollary 7.2.6 and choosing $D = H^1(M)$. □

We can now finish the proof of Theorem 7.2.1.

Proof of Theorem 7.2.1. It follows from Theorem 7.2.9 that

(7.2.15) $$\mathcal{N}_M \left((\sigma - C)^2 \right) \leq \mathcal{N}^S(\sigma) \leq \mathcal{N}_M(\sigma^2 + C\sigma),$$

where $\mathcal{N}_M(\cdot)$ is the eigenvalue counting function of the Laplace–Beltrami operator on M. Indeed, to prove the left inequality (7.2.15) we observe that if $\nu_k \leq (\sigma - C)^2$, then $\sigma \geq \sqrt{\nu_k} + C \geq \sigma_k$ by (7.2.14). To prove the right inequality (7.2.15), we note that if $\sigma_k \leq \sigma$, then $\nu_k \leq \sigma_k^2 + C\sigma_k \leq \sigma^2 + C\sigma_k$, once more using (7.2.14). An application of Theorem 6.1.9 to both sides of (7.2.15) then yields the result. □

Remark 7.2.10. We have stated Theorem 6.1.9 in the Riemannian setting but have proved it in the Euclidean case only. The Riemannian argument goes through identically, with the only modification required being in (7.2.5) where $\mathrm{Jac}_\mathbf{F}[\nabla U, \nabla U]$ in the last integral should be replaced by $(\nabla_{\nabla U}, \mathbf{F}) \nabla U$, where $\nabla_{\nabla U}$ denotes a covariant derivative in the direction ∇U; see [**GirKLP22**] for details.

Remark 7.2.11. There exist various improvements and extensions of the results presented in this subsection. In particular, Theorem 7.2.4 can be proved verbatim under the assumption that Ω has Lipschitz boundary and F is a Lipschitz vector field. Consequently, Theorem 7.2.5 holds if F is a Lipschitz vector field and Ω has $C^{1,1}$ boundary, so that the normal field on the boundary is Lipschitz. As a result, the regularity assumptions in Theorem 7.2.1 can be significantly relaxed; moreover, the error term estimate in (7.2.1) can be improved to $O\left(\sigma^{d-2}\right)$ using the sharp Weyl law for the boundary Laplacian. This improvement holds for domains with $C^{2,\alpha}$ boundary for some $\alpha > 0$ in arbitrary dimension and for domains with $C^{1,1}$ boundary in dimension two. We refer to [**GirKLP22**] for a detailed exposition of these results.

Remark 7.2.12 (The two-dimensional case). Let Ω be a smooth simply connected planar domain. Then $M = \partial\Omega$ is one dimensional and hence locally isometric to a circle. Hence, by Remark 7.1.5, $\sqrt{\nu_k(M)} = \sigma_k(\Omega^\star)$, where $\Omega^\star$ is a disk of the same *perimeter* as Ω. Therefore, in this case (7.2.14) yields

$$|\sigma_k(\Omega) - \sigma_k(\Omega^\star)| < C, \quad k \in \mathbb{N}.$$

This bound admits a significant asymptotic improvement:

$$(7.2.16) \qquad\qquad |\sigma_k(\Omega) - \sigma_k(\Omega^\star)| = o\left(k^{-N}\right),$$

for any $N > 0$, see [**Roz86**], [**Edw93a**], which is proved using pseudodifferential techniques. In particular, this implies that in this case the remainder estimate in (7.2.1) can be replaced by $o\left(\sigma^{-N}\right)$ for any $N > 0$.

We refer also to [**GirPPS14**] for a generalisation of (7.2.16) to arbitrary Riemannian surfaces with boundary.

Remark 7.2.13 (Steklov isospectrality). Similarly to Definition 6.2.1, we can say that two Riemannian manifolds with boundary, or two Euclidean domains, are *Steklov isospectral* if their Steklov spectra coincide. For some examples of Steklov isospectral manifolds, see, e.g., [**GirPPS14**]. It is immediately clear from (7.2.1) that the volume $\mathrm{Vol}(M)$ of the boundary $M = \partial\Omega$ of a complete Riemannian manifold is a Steklov spectral invariant.

Interestingly enough, no examples of nonisometric Steklov isospectral planar domains are presently known [**GirPol17**, Open problem 6]; we refer also to [**Edw93b**, **MalSha15**, **JolSha14**, **JolSha18**] for some related results and conjectures. At the same time, Steklov spectral invariants of planar domains are also quite scarce — we know, in addition to the perimeter, that if the boundary of $\Omega \subset \mathbb{R}^2$ is smooth, its Steklov spectrum determines the number of boundary components of $M = \partial\Omega$ and their lengths [**GirPPS14**]. However, for smooth simply connected planar domains, extracting further geometric information from the Steklov problem is quite difficult. In part, the reason for that lies in formula (7.2.16): any two smooth simply connected planar domains Ω_I and Ω_II of the same perimeter will have Steklov eigenvalues which differ as $|\sigma_k(\Omega_\mathrm{I}) - \sigma_k(\Omega_\mathrm{II})| = O\left(k^{-\infty}\right)$ as $k \to \infty$. As a result, no other spectral invariants except the perimeter can be obtained from the eigenvalue asymptotics on the polynomial scale.

We will say that two (not necessarily smooth) planar domains Ω_{I} and Ω_{II} are *asymptotically Steklov isospectral* if

$$|\sigma_k(\Omega_{\mathrm{I}}) - \sigma_k(\Omega_{\mathrm{II}})| = o\,(1) \qquad \text{as } k \to \infty.$$

We will consider some further examples of asymptotically Steklov isospectral planar domains and corresponding Steklov spectral invariants in §7.3.6.

7.3. Steklov spectra on domains with corners

7.3.1. Asymptotics of Steklov eigenvalues for curvilinear polygons.
In this section, we mostly follow [**LevPPS22b**]. Let $\mathcal{P} = \mathcal{P}_{\boldsymbol{\alpha},\boldsymbol{\ell}}$ be a (simply connected) curvilinear polygon in $\mathbb{R}^2$ with n vertices $V_1, \ldots, V_n$ numbered clockwise, corresponding internal angles $0 < \alpha_j < \pi$ at V_j, and smooth sides I_j of length ℓ_j joining V_{j-1} and V_j. Here, $\boldsymbol{\alpha} = (\alpha_1, \ldots, \alpha_n) \in (0, \pi)^n$, $\boldsymbol{\ell} = (\ell_1, \ldots, \ell_n) \in \mathbb{R}_+^n$, and we will use cyclic subscript identification $n + 1 \equiv 1$. Our choice of orientation ensures that an internal angle α_j is measured from I_j to I_{j+1} in the counterclockwise direction, as in Figure 7.4. The perimeter of $\mathcal{P}$ is $L(\partial\mathcal{P}) = L = \ell_1 + \cdots + \ell_n$.

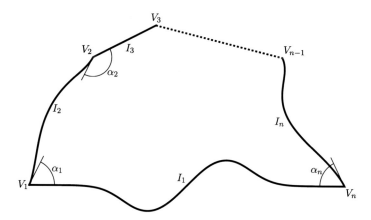

Figure 7.4. A curvilinear polygon.

We will give an improved asymptotics of the Steklov eigenvalues $\sigma_m(\mathcal{P})$ as $m \to +\infty$, which takes into account not just the perimeter of a curvilinear polygon but also the lengths of individual sides and the angles between them. The philosophy behind this result is somewhat similar to the principle of Theorem 7.2.9: we will compare (asymptotically only, and using a completely different set of techniques) the Steklov eigenvalues of $\mathcal{P}$ to the eigenvalues of a particular "boundary Laplacian" on $M = \partial\mathcal{P}$. More precisely, the role of this boundary Laplacian is played here by a certain *quantum graph Laplacian* (see [**BerKuc13**] and references therein for a comprehensive spectral theory of quantum graphs).

Let us associate with the boundary of a curvilinear polygon $\mathcal{P}_{\alpha,\ell}$ a cyclic metric graph $\mathcal{M}_{\alpha,\ell}$ with n vertices $V_1, \ldots, V_n$ and n edges I_j (joining V_{j-1} and V_j, with V_0 identified with V_n) of length ℓ_j, $j = 1, \ldots, n$. Let s be the arc-length parameter on $\mathcal{M}_{\alpha,\ell}$ starting at V_1 and going in the clockwise direction; see Figure 7.5.

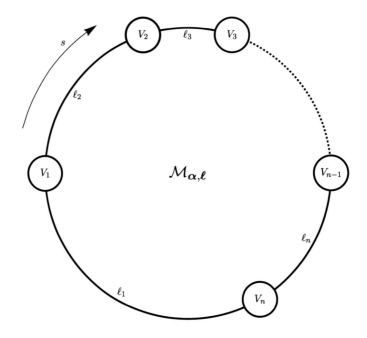

Figure 7.5. A quantum graph.

Consider the spectral problem for a quantum graph Laplacian on $\mathcal{M}_{\alpha,\ell}$,

$$(7.3.1) \qquad -\frac{\mathrm{d}^2 f}{\mathrm{d}s^2} = \nu f,$$

with matching conditions

$$(7.3.2) \qquad \begin{aligned} \sin\left(\frac{\pi^2}{4\alpha_j}\right) f|_{V_j+0} &= \cos\left(\frac{\pi^2}{4\alpha_j}\right) f|_{V_j-0}, \\ \cos\left(\frac{\pi^2}{4\alpha_j}\right) f'|_{V_j+0} &= \sin\left(\frac{\pi^2}{4\alpha_j}\right) f'|_{V_j-0}. \end{aligned}$$

Hereinafter at each vertex V_j, $j = 1, \ldots, n$, $g|_{V_j-0}$ and $g|_{V_j+0}$ denote the limiting values of a quantity $g(s)$ as s approaches the vertex V_j from the left and from the right, respectively, in the direction of s.

We will denote the operator $f \mapsto -\frac{\mathrm{d}^2 f}{\mathrm{d}s^2}$ subject to matching conditions (7.3.2) by $-\Delta_{\mathcal{M}}$. It is easy to check that $-\Delta_{\mathcal{M}}$ is selfadjoint and nonnegative. Therefore, its spectrum is given by a sequence of nonnegative real eigenvalues

$$0 \leq \nu_1 \leq \nu_2 \leq \cdots \nu_m \leq \cdots \nearrow +\infty,$$

listed with multiplicity.

Remark 7.3.1. The eigenvalues ν_m also satisfy a standard variational principle: if

$$\mathrm{Dom}(\mathcal{Q}_\mathcal{M}) := \left\{ f \in \bigoplus_{j=1}^{n} H^1(I_j) : \sin\left(\frac{\pi^2}{4\alpha_j}\right) f|_{V_j+0} = \cos\left(\frac{\pi^2}{4\alpha_j}\right) f|_{V_j-0} \right\}$$

denotes the domain of the quadratic form

$$\mathcal{Q}_\mathcal{M}[f] := \sum_{j=1}^{n} \int_{I_j} (f'(s))^2 \, \mathrm{d}s$$

of $-\Delta_\mathcal{M}$, then

$$\nu_m = \min_{\substack{S \subset \mathrm{Dom}(\mathcal{Q}_\mathcal{M}) \\ \dim S = m}} \max_{0 \neq f \in S} \frac{\mathcal{Q}_\mathcal{M}[f]}{\sum\limits_{j=1}^{n} \int_{I_j} (f(s))^2 \, \mathrm{d}s}.$$

We now have

Theorem 7.3.2 (Eigenvalue asymptotics for curvilinear polygons [**LevPPS22b**, Theorem 1.4]). Let $\mathcal{P} = \mathcal{P}_{\boldsymbol{\alpha}, \boldsymbol{\ell}}$ be a curvilinear polygon defined above, let σ_m, $m \in \mathbb{N}$, be its Steklov eigenvalues, and let ν_m, $m \in \mathbb{N}$, be the eigenvalues of the associated quantum graph problem (7.3.1), (7.3.2). Then there exists $\varepsilon > 0$ such that we have

$$\sigma_m = \sqrt{\nu_m} + O\left(m^{-\varepsilon}\right) \qquad \text{as } m \to +\infty.$$

From now on, we will call the numbers[12]

$$\tau_m := \sqrt{\nu_m}$$

the *quasi-eigenvalues* of the Steklov problem on $\mathcal{P}$.

Theorem 7.3.2 immediately implies

Corollary 7.3.3. Let $\mathcal{P}^{\mathrm{I}}_{\boldsymbol{\alpha}, \boldsymbol{\ell}}$ and $\mathcal{P}^{\mathrm{II}}_{\boldsymbol{\alpha}, \boldsymbol{\ell}}$ be two curvilinear polygons with the same angles $\boldsymbol{\alpha}$ and the same side lengths $\boldsymbol{\ell}$. Then there exists $\varepsilon > 0$ such that

$$\sigma_m\left(\mathcal{P}^{\mathrm{I}}\right) - \sigma_m\left(\mathcal{P}^{\mathrm{II}}\right) = O\left(m^{-\varepsilon}\right) \qquad \text{as } m \to +\infty.$$

---

[12]We emphasise that in [**LevPPS22b**] and a related paper [**LevPPS22a**], the Steklov eigenvalues are denoted by λ (rather than σ) and the quasi-eigenvalues are denoted by σ (rather than τ).

As it turns out, the Steklov quasi-eigenvalues τ_m can be determined as the roots of a particular trigonometric function which depends only on the side lengths $\boldsymbol{\ell}$ and angles $\boldsymbol{\alpha}$ of the curvilinear polygon $\mathcal{P}$. To define this trigonometric function, we need to introduce some combinatorial notation.

Let

$$3^n = \{\pm 1\}^n,$$

and for a vector $\boldsymbol{\zeta} = (\zeta_1, \ldots, \zeta_n) \in 3^n$ with cyclic identification $\zeta_{n+1} \equiv \zeta_1$, let

$$\mathbf{Ch}(\boldsymbol{\zeta}) := \{j \in \{1, \ldots, n\} \mid \zeta_j \neq \zeta_{j+1}\}$$

denote the set of indices of sign change in $\boldsymbol{\zeta}$; e.g.,

$$\mathbf{Ch}((1,1,1)) = \varnothing; \quad \mathbf{Ch}((-1,-1,1,1)) = \{2,4\}.$$

Given a curvilinear polygon $\mathcal{P}_{\boldsymbol{\alpha},\boldsymbol{\ell}}$, we now define the following trigonometric function in real variable σ:

$$(7.3.3) \qquad F_{\boldsymbol{\alpha},\boldsymbol{\ell}}(\tau) := \sum_{\substack{\boldsymbol{\zeta} \in 3^n \\ \zeta_1 = 1}} \mathfrak{p}_{\boldsymbol{\zeta}} \cos(\langle \boldsymbol{\ell}, \boldsymbol{\zeta} \rangle \tau) - \prod_{j=1}^n \sin\left(\frac{\pi^2}{2\alpha_j}\right),$$

where

$$\mathfrak{p}_{\boldsymbol{\zeta}} = \mathfrak{p}_{\boldsymbol{\zeta}}(\boldsymbol{\alpha}) := \prod_{j \in \mathbf{Ch}(\boldsymbol{\zeta})} \cos\left(\frac{\pi^2}{2\alpha_j}\right),$$

and we assume the convention $\prod_\emptyset = 1$.

We can now state

Theorem 7.3.4 ([**LevPPS22b**, Theorem 2.16]). Let $\mathcal{P}_{\boldsymbol{\alpha},\boldsymbol{\ell}}$ be a curvilinear polygon. Then $\tau \geq 0$ is its quasi-eigenvalue if and only if it is a root of the trigonometric function $F_{\boldsymbol{\alpha},\boldsymbol{\ell}}(\tau)$. The multiplicity of a quasi-eigenvalue $\tau > 0$ coincides with its multiplicity as a root of (7.3.3), and the multiplicity of the quasi-eigenvalue $\tau = 0$ (if present) is half its multiplicity as a root of (7.3.3).

Theorem 7.3.4 is proved by a rather complicated but straightforward computation of the *secular equation* of the quantum graph problem (7.3.1), (7.3.2) using the methods of [**KotSmi99**, **KurNow10**, **BerKuc13**, **Ber17**], which shows that $F_{\boldsymbol{\alpha},\boldsymbol{\ell}}(\sqrt{\nu_k}) = 0$ with the same multiplicities as in Theorem 7.3.4.

Example 7.3.5. Let $\mathcal{P}$ be the isosceles right-angled triangle with $\boldsymbol{\alpha} = \left(\frac{\pi}{4}, \frac{\pi}{4}, \frac{\pi}{2}\right)$ and $\boldsymbol{\ell} = (1, \sqrt{2}, 1)$. For each $\boldsymbol{\zeta} \in 3^3$ with $\zeta_1 = 1$ we list the corresponding set $\mathbf{Ch}(\boldsymbol{\zeta})$ and the quantities $\langle \boldsymbol{\ell}, \boldsymbol{\zeta} \rangle$ and $\mathfrak{p}_{\boldsymbol{\zeta}}$ in the table below:

$\boldsymbol{\zeta}$	$\langle \boldsymbol{\ell}, \boldsymbol{\zeta} \rangle$	$\mathbf{Ch}(\boldsymbol{\zeta})$	$\mathfrak{p}_{\boldsymbol{\zeta}}$
$(1,1,1)$	$2 + \sqrt{2}$	$\emptyset$	1
$(1,1,-1)$	$\sqrt{2}$	$\{2,3\}$	-1
$(1,-1,1)$	$2 - \sqrt{2}$	$\{1,2\}$	1
$(1,-1,-1)$	$-\sqrt{2}$	$\{1,3\}$	-1

Since in this case we also have $\prod\limits_{j=1}^{3} \sin\left(\frac{\pi^2}{2\alpha_j}\right) = 0$, the definition (7.3.3) yields

$$F_{\boldsymbol{\alpha}, \boldsymbol{\ell}}(\tau) = \cos\left(\left(2 + \sqrt{2}\right)\tau\right) - 2\cos\left(\sqrt{2}\tau\right) + \cos\left(\left(2 - \sqrt{2}\right)\tau\right)$$
$$= -4\left(\cos^2 \tau - 1\right)\cos\left(\sqrt{2}\tau\right),$$

where the second equality follows from some elementary trigonometry. Therefore, by solving $F_{\boldsymbol{\alpha}, \boldsymbol{\ell}}(\tau) = 0$ and using Theorem 7.3.4, we deduce that we have a single quasi-eigenvalue $\tau = 0$, a subsequence of quasi-eigenvalues $\tau = \pi m$, $m \in \mathbb{N}$, of multiplicity two, and another subsequence of quasi-eigenvalues $\tau = \frac{\pi}{\sqrt{2}}\left(m - \frac{1}{2}\right)$, $m \in \mathbb{N}$, of multiplicity one. See also Remark 7.3.7.

■ **Exercise 7.3.6.** For each of the following polygons, write down the trigonometric function $F_{\boldsymbol{\alpha}, \boldsymbol{\ell}}(\tau)$ and hence find the quasi-eigenvalues, with multiplicities:

(i) The equilateral triangle with $\boldsymbol{\alpha} = \left(\frac{\pi}{3}, \frac{\pi}{3}, \frac{\pi}{3}\right)$ and $\boldsymbol{\ell} = (1,1,1)$.

(ii) The right-angled triangle with $\boldsymbol{\alpha} = \left(\frac{\pi}{3}, \frac{\pi}{6}, \frac{\pi}{2}\right)$ and $\boldsymbol{\ell} = \left(1, 2, \sqrt{3}\right)$.

(iii) The square with $\boldsymbol{\alpha} = \left(\frac{\pi}{2}, \frac{\pi}{2}, \frac{\pi}{2}, \frac{\pi}{2}\right)$ and $\boldsymbol{\ell} = (2,2,2,2)$. In this case, additionally compare the quasi-eigenvalues with those implied by Exercise 7.1.11.

Remark 7.3.7. Although it is not immediately transparent from the statements of Theorems 7.3.2 and 7.3.4, the asymptotics of the Steklov eigenvalues and eigenfunctions of a curvilinear polygon is strongly affected by the arithmetic properties of its angles, in particular by the presence or absence of the so-called *exceptional angles* of the form $\frac{\pi}{2k}$, $k \in \mathbb{N}$, and *special angles* of the form $\frac{\pi}{2k-1}$, $k \in \mathbb{N}$. Firstly, in the absence of exceptional angles a multiplicity of every quasi-eigenvalue is either one or two, whereas in the presence of K exceptional angles a multiplicity of a quasi-eigenvalue may be as high as K (compare with the results of Exercise 7.3.6(iii): the square has four exceptional angles, and the multiplicity of every quasi-eigenvalue is in fact four). Secondly, the asymptotic behaviour of the eigenfunctions u_m of the Dirichlet-to-Neumann map (that is, of the boundary traces $U_m|_{\partial\mathcal{P}}$ of the Steklov eigenfunctions) as $m \to \infty$ may also be different: if all angles are special, then the Dirichlet-to-Neumann eigenfunctions u_m are *asymptotically equidistributed* on the boundary in the sense that for any arc $I \subset \partial\mathcal{P}$ (not necessarily a side),

$$\lim_{m\to\infty} \frac{\|u_m\|_{L^2(I)}}{\|u_m\|_{L^2(\partial\mathcal{P})}} = \frac{\text{Length}(I)}{\text{Length}(\partial\mathcal{P})},$$

whilst in the presence of exceptional angles the eigenfunctions tend to *concentrate* on the *exceptional components* of $\partial\mathcal{P}$: the parts of the boundary between two consecutive exceptional angles. For an illustration of this phenomenon see Figures 7.6 and 7.7, which show some numerically computed eigenfunctions u_m for the equilateral triangle from Exercise 7.3.6(i) and for the isosceles right-angled triangle from Example 7.3.5. In the former case all angles are special, and one observes that the eigenfunctions are more or less equally distributed on all sides, whereas in the latter case there are three exceptional angles, and the eigenfunction u_{18} is mostly concentrated on the union of two sides, and the eigenfunction u_{19} is mostly concentrated on the hypothenuse.

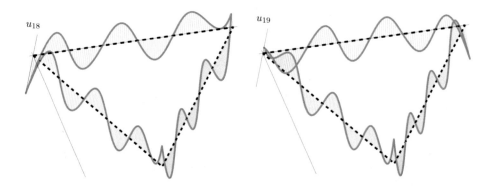

Figure 7.6. Numerically computed eigenfunctions of the Dirichlet-to-Neumann map on the equilateral triangle corresponding to the eigenvalues $\sigma_{18} \approx 17.8023$ and $\sigma_{19} \approx 16.6608$, which in turn correspond to the quasi-eigenvalues $\tau_{18} = 5\pi$ and $\tau_{19} = \frac{19\pi}{3}$ (both of which are in fact double, $\tau_{17} = \tau_{18}$ and $\tau_{19} = \tau_{20}$).

The complete proofs of Theorems 7.3.2 and 7.3.4 are rather difficult and lie well beyond the scope of this book. In the next subsections, we

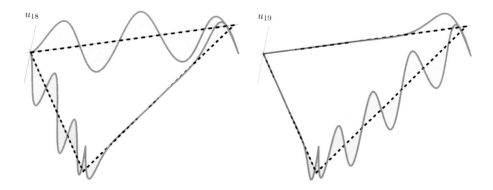

Figure 7.7. Numerically computed eigenfunctions of the Dirichlet-to-Neumann map on the isosceles right-angled triangle from Example 7.3.5, corresponding to the eigenvalues $\sigma_{18} \approx 15.708$ and $\sigma_{19} \approx 19.8968$, which in turn correspond to the quasi-eigenvalues $\tau_{18} = \frac{17\pi}{3}$ (which is in fact double, $\tau_{17} = \tau_{18}$) and $\tau_{19} = \frac{15\pi}{2\sqrt{2}}$ (which is single).

explain some main ideas underlying these proofs and their links to some classical problems in hydrodynamics, including the sloshing problem we have mentioned already.

7.3.2. Sloping beach problems. Let (x, y) be Cartesian coordinates in $\mathbb{R}^2$, and let (ρ, θ) denote the polar coordinates. Let

$$\mathfrak{S}_\alpha = \{(r, \theta) : r > 0, -\alpha < \theta < 0\}$$

denote an infinite sector of angle α with the vertex at the origin, where $0 < \alpha \leq \pi$. For future use, we denote its sides as

$$I_{\text{in}} := \{(r, -\alpha) : r > 0\}, \qquad I_{\text{out}} := \{(r, 0) : r > 0\}$$

and call them the *incoming* and *outgoing* side, respectively, so that the angle α is measured counterclockwise from I_{in} to I_{out}. We also denote the bisector by $I_{\text{b}} := \{(r, -\alpha/2) : r > 0\}$ and introduce the boundary coordinate s on $\partial \mathfrak{S}_\alpha = I_{\text{in}} \cup \{(0, 0)\} \cup I_{\text{out}}$ as shown in Figure 7.8, with $s = 0$ at the vertex, s negative on I_{in}, and s positive on I_{out}.

Restricting for the moment our attention to the half-sector $\mathfrak{S}_{\frac{\alpha}{2}}$, we consider two problems there: a mixed Robin–Neumann problem

$$(7.3.4) \qquad \Delta\Phi = 0 \quad \text{in } \mathfrak{S}_{\frac{\alpha}{2}}, \qquad \left(\frac{\partial\Phi}{\partial y} - \Phi\right)\Bigg|_{I_{\text{out}}} = 0, \qquad \partial_n\Phi|_{I_{\text{b}}} = 0,$$

and a similar mixed Robin–Dirichlet problem

$$(7.3.5) \qquad \Delta\Phi = 0 \quad \text{in } \mathfrak{S}_{\frac{\alpha}{2}}, \qquad \left(\frac{\partial\Phi}{\partial y} - \Phi\right)\Bigg|_{I_{\text{out}}} = 0, \qquad \Phi|_{I_{\text{b}}} = 0.$$

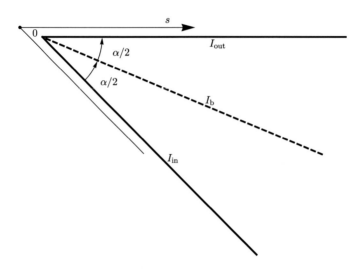

Figure 7.8. Infinite sectors $\mathfrak{S}_\alpha$ and $\mathfrak{S}_{\frac{\alpha}{2}}$.

We are particularly interested, in each case, in the existence of solutions which are bounded in the closed sector $\overline{\mathfrak{S}_{\frac{\alpha}{2}}}$ and behave far from the origin as $\cos(x-\xi)e^y$, with some constant ξ. More precisely, we additionally impose the conditions

$$(7.3.6) \qquad \Phi(x,y) = \cos(x-\xi)e^y + R(x,y),$$

where

$$(7.3.7) \qquad R(x,y) + |\rho\nabla R(x,y)| = O\left(\rho^{-r}\right) \qquad \text{as } \rho \to \infty,$$

with some constant $r > 0$ (which may depend on the angle of the sector) to be determined.

The Robin–Neumann problem (7.3.4), (7.3.6), (7.3.7) is known as the *sloping beach* or the *floating mat* problem and has a long and storied history in hydrodynamics; see [**Lew46**] and references therein[13]. We will also refer to the Robin–Dirichlet problem (7.3.5)–(7.3.7) as a sloping beach problem, somewhat abusing terminology. In particular, each of these problems has a solution of the required form if the parameter ξ takes a specific value which depends on the angle $\frac{\alpha}{2}$; in the Robin–Neumann case, one needs to take

$$(7.3.8) \qquad \xi = \xi_{\frac{\alpha}{2},\mathrm{N}} = \frac{\pi}{4} - \frac{\pi^2}{4\alpha},$$

and in the Robin–Dirichlet case,

$$(7.3.9) \qquad \xi = \xi_{\frac{\alpha}{2},\mathrm{D}} = \frac{\pi}{4} + \frac{\pi^2}{4\alpha}.$$

---

[13]Remarkably, Lewy's paper also recovers an elementary proof of the number-theoretical quadratic reciprocity law as a corollary of his hydrodynamics results.

The first result is due to A. S. Peters [**Pet50**], which was extended to the second problem in [**LevPPS22a**, Theorem 2.1]; in both cases one can take $r = \frac{\pi}{\alpha}$ in (7.3.7). We denote the corresponding solutions of (7.3.4), (7.3.6), (7.3.7) and (7.3.5)–(7.3.7) by $\Phi_{\frac{\alpha}{2},\mathrm{N}}(x, y)$ and $\Phi_{\frac{\alpha}{2},\mathrm{D}}(x, y)$, respectively.

In §7.3.5, we will outline how to use the solutions $\Phi_{\frac{\alpha}{2},\mathrm{N}}(x, y)$ and $\Phi_{\frac{\alpha}{2},\mathrm{D}}(x, y)$ of the sloping beach problems to obtain the asymptotics of the eigenvalues of the sloshing problem (7.1.12), $\mathcal{W}_\mathrm{D} = \emptyset$.

7.3.3. Peters solutions of the Robin problem in an infinite sector.

We will now use the sloping beach solutions $\Phi_{\frac{\alpha}{2},\mathrm{N}}(x, y)$ and $\Phi_{\frac{\alpha}{2},\mathrm{D}}(x, y)$ in the "half"-sector $\mathfrak{S}_{\frac{\alpha}{2}}$ to construct some specific solutions of the Robin problem

$$(7.3.10) \qquad \Delta\widetilde{\Phi} = 0 \quad \text{in } \mathfrak{S}_\alpha, \qquad \partial_n\widetilde{\Phi} = \tau\widetilde{\Phi} \quad \text{on } I_\text{in} \cup I_\text{out},$$

in the "full" sector $\mathfrak{S}_\alpha$ for large values of the Robin parameter τ. To do so, we start by extending the rescaled Robin–Neumann solution $\Phi_{\frac{\alpha}{2},\mathrm{N}}(\tau x, \tau y)$ symmetrically across I_b to a *symmetric Peters* solution $\widetilde{\Phi}_\mathrm{s}(x, y)$ of (7.3.10). Similarly, we extend the rescaled Robin–Dirichlet solution $\Phi_{\frac{\alpha}{2},\mathrm{D}}(\tau x, \tau y)$ antisymmetrically across I_b to an *antisymmetric Peters* solution $\widetilde{\Phi}_\mathrm{a}(x, y)$ of (7.3.10).

Let us now consider an arbitrary nontrivial linear combination $\widetilde{\Phi}(x, y)$ of $\widetilde{\Phi}_\mathrm{s}(x, y)$ and $\widetilde{\Phi}_\mathrm{a}(x, y)$ with constant complex coefficients. It is a solution of the Robin problem (7.3.10) which we call its *Peters solution*. It is also clear from (7.3.6), (7.3.7), by converting the cosines into the complex exponentials, that, as $\tau \to +\infty$, the leading terms of the traces of $\widetilde{\Phi}(x, y)$ on the boundary rays I_in and I_out are oscillatory in the variable s,

$$(7.3.11)$$

$$\widetilde{\Phi}\Big|_{I_\text{in}}(s) = h_{\text{in},1}\mathrm{e}^{\mathrm{i}\tau s} + h_{\text{in},2}\mathrm{e}^{-\mathrm{i}\tau s} + o(1) = \left\langle \mathbf{h}_\text{in}, \begin{pmatrix} \mathrm{e}^{-\mathrm{i}\tau s} \\ \mathrm{e}^{\mathrm{i}\tau s} \end{pmatrix} \right\rangle_{\mathbb{C}^2} + o(1),$$

$$\widetilde{\Phi}\Big|_{I_\text{out}}(s) = h_{\text{out},1}\mathrm{e}^{\mathrm{i}\tau s} + h_{\text{out},2}\mathrm{e}^{-\mathrm{i}\tau s} + o(1) = \left\langle \mathbf{h}_\text{out}, \begin{pmatrix} \mathrm{e}^{-\mathrm{i}\tau s} \\ \mathrm{e}^{\mathrm{i}\tau s} \end{pmatrix} \right\rangle_{\mathbb{C}^2} + o(1),$$

with some vectors

$$\mathbf{h}_\text{in} := \begin{pmatrix} h_{\text{in},1} \\ h_{\text{in},2} \end{pmatrix} \quad \text{and} \quad \mathbf{h}_\text{out} := \begin{pmatrix} h_{\text{out},1} \\ h_{\text{out},2} \end{pmatrix} \in \mathbb{C}^2.$$

We will denote such a Peters solution by

$$\widetilde{\Phi}_\tau(x, y; \mathbf{h}_\text{in}, \mathbf{h}_\text{out}).$$

We now ask what should be the relations (if any) between vectors $\mathbf{h}^+$ and $\mathbf{h}^-$ for the existence of a Peters solution $\widetilde{\Phi}_\tau(x, y; \mathbf{h}_\text{in}, \mathbf{h}_\text{out})$ of (7.3.10) with asymptotics (7.3.11). The equations (7.3.6), (7.3.8), and (7.3.9) imply, after some linear algebra, that the relations we seek in fact depend upon the arithmetic properties of the angle α; more precisely, they depend upon whether or not the angle is *exceptional*; see Remark 7.3.7.

Theorem 7.3.8 ([**LevPPS22b**, Theorem 3.1]).

(i) Let α be a nonexceptional angle. Then for every $\mathbf{h}_{\mathrm{in}} \in \mathbb{C}^2$ there exists a Peters solution $\widetilde{\Phi}_\tau(x, y; \mathbf{h}_{\mathrm{in}}, \mathbf{h}_{\mathrm{out}})$ of (7.3.10) satisfying (7.3.11) with

$$(7.3.12) \qquad \mathbf{h}_{\mathrm{out}} = \mathsf{A}(\alpha)\mathbf{h}_{\mathrm{in}},$$

where

$$(7.3.13) \qquad \mathsf{A}(\alpha) := \begin{pmatrix} \operatorname{cosec} \frac{\pi^2}{2\alpha} & -\mathrm{i}\cot \frac{\pi^2}{2\alpha} \\ \mathrm{i}\cot \frac{\pi^2}{2\alpha} & \operatorname{cosec} \frac{\pi^2}{2\alpha} \end{pmatrix}.$$

(ii) Let $\alpha = \frac{\pi}{2k}$, $k \in \mathbb{N}$, be an exceptional angle. Then a Peters solution $\widetilde{\Phi}_\tau(x, y; \mathbf{h}_{\mathrm{in}}, \mathbf{h}_{\mathrm{out}})$ of (7.3.10) satisfying (7.3.11) exists if the vectors $\mathbf{h}_{\mathrm{in}}, \mathbf{h}_{\mathrm{out}}$ satisfy

$$(7.3.14) \qquad \langle \mathbf{h}_{\mathrm{in}}, \mathbf{X} \rangle_{\mathbb{C}^2} = \langle \mathbf{h}_{\mathrm{out}}, \overline{\mathbf{X}} \rangle_{\mathbb{C}^2} = 0,$$

where

$$\mathbf{X} := \begin{pmatrix} \mathrm{e}^{(-1)^{k+1}\mathrm{i}\pi/4} \\ \mathrm{e}^{(-1)^k\mathrm{i}\pi/4} \end{pmatrix}.$$

Remark 7.3.9. In both cases in Theorem 7.3.8, we obtain the existence of a Peters solution $\widetilde{\Phi}_\tau(x, y; \mathbf{h}_{\mathrm{in}}, \mathbf{h}_{\mathrm{out}})$ by fixing two out of the four components of the vectors $\mathbf{h}_{\mathrm{in}}, \mathbf{h}_{\mathrm{out}}$. The difference is that in the nonexceptional case we fix the two components of the same vector and find the other vector from (7.3.12) (it does not in fact matter whether we fix either of the two vectors as the matrix $\mathsf{A}(\alpha)$ is invertible), whereas in the exceptional case we fix exactly one component of each of $\mathbf{h}_{\mathrm{in}}$ and $\mathbf{h}_{\mathrm{out}}$ and recover the other ones from (7.3.14).

Remark 7.3.10. It may be shown that the conditions on $\mathbf{h}_{\mathrm{in}}, \mathbf{h}_{\mathrm{out}}$ in Theorem 7.3.8 are not only sufficient but also necessary for the existence of Peters solutions.

7.3.4. Quasi-mode construction for the Steklov problem in a curvilinear polygon. We now outline the main ideas behind the proofs of Theorems 7.3.2 and 7.3.4 following the exposition in [**LevPPS22b**]. As in the sloshing problem, we start by describing the construction of the corresponding quasi-modes.

Assume for simplicity that the polygon $\mathcal{P} = \mathcal{P}(\boldsymbol{\alpha}, \boldsymbol{\ell})$ has straight sides and that all angles are nonexceptional. We introduce on ∂P near each vertex V_j the local coordinate s_j such that s_j is zero at V_j, negative on the side

I_j, and positive on the side I_{j+1}; see Figure 7.9. Note that on each side I_j joining V_{j-1} and V_j we have effectively two coordinates: the coordinate s_j running from $-\ell_j$ to 0 and the coordinate s_{j-1} running from 0 to ℓ_j, related as

$$(7.3.15) \qquad s_j = s_{j-1} - \ell_j.$$

This emphasises the fact that I_j is the outgoing side of the sector with the vertex at V_{j-1} and the incoming side of the sector with the vertex at V_j.

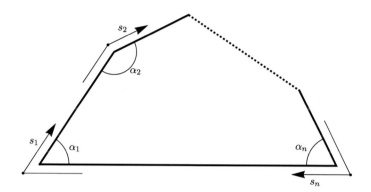

Figure 7.9. A straight polygon with local boundary coordinates.

Let $\mathcal{V}_j$ be the orientation-preserving isometry of the plane which maps the sector $V_{j-1}V_jV_{j+1}$ into the sector $\mathfrak{S}_{\alpha_j}$ with the vertex at the origin, and let $(x'_j, y'_j) := \mathcal{V}_j(x, y)$ be the local Cartesian coordinates with the origin at V_j. We will seek the quasi-modes $\widetilde{U}_\tau(z)$ of the Steklov problem on $\mathcal{P}$ which coincide, in the vicinity of each vertex V_j, with a Peters solution

$$\widetilde{\Phi}_\tau(x'_j, y'_j; \mathbf{h}_{j,\mathrm{in}}, \mathbf{h}_{j,\mathrm{out}}),$$

where suitable values of the quasi-eigenvalues τ and the coefficient vectors $\mathbf{h}_{j,\mathrm{in}}, \mathbf{h}_{j,\mathrm{out}} \in \mathbb{C}^2$ are to be determined. By Theorem 7.3.8(i), these vectors should be related by

$$(7.3.16) \qquad \mathbf{h}_{j,\mathrm{out}} := \mathsf{A}(\alpha_j)\mathbf{h}_{j,\mathrm{in}}$$

to ensure the existence of the Peters solutions.

As a consequence of (7.3.11),

$$\widetilde{U}_\tau\Big|_{\partial\mathcal{P}} = \widetilde{u} + o(1) \qquad \text{as } \tau \to \infty,$$

where we can write $\widetilde{u}|_{I_j}$ as a trigonometric function of the variable s_j involving the vectors $\mathbf{h}_{j,\mathrm{in}}, \mathbf{h}_{j,\mathrm{out}}$ (using $\widetilde{\Phi}_\tau(x'_j, y'_j; \mathbf{h}_{j,\mathrm{in}}, \mathbf{h}_{j,\mathrm{out}})$) or as a trigonometric function of the variable s_{j-1} involving the vectors $\mathbf{h}_{j-1,\mathrm{in}}, \mathbf{h}_{j-1,\mathrm{out}}$ (using $\widetilde{\Phi}_\tau(x'_{j-1}, y'_{j-1}; \mathbf{h}_{j-1,\mathrm{in}}, \mathbf{h}_{j-1,\mathrm{out}})$).

These expressions should match, so an easy computation shows that we must have, due to (7.3.15),

(7.3.17) $$\mathbf{h}_{j,\text{in}} = \mathsf{B}(\ell_j, \tau)\mathbf{h}_{j-1,\text{out}},$$

where the *side transfer matrices* $\mathsf{B}(\ell_j, \tau)$ are defined by

(7.3.18) $$\mathsf{B}(\ell, \tau) := \begin{pmatrix} \exp(i\ell\tau) & 0 \\ 0 & \exp(-i\ell\tau) \end{pmatrix}$$

(the relations (7.3.17) and (7.3.18) essentially manifest just a change of variables on I_j). We will call the vector $\mathbf{h}_{j,\text{in}}$ the *boundary quasi-wave incoming into* V_j (from V_{j-1}), and the vector $\mathbf{h}_{j-1,\text{out}}$ the *boundary quasi-wave outgoing from* V_{j-1} (towards V_j). In order for our Peters solutions on I_j to match, these must be related by (7.3.17).

This formulation allows us to think of our problem as a transfer problem. Consider a boundary quasi-wave $\mathbf{b} := \mathbf{h}_{n,\text{out}}$ outgoing from the vertex V_n towards V_1. It arrives at the vertex V_1 as an incoming quasi-wave $\mathbf{h}_{1,\text{in}} = \mathsf{B}(\ell_1, \tau)\mathbf{b}$, and, according to (7.3.16), it leaves V_1 towards V_2 as an outgoing boundary quasi-wave

$$\mathbf{h}_{1,\text{out}} = \mathsf{A}(\alpha_1)\mathbf{h}_{1,\text{in}} = \mathsf{A}(\alpha_1)\mathsf{B}(\ell_1, \tau)\mathbf{b}.$$

It then arrives at V_2 as an incoming boundary quasi-wave

$$\mathbf{h}_{2,\text{in}} = \mathsf{B}(\ell_2, \tau)\mathsf{A}(\alpha_1)\mathsf{B}(\ell_1, \tau)\mathbf{b}$$

and leaves V_2 towards V_3 as an outgoing boundary quasi-wave

$$\mathbf{h}_{2,\text{out}} = \mathsf{A}(\alpha_2)\mathsf{B}(\ell_2, \tau)\mathsf{A}(\alpha_1)\mathsf{B}(\ell_1, \tau)\mathbf{b}.$$

Continuing the process, we conclude that it arrives at V_n from V_{n-1} as an incoming boundary quasi-wave

$$\mathbf{h}_{n,\text{in}} = \mathsf{B}(\ell_n, \tau)\mathsf{A}(\alpha_{n-1})\mathsf{B}(\ell_{n-1}, \tau) \cdots \mathsf{A}(\alpha_1)\mathsf{B}(\ell_1, \tau)\mathbf{b}$$

and leaves V_n towards V_1 as an outgoing boundary quasi-wave

$$\mathbf{h}_{n,\text{out}} = \mathsf{A}(\alpha_n)\mathsf{B}(\ell_n, \tau)\mathsf{A}(\alpha_{n-1})\mathsf{B}(\ell_{n-1}, \tau) \cdots \mathsf{A}(\alpha_1)\mathsf{B}(\ell_1, \tau)\mathbf{b} = \mathsf{T}(\boldsymbol{\alpha}, \boldsymbol{\ell})\mathbf{b},$$

where we have denoted

$$\mathsf{T}(\boldsymbol{\alpha}, \boldsymbol{\ell}) := \mathsf{A}(\alpha_n)\mathsf{B}(\ell_n, \tau)\mathsf{A}(\alpha_{n-1})\mathsf{B}(\ell_{n-1}, \tau) \cdots \mathsf{A}(\alpha_1)\mathsf{B}(\ell_1, \tau).$$

The boundary quasi-wave $\mathbf{h}_{n,\text{out}}$ must match the original outgoing boundary quasi-wave $\mathbf{b}$, which imposes the following quantisation condition on τ:

(7.3.19)   the matrix $\mathsf{T}(\boldsymbol{\alpha}, \boldsymbol{\ell})$ has an eigenvalue 1.

Using the explicit definitions (7.3.13) of the matrices $\mathsf{A}(\alpha_j)$ and (7.3.18) of the matrices $\mathsf{B}(\ell_j, \tau)$, it is easily seen that (7.3.19) is equivalent to

(7.3.20) $$\operatorname{Tr}(\mathsf{T}(\boldsymbol{\alpha}, \boldsymbol{\ell})) = 2.$$

Some rather elaborate calculations then demonstrate that every nonnegative solution τ of (7.3.20) is a root of the trigonometric equation $F_{\alpha,\ell}(\tau) = 0$ and vice versa, with $F_{\alpha,\ell}$ defined by (7.3.3) and with multiplicities as stated in Theorem 7.3.4, thus giving the first hint of the validity of that theorem.

The full proof of Theorem 7.3.4 is highly nontrivial, and we only mention the remaining steps briefly. First, after a rigorous construction of quasi-modes $\tilde{U}_m$ using appropriate cut-offs and with τ_m being the roots of (7.3.20), it is relatively easy to see that U_m approximately satisfy the Laplace equation and the Steklov boundary condition with suitably diminishing errors as $m \to \infty$. That allows us to conclude, in a standard manner, that τ_m are indeed the approximate eigenvalues of the Steklov problem on the curvilinear polygon in a sense that there exists a subsequence of exact Steklov eigenvalues σ_{i_m} such that $|\tau_m - \sigma_{i_m}| = o(1)$ as $m \to \infty$.

The most difficult part of the proof consists in establishing the correct enumeration of quasi-eigenvalues by showing that $i_m = m$. This is done with the help of Dirichlet–Neumann bracketing: a suitably chosen sequence of cuts perpendicular to the boundary is added to $\partial \mathcal{P}$, on which either the Dirichlet or Neumann conditions are imposed; see Figure 7.10. These cuts are introduced not simultaneously but in a particular order, allowing at each step a quantitative comparison with the known asymptotics of sloshing problems (mixed Steklov–Neumann problems) and other mixed Steklov–Dirichlet and Steklov–Neumann–Dirichlet problems obtained in Theorem 7.3.11 and Remark 7.3.12 below, either directly or using transplantation tricks similar to those used in the proof of Theorem 6.2.17.

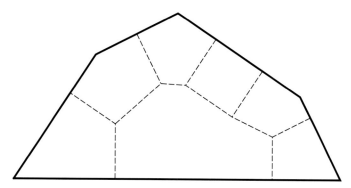

Figure 7.10. An example of a polygon with cuts.

Then all the results are extended from straight polygons to curvilinear polygons; here the curvature of the boundary at the vertices requires special treatment using potential theory. Finally, each step should be adjusted for the case of polygons with exceptional angles which need to be analysed separately.

7.3.5. Asymptotics of the sloshing eigenvalues. We are now able to outline, following [**LevPPS22a**], how to use the solutions $\Phi_{\frac{\alpha}{2},\mathrm{N}}(x,y)$ and $\Phi_{\frac{\alpha}{2},\mathrm{D}}(x,y)$ of the sloping beach problem to obtain the asymptotics of eigenvalues of the sloshing problem (7.1.12), $\mathcal{W}_{\mathrm{D}} = \emptyset$ — this does not require the full machinery of §7.3.4 and is, in fact, a preliminary step for that. For simplicity, we assume that Ω is a triangle, the sloshing surface $\mathcal{S}$ coincides with the interval $(A, B) = (0, L)$ of the horizontal axis, and that the walls $\mathcal{W}$ form the angles $\frac{\alpha}{2}$ and $\frac{\beta}{2}$ with the sloshing surface at the points A and B, respectively; see Figure 7.11.

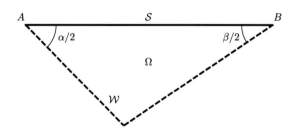

Figure 7.11. A sloshing problem in a triangular domain.

We are looking for *quasi-modes* (approximate solutions) of (7.1.12), $\mathcal{W}_{\mathrm{D}} = \emptyset$, which are constructed, in the first approximation, by gluing together a sloping beach solution $\pm\Phi_{\frac{\alpha}{2},\mathrm{N}}(\sigma x, \sigma y)$ near the corner point A and a sloping beach solution $\pm\Phi_{\frac{\beta}{2},\mathrm{N}}(\sigma(L-x), \sigma y)$ near the corner point B. As the traces of these two solutions on $\mathcal{S}$ behave asymptotically as $\pm\cos\left(\sigma x - \xi_{\frac{\alpha}{2},\mathrm{N}}\right)$ and $\pm\cos\left(\sigma x - \sigma L + \xi_{\frac{\beta}{2},\mathrm{N}}\right)$ for $\sigma \to +\infty$, cf. (7.3.6) and (7.3.8), the phases of the cosines should match. This matching condition yields an asymptotic quantisation condition for the eigenvalues σ_m subject to which the quasi-modes can be rigorously constructed. The quasi-mode analysis can be extended to the more general (no longer triangular and with possibly curved walls) sloshing domains, such as the one shown in Figure 7.1, which eventually leads to

Theorem 7.3.11 ([**LevPPS22a**, Theorem 1.1]). *Let $\Omega \subset \mathbb{R}^d$ be a bounded simply connected domain with a Lipschitz boundary $M = \partial\Omega$, the sloshing surface $\mathcal{S} \subset M$ which is a straight line interval (A, B) of length L, and the walls $\mathcal{W} = M \setminus \overline{\mathcal{S}}$ which form the interior angles $0 < \frac{\alpha}{2}, \frac{\beta}{2} < \frac{\pi}{2}$ with $\mathcal{S}$ at the points A and B. Then the eigenvalues σ_m, $m \in \mathbb{N}$, of the sloshing problem (7.1.12), $\mathcal{W}_{\mathrm{D}} = \emptyset$, have the asymptotics*

$$(7.3.21) \quad L\sigma_m = \pi\left(m - \frac{1}{2}\right) - \frac{\pi^2}{8}\left(\frac{2}{\alpha} + \frac{2}{\beta}\right) + o(1) \qquad \text{as } m \to \infty.$$

Remark 7.3.12. A similar method of constructing the quasi-modes can also be applied in a full mixed Steklov–Neumann–Dirichlet problem (7.1.12), with the following modifications: if the Dirichlet condition is imposed on $\mathcal{W}$ near the corner A, we use a sloping beach solution $\pm\Phi_{\frac{\alpha}{2},\mathrm{D}}(\sigma x, \sigma y)$ there, and similarly near corner B. The result is the asymptotic formula for the eigenvalues similar to (7.3.21), see [**LevPPS22a**, Theorem 1.8],

$$(7.3.22) \qquad L\sigma_m = \pi\left(m - \frac{1}{2}\right) + \frac{\pi^2}{8}\left(\pm\frac{2}{\alpha} \pm \frac{2}{\beta}\right) + o(1) \qquad \text{as } m \to \infty,$$

where the contributions from the angles $\frac{\alpha}{2}$ and $\frac{\beta}{2}$ appear with the plus sign if a Dirichlet condition is imposed on $\mathcal{W}$ adjacently to the corner points A, B, respectively, and with a minus sign in the case of a Neumann condition.

Remark 7.3.13. As was additionally shown in [**LevPPS22b**], the remainder estimates in (7.3.21) and (7.3.22) can be improved if the walls are straight near the corner. The formula (7.3.21) is also applicable if the walls form right angles with the sloshing surface subject to some additional geometric constrains.

Remark 7.3.14. Numerical evidence suggests that asymptotics (7.3.21) and (7.3.22) remain valid for angles $\frac{\alpha}{2}, \frac{\beta}{2} \in \left[\frac{\pi}{2}, \pi\right]$. In the same vein, numerics suggest that Theorem 7.3.2 also remains valid if the restriction $\alpha_j < \pi$ on the angles of a curvilinear polygon is replaced by $\alpha_j < 2\pi$. However, there is no proof of that in either case as the exponent r in the error estimate (7.3.7) is not good enough to implement the quasi-mode argument.

■ **Exercise 7.3.15.** Verify the asymptotics (7.3.21) and (7.3.22) for the sloshing problems allowing separation of variables: the rectangle $(0,1) \times (-h, 0)$ from Exercise 7.1.10 (in which $L = 1$ and $\frac{\alpha}{2} = \frac{\beta}{2} = \frac{\pi}{2}$) and the mixed problems I–IV on the triangular domains from Figure 7.3 (in which $L = 2$ and $\frac{\alpha}{2} = \frac{\beta}{2} = \frac{\pi}{4}$; see also the discussion at the end of §7.1.2).

Remark 7.3.16. Very little is known about the spectral asymptotics for sloshing eigenvalues in higher dimensions beyond the leading term. We refer to [**MaySenStA22**] for some partial results in that direction, as well as to [**GirLPS19**], [**Ivr19**] for related developments in the case of Steklov eigenvalues when the boundary has edges.

7.3.6. Inverse spectral problem for curvilinear polygons. Here we follow [**KryLPPS21**]. Recalling, first of all, the definition of asymptotically Steklov isospectral domains from Remark 7.2.13, we note that two curvilinear polygons with the same side lengths ℓ and angles $\boldsymbol{\alpha}$ are asymptotically Steklov isospectral by Corollary 7.3.3 (of course, at the same time they need not be isospectral).

We further have

Theorem 7.3.17. Two curvilinear polygons are asymptotically Steklov isospectral if and only if their trigonometric characteristic functions (7.3.3) coincide. Moreover, the trigonometric characteristic functions of two curvilinear polygons coincide if and only if their nonnegative real roots (that is, the quasi-eigenvaluies τ_m of the polygons) coincide with account of multiplicities. Additionally, the trigonometric characteristic function $F_{\alpha,\ell}(\tau)$ of a curvilinear polygon $\mathcal{P}(\alpha,\ell)$ can be uniquely reconstructed from the Steklov spectrum of $\mathcal{P}(\alpha,\ell)$.

The proof of Theorem 7.3.17 is based on the application of the Hadamard–Weierstrass factorisation theorem for entire functions and the property of almost periodic real functions with all real zeros: if two such functions have asymptotically close zeros, they have exactly the same zeros [**KurSuh20**, Theorem 6].

We will now describe what information on the geometry of a curvilinear polygon $\mathcal{P}(\alpha,\ell)$ may be deduced from its Steklov spectrum (or, equivalently, in accordance with Theorem 7.3.17, from a characteristic trigonometric function $F(\tau)$). To do so, we need to work within a generic class of curvilinear polygons, which we call *admissible polygons* and which satisfy the following two conditions:

(7.3.23)   the lengths $\ell_1,\ldots,\ell_n$ are incommensurable over $\{-1,0,+1\}$

(that is, only the trivial linear combination of $\ell_1,\ldots,\ell_n$ with these coefficients vanishes), and

(7.3.24)   all angles $\alpha_1,\ldots,\alpha_n$ are not special

(see Remark 7.3.7 for the definition).

Theorem 7.3.18. Given a characteristic trigonometric function $F(\tau) = F_{\alpha,\ell}(\tau)$ of an admissible curvilinear polygon, we can constructively recover, in a finite number of steps:

(i) The number of vertices n and the number of exceptional angles K.

(ii) If $K = 0$, then the vector of side lengths ℓ, in the correct order, subject to a cyclic shift and a change of orientation, and further, once the enumeration of ℓ is fixed, the vector

$$\mathbf{c}(\boldsymbol{\alpha}) = \left(\cos \frac{\pi^2}{2\alpha_1}, \ldots, \cos \frac{\pi^2}{2\alpha_n} \right),$$

modulo a global change of sign.

(iii) If $K > 0$, then we can recover the same information as in (ii) for each exceptional component of $\partial\mathcal{P}$ (a part of the boundary between two exceptional angles) but not the order in which the exceptional components are joined together.

If either (or both) of the admissibility conditions (7.3.23) and (7.3.24) is not satisfied, then Theorem 7.3.18 is no longer applicable.

Example 7.3.19.

(i) Consider a family of straight parallelograms P_a depending on a parameter $a \in (0,1)$, with angles $\frac{\pi}{5}$ (which is special) and $\frac{4\pi}{5}$, and side lengths a and $1-a$. In this case the characteristic function

$$F(\tau) = \cos(2\tau) - \frac{1}{\sqrt{2}}$$

is independent of a, and we therefore cannot reconstruct side lengths from it — all these parallelograms are asymptotically Steklov isospectral. In this example, both conditions (7.3.23) and (7.3.24) are not satisfied.

(ii) Two straight triangles of the same perimeter and with angles $\boldsymbol{\alpha} = \left(\frac{\pi}{7}, \frac{\pi}{63}, \frac{53\pi}{63}\right)$ and $\widetilde{\boldsymbol{\alpha}} = \left(\frac{\pi}{9}, \frac{\pi}{21}, \frac{53\pi}{63}\right)$, respectively (in each case there are two special angles), have the same characteristic function $F(\tau)$ and are therefore asymptotically Steklov isospectral.

7.4. The Dirichlet-to-Neumann map for the Helmholtz equation

7.4.1. Definition and basic properties. Let, as in §7.1, Ω be a bounded domain in a complete Riemannian manifold of dimension $d \geq 2$, with a Lipschitz boundary $M := \partial\Omega$. Let us choose a real parameter $\Lambda \notin \mathrm{Spec}(-\Delta_\Omega^{\mathrm{D}})$ and consider, for a given $u \in H^{1/2}(\Omega)$, a nonhomogeneous Dirichlet problem

(7.4.1)
$$\begin{cases} -\Delta U = \Lambda U & \text{in } \Omega, \\ U = u & \text{on } M. \end{cases}$$

This problem has a unique solution $U \in H^1(\Omega)$ which we will call the Λ-*Helmholtz extension* of u and which we denote as

$$U := \mathcal{E}_\Lambda u \in \mathcal{H}_\Lambda(\Omega),$$

where by analogy with (7.1.4) we define

$$(7.4.2) \quad \mathcal{H}_\Lambda(\Omega) := \{U \in H^1(\Omega) : -\Delta U = \Lambda U\} = \{\mathcal{E}_\Lambda u : u \in H^{1/2}(M)\}$$

to be the subspace of Λ-harmonic functions in $H^1(\Omega)$.

Definition 7.4.1. Let $\Lambda \notin \mathrm{Spec}(-\Delta_\Omega^D)$. The linear operator

$$\mathcal{D}_\Lambda : H^{1/2}(\Omega) \to H^{-1/2}(\Omega), \qquad \mathcal{D}_\Lambda : u \mapsto \partial_n(\mathcal{E}_\Lambda u)|_M,$$

which maps u into the trace of the normal derivative of its Λ-Helmholtz extension, is called the *Dirichlet-to-Neumann map* for the Helmholtz equation.

We want to extend Definition 7.4.1 to the case when $\Lambda \in \mathrm{Spec}(-\Delta_\Omega^D)$. We only do it briefly, outlining the major steps; for the full rigorous definition in terms of the so-called *linear relations*, see [**BehtEl15**] and also [**AreMaz12**]. Let

$$\mathcal{K}_\Lambda := \{\partial_n U : U \in \mathcal{H}_\Lambda(\Omega) \cap H_0^1(\Omega)\}$$

be the finite-dimensional linear space of the Neumann boundary traces of eigenfunctions of $-\Delta^D$ corresponding to a Dirichlet eigenvalue Λ. The non-homogenous problem (7.4.1) is solvable if and only if u is orthogonal in $L^2(M)$ to $\mathcal{K}_\Lambda$; see [**McL00**, Theorem 4.10]. The necessity of this condition is immediate by Green's formula: if U^D is an eigenfunction of $-\Delta^D$ corresponding to Λ, then from (7.4.1)

$$\Lambda \left(U, U^D\right)_{L^2(\Omega)} = \left(-\Delta U, U^D\right)_{L^2(\Omega)} = \left(U, -\Delta U^D\right)_{L^2(\Omega)} + \left(u, \partial_n U^D\right)_{L^2(M)}$$
$$= \Lambda \left(U, U^D\right)_{L^2(\Omega)} + \left(u, \partial_n U^D\right)_{L^2(M)},$$

implying $\left(u, \partial_n U^D\right)_{L^2(M)} = 0$.

Let $\mathcal{K}_\Lambda^\perp$ denote an orthogonal complement to $\mathcal{K}_\Lambda$ in $L^2(M)$, and let $\Pi_{\mathcal{K}_\Lambda^\perp}$ denote the orthogonal projection onto it. For any $u \in H^1(M) \cap \mathcal{K}_\Lambda^\perp$, a solution to (7.4.1) exists but is not unique as it is defined modulo an addition of an eigenfunction of $-\Delta^D$ corresponding to the eigenvalue Λ. If we however treat $\mathcal{E}_\Lambda u$ as a multi-valued map, then the map $u \mapsto \Pi_{\mathcal{K}_\Lambda^\perp} \partial_n \mathcal{E}_\Lambda u$ is still uniquely defined for $u \in H^1(M) \cap \mathcal{K}_\Lambda^\perp$, and we will call it the Dirichlet-to-Neumann map for the Helmholtz equation for $\Lambda \in \mathrm{Spec}(-\Delta_\Omega^D)$. We note that this construction relies on the fact that the eigenfunctions of the Dirichlet Laplacian on Ω belong to the space $H^{3/2}(\Omega)$; see Remark 2.2.20.

For any fixed $\Lambda \in \mathbb{R}$, the Dirichlet-to-Neumann map $\mathcal{D}_\Lambda$ is a selfadjoint operator in $L^2(M)$ with a discrete spectrum of real eigenvalues

$$\sigma_1^\Lambda \le \sigma_2^\Lambda \le \cdots;$$

see [**BehtEl15**], [**AreMaz12**], and also [**GréNédPla76**]. The eigenvalues and the corresponding eigenfunctions u_j^Λ, $j = 1, \dots, \infty$, satisfy

(7.4.3)
$$\begin{cases} -\Delta U = \Lambda U & \text{in } \Omega, \\ \partial_n U = \sigma_j^\Lambda u_j & \text{on } M, \end{cases}$$

with $U := \mathcal{E}_\Lambda u_j$, and the basis of eigenfunctions may be chosen to be orthogonal in $L^2(M)$. The analogue of the weak Steklov spectral problem (7.1.5) for (7.4.3) is

(7.4.4)
$$(\nabla U, \nabla V)_{L^2(\Omega)} - \Lambda (U, V)_{L^2(\Omega)} = \sigma(U, V)_{L^2(M)} \qquad \text{for all } V \in H^1(\Omega).$$

Let now
$$u \in \text{Dom}(\mathcal{D}_\Lambda) = \begin{cases} H^{1/2}(M) & \text{if } \Lambda \notin \text{Spec}(-\Delta_\Omega^D), \\ H^1(M) \cap \mathcal{K}_\Lambda^\perp & \text{if } \Lambda \in \text{Spec}(-\Delta_\Omega^D). \end{cases}$$

The quadratic form of the Dirichlet-to-Neumann map $\mathcal{D}_\Lambda$ is given by

(7.4.5)
$$(\mathcal{D}_\Lambda u, u)_{L^2(\Omega)} = (\partial_n U, u)_{L^2(M)} = \|\nabla U\|_{L^2(\Omega)}^2 - \Lambda \|U\|_{L^2(\Omega)}^2;$$

cf. (7.1.6). We have the following analogue of (7.1.11) and Theorem 7.1.9.

Theorem 7.4.2 (The variational principle for the eigenvalues of the Dirichlet-to-Neumann map). Let Ω be a bounded open set in $\mathbb{R}^d$, with a Lipschitz boundary $M = \partial\Omega$, let $\Lambda \in \mathbb{R}$, and let σ_k^Λ be the eigenvalues of the Dirichlet-to-Neumann map for the Helmholtz equation in Ω. Then

(7.4.6)
$$\sigma_k^\Lambda = \min_{\substack{\widetilde{\mathcal{L}} \subset \text{Dom}(\mathcal{D}_\Lambda) \\ \dim \widetilde{\mathcal{L}} = k}} \max_{u \in \widetilde{\mathcal{L}} \backslash \{0\}} \frac{\|\nabla \mathcal{E}_\Lambda u\|_{L^2(\Omega)}^2 - \Lambda \|\mathcal{E}_\Lambda u\|_{L^2(\Omega)}^2}{\|u\|_{L^2(M)}^2}$$

$$= \min_{\substack{\mathcal{L} \subset \mathcal{H}_\Lambda(\Omega) \\ \dim \mathcal{L} = k}} \max_{\substack{U \in \mathcal{L} \\ U \neq 0}} \frac{\|\nabla U\|_{L^2(\Omega)}^2 - \Lambda \|U\|_{L^2(\Omega)}^2}{\|U|_M\|_{L^2(M)}^2}, \qquad k \in \mathbb{N}.$$

Moreover, if $\Lambda < \lambda_1^D(\Omega)$, then $\mathcal{H}_\Lambda(\Omega)$ in the right-hand side of (7.4.6) may be replaced by $H^1(\Omega)$, and we have

(7.4.7) $\quad \sigma_k^\Lambda(\Omega) = \min_{\substack{\mathcal{L} \subset H^1(\Omega) \\ \dim \mathcal{L} = k}} \max_{\substack{W \in \mathcal{L} \\ W \neq 0}} \dfrac{\|\nabla W\|_{L^2(\Omega)}^2 - \Lambda \|W\|_{L^2(\Omega)}^2}{\|W|_M\|_{L^2(M)}^2}, \qquad k \in \mathbb{N}.$

Proof. The formula (7.4.6) is just the standard variational principle taking into account (7.4.5), (7.4.2), and the definition of $\mathcal{E}_\Lambda$. In order to prove the validity of (7.4.7) we first need, assuming $\Lambda < \lambda_1^D(\Omega)$, the following analogue of Proposition 7.1.8: we have $H^1(\Omega) = \mathcal{H}_\Lambda(\Omega) \oplus H_0^1(\Omega)$ and

$$(\nabla U, \nabla V)_{L^2(\Omega)} = \Lambda (U, V)_{L^2(\Omega)} \qquad \text{for any } U \in \mathcal{H}_\Lambda(\Omega), V \in H_0^1(\Omega).$$

Taking now in (7.4.7) $H^1(\Omega) \ni W = U + V$, with $U \in \mathcal{H}_\Lambda(\Omega)$, $V \in H^1_0(\Omega)$, we obtain

$$\|\nabla W\|^2_{L^2(\Omega)} - \Lambda \|W\|^2_{L^2(\Omega)} \geq \|\nabla U\|^2_{L^2(\Omega)} - \Lambda \|U\|^2_{L^2(\Omega)} + (\lambda^D_1(\Omega) - \Lambda)\|V\|^2_{L^2(\Omega)},$$

and the minimisation procedure now requires taking $V = 0$.   □

■ **Exercise 7.4.3.** By separating variables in polar coordinates (r, θ), show that the spectrum of the Dirichlet-to-Neumann map $\mathcal{D}_\Lambda$ in the unit disk consists of the single eigenvalues

$$\begin{cases} \dfrac{\sqrt{-\Lambda}I'_0(\sqrt{-\Lambda})}{I_0(\sqrt{-\Lambda})} & \text{if } \Lambda < 0, \\ 0 & \text{if } \Lambda = 0, \\ \dfrac{\sqrt{\Lambda}J'_0(\sqrt{\Lambda})}{J_0(\sqrt{\Lambda})} & \text{if } \Lambda > 0, \end{cases}$$

with the corresponding eigenfunction $u(\theta) = 1$, and the double eigenvalues

$$\begin{cases} \dfrac{\sqrt{-\Lambda}I'_m(\sqrt{-\Lambda})}{I_m(\sqrt{-\Lambda})} & \text{if } \Lambda < 0, \\ m & \text{if } \Lambda = 0, \qquad m \in \mathbb{N}, \\ \dfrac{\sqrt{\Lambda}J'_m(\sqrt{\Lambda})}{J_m(\sqrt{\Lambda})} & \text{if } \Lambda > 0, \end{cases}$$

with the corresponding eigenfunctions $u(\theta) = \cos m\theta$ and $u(\theta) = \sin m\theta$, where J_m and I_m are the Bessel functions and the modified Bessel functions, respectively. Use these expressions to reproduce Figure 7.12, and compare it to Figure 3.1; cf. also Numerical Exercise 3.1.17.

7.4.2. Dependence of the eigenvalues of the Dirichlet-to-Neumann map on the parameter. The behaviour of eigenvalues of the Dirichlet-to-Neumann map $\mathcal{D}_\Lambda$ as functions of Λ shown in Figure 7.12 for the unit disk is in fact typical (except for the multiplicities of the eigenvalues) for a generic Lipschitz domain $\Omega \subset \mathbb{R}^d$. We start by revisiting Remark 3.1.19 and restating it rigorously.

Proposition 7.4.4 (Robin–Dirichlet-to-Neumann duality [**AreMaz12**, Theorem 3.1], [**HasShe22**]). Let $\Omega \subset \mathbb{R}^d$ be a Lipschitz domain, and let $\Lambda, \sigma \in \mathbb{R}$. Then σ is an eigenvalue of the Dirichlet-to-Neumann map $\mathcal{D}_\Lambda$ if and only if Λ is an eigenvalue of the Robin Laplacian $-\Delta^{R,-\sigma}$. Moreover, the multiplicities of σ as an eigenvalue of $\mathcal{D}_\Lambda$ and of Λ as an eigenvalue of $-\Delta^{R,-\sigma}$ coincide.

Proposition 7.4.4 is almost immediately obvious (at least when $\Lambda \notin$ Spec$(-\Delta^D)$) from the fact that the mapping $T : \mathcal{H}_\Lambda(\Omega) \to H^{1/2}(M)$ which

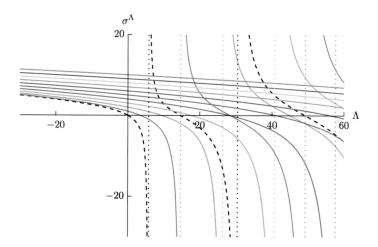

Figure 7.12. Some eigenvalues of the Dirichlet-to-Neumann map $\mathcal{D}_\Lambda$ for the unit disk as functions of Λ. The dashed black curves correspond to single eigenvalues, and the solid curves to double eigenvalues. The vertical dotted lines are placed at the abscissae coinciding with the Dirichlet eigenvalues of the unit disk.

acts as $T : U \to U|_M$ is an isomorphism between the corresponding eigenspaces (as well as its inverse $\mathcal{E}_\Lambda : H^{1/2}(M) \to \mathcal{H}_\Lambda(\Omega)$).

We now go back to the Robin problem and state the following extension of (3.1.19).

Proposition 7.4.5 ([**AreMaz12**, Proposition 3]). Let $\Omega \subset \mathbb{R}^d$ be a Lipschitz domain. For a fixed $k \in \mathbb{N}$, the eigenvalues $\lambda_k^{\mathrm{R},\gamma}(\Omega)$ of the Robin Laplacian on Ω are continuous strictly monotone increasing functions of $\gamma \in \mathbb{R}$ and satisfy

$$(7.4.8) \qquad \lim_{\gamma \to +\infty} \lambda_k^{\mathrm{R},\gamma}(\Omega) = \sup \left\{ \lambda_k^{\mathrm{R},\gamma}(\Omega) : \gamma \in \mathbb{R} \right\} = \lambda_k^{\mathrm{D}}(\Omega),$$

$$(7.4.9) \qquad \lim_{\gamma \to -\infty} \lambda_k^{\mathrm{R},\gamma}(\Omega) = -\infty.$$

Proof. For an illustration, see once more Figure 3.1. We have already established the (nonstrict) monotonicity of the Robin eigenvalues as functions of γ in Theorem 3.2.9. To prove the strict monotonicity, assume for contradiction that for some $k \in \mathbb{N}$ we have $\lambda_k^{\mathrm{R},\gamma_2} = \lambda_k^{\mathrm{R},\gamma_1} =: \Lambda$ with some $\gamma_1 < \gamma_2$. Then by Proposition 7.4.4,

$$[-\gamma_2, -\gamma_1] \subset \mathrm{Spec}\left(\mathcal{D}_\Lambda\right),$$

which contradicts the fact that the spectrum of the Dirichlet-to-Neumann map $\mathcal{D}_\Lambda$ is discrete.

The limiting behaviour (7.4.8) of the Robin eigenvalues as $\gamma \to +\infty$ has already been discussed in §3.1.3. To prove the limiting identity (7.4.9), assume for contradiction that for some $k \in \mathbb{N}$, the eigenvalue $\lambda_k^{\gamma,\mathrm{R}}$ is bounded below by $\Lambda := \inf_{\gamma \in \mathbb{R}} \lambda_k^{\gamma,\mathrm{R}} > -\infty$. Then by Proposition 7.4.4,

$$\mathrm{Spec}\,(\mathcal{D}_\Lambda) \subseteq \left\{-\gamma : \lambda_j^{\gamma,\mathrm{R}} = \Lambda, j = 1, \ldots, k\right\},$$

which is a finite set and therefore impossible, thus proving (7.4.9). □

Remark 7.4.6. As can be seen from Figure 3.1, the kth Robin eigenvalue $\lambda_k^{\mathrm{R},\gamma}$ is only continuous in γ and not necessarily smooth. If however we follow the eigenvalue branches correctly through their crossings, forsaking the ordering of eigenvalues, then the union over γ of spectra of the Robin Laplacians $-\Delta^{\mathrm{R},\gamma}$ may be decomposed into the union of analytic eigencurves; see [**BucFreKen17**, §4.4.2] for details. Moreover, if $\lambda_k^{\mathrm{R},\gamma}$ is a simple eigenvalue of the Robin Laplacian $-\Delta^{\mathrm{R},\gamma}$ and U_k is the corresponding eigenfunction, then

$$\frac{\mathrm{d}}{\mathrm{d}\gamma}\lambda_k^{\mathrm{R},\gamma} = \frac{\|U_k|_M\|_{L^2(M)}^2}{\|U_k\|_{L^2(\Omega)}^2}.$$

Let us now consider the functions $\gamma_k : (-\infty, \lambda_k^{\mathrm{D}}) \to \mathbb{R}$ which are the inverses of $\lambda_k^{\mathrm{R},\gamma}$ viewed as functions of γ. These inverses are well-defined due to the strict monotonicity of the Robin eigenvalues established in Proposition 7.4.5. The functions $-\gamma_k(\Lambda)$ are continuous and strictly monotone decreasing for $\Lambda \in \left(-\infty, \lambda_k^{\mathrm{D}}\right)$ and satisfy

$$\lim_{\Lambda \to -\infty} (-\gamma_k(\Lambda)) = +\infty, \qquad \lim_{\Lambda \to (\lambda_k^{\mathrm{D}})^-} (-\gamma_k(\Lambda)) = -\infty.$$

Using Proposition 7.4.4, we can now explicitly find the spectrum of the Dirichlet-to-Neumann map $\mathcal{D}_\Lambda$ in terms of the functions $-\gamma_k(\Lambda)$.

Proposition 7.4.7 ([**AreMaz12**, Proposition 5]). Let $\Omega \subset \mathbb{R}^d$ be a Lipschitz domain, and let $\Lambda \in \mathbb{R}$. Choose $m \in \mathbb{N}$ such that $\lambda_{m-1}^{\mathrm{D}} \leq \Lambda < \lambda_m^{\mathrm{D}}$, where we assume the convention $\lambda_0^{\mathrm{D}} := -\infty$. Then

$$\mathrm{Spec}\,(\mathcal{D}_\Lambda) = \{-\gamma_k(\Lambda) : k \geq m\}.$$

Using Proposition 7.4.7 we immediately deduce

Theorem 7.4.8. Let $\Omega \subset \mathbb{R}^d$ be a Lipschitz domain. The eigenvalues $\sigma_k^\Lambda(\Omega)$ of the Dirichlet-to-Neumann map $\mathcal{D}_\Lambda$ are continuous and strictly monotone decreasing functions of Λ on each interval of the real line not containing the points of $\mathrm{Spec}\left(-\Delta_\Omega^D\right)$. As Λ approaches from below a Dirichlet eigenvalue λ^D of multiplicity m, the first m eigenvalues $\sigma_1^\Lambda, \ldots, \sigma_m^\Lambda$ of $\mathcal{D}_\Lambda$ tend to $-\infty$.

Remark 7.4.9. In the smooth case, Theorem 7.4.8 was first stated in [**Fri91**, Lemma 2.3]. Further results on the asymptotics of eigenvalues $\sigma_1^\Lambda, \ldots, \sigma_m^\Lambda$ as $\Lambda \to \left(\lambda^D\right)^-$ can be deduced from [**BelBBT18**]; see also [**GirKLP22**, §4.4].

Remark 7.4.10. As we already know that the eigenvalues of the Steklov problem (or the operator $\mathcal{D}_0$) are nonnegative, Theorem 7.4.8 immediately implies that

$$\sigma_k^\Lambda > 0 \qquad \text{for all } \Lambda < 0 \text{ and all } k \in \mathbb{N}.$$

The behaviour of the Dirichlet-to-Neumann eigenvalues σ_k^Λ as functions of Λ will also be discussed below in §7.4.3. We now concentrate on the analogue of Theorem 7.2.9 in order to compare the eigenvalues of the Dirichlet-to-Neumann map $\mathcal{D}_\Lambda$ with $\Lambda \leq 0$ with those of the boundary Laplace–Beltrami operator $-\Delta_M$. Namely, we state

Theorem 7.4.11 ([**GirKLP22**, Theorem 4.2]). *Let $\Omega \subset \mathbb{R}^d$ be a bounded domain with a smooth boundary $M = \partial\Omega$, and let σ_k^Λ and ν_k, $k \in \mathbb{N}$, be the eigenvalues of the Dirichlet-to-Neumann map $\mathcal{D}^\Lambda$ on Ω and of the Laplace–Beltrami operator on M, respectively. Then there exists a constant $C > 0$ such that*

$$(7.4.10) \qquad \left| \sigma_k^\Lambda - \sqrt{\nu_k - \Lambda} \right| \leq C$$

uniformly over all $\Lambda \in (-\infty, 0]$ and all $k \in \mathbb{N}$.

We note that in two dimensions, much more precise results are available as $k \to \infty$ [**LagStA21**]; cf. Remark 7.2.12 in the case $\Lambda = 0$.

The proof of Theorem 7.4.11 relies on the following generalisation of Hörmander's identity of Theorem 7.2.5.

Theorem 7.4.12 (The generalised Hörmander's identity [**GirKLP22**, Theorem 4.3]). *Let $\Omega \subset \mathbb{R}^d$ be a bounded domain with a smooth boundary $M = \partial\Omega$. Let $\mathbf{F}$ be a smooth vector field on $\overline{\Omega}$ which on the boundary of Ω coincides with the exterior unit normal, $\mathbf{F}|_M = \mathbf{n}$. Let $u \in H^1(M)$, let $\Lambda \leq 0$, and let $U = \mathcal{E}_\Lambda u$ be the unique Λ-Helmholtz extension of u onto Ω. Then*

$$(\mathcal{D}_\Lambda u, \mathcal{D}_\Lambda u)_{L^2(M)} - (-\Delta_M u, u)_{L^2(M)} + \Lambda (u, u)_{L^2(M)}$$

$$(7.4.11) \qquad = \int_\Omega \left(2\mathrm{Jac}_{\mathbf{F}}\left[\nabla U, \nabla U\right] - |\nabla U|^2 \operatorname{div} \mathbf{F} + \Lambda U^2 \operatorname{div} \mathbf{F} \right) \, \mathrm{dx}.$$

■ **Exercise 7.4.13.** Prove Theorem 7.4.12 by first showing that after replacing the harmonic extension $U = \mathcal{E}_0 u$ by the Λ-Helmholtz extension $U = \mathcal{E}_\Lambda u$ in Theorem 7.2.4 the formula (7.2.5) becomes

(7.4.12)
$$\int_M \langle \mathbf{F}, \nabla U \rangle \, \partial_n U \, \mathrm{d}s - \frac{1}{2} \int_M |\nabla U|^2 \langle \mathbf{F}, \mathbf{n} \rangle \, \mathrm{d}s$$
$$+ \frac{\Lambda}{2} \int_M u^2 \langle \mathbf{F}, \mathbf{n} \rangle \, \mathrm{d}s + \frac{1}{2} \int_\Omega |\nabla U|^2 \operatorname{div} F \, \mathrm{d}\mathbf{x}$$
$$- \int_\Omega \operatorname{Jac}_{\mathbf{F}} [\nabla U, \nabla U] \, \mathrm{d}\mathbf{x} - \frac{\Lambda}{2} \int_\Omega U^2 \operatorname{div} F \, \mathrm{d}\mathbf{x} = 0$$

(see [**HasSif20**, Theorem 3.1]) and then using (7.4.12) and repeating the arguments in the proof of Theorem 7.2.5, keeping track of Λ-dependent terms. See also Exercise 7.4.15 for further applications of (7.4.12).

Proof of Theorem 7.4.11. We first note that under the conditions of Theorem 7.4.12 there exists a constant $C > 0$ such that

(7.4.13) $\left| (\mathcal{D}_\Lambda u, \mathcal{D}_\Lambda u)_{L^2(M)} - ((-\Delta_M - \Lambda) u, u)_{L^2(M)} \right| \leq C \, (\mathcal{D}_\Lambda u, u)_{L^2(M)} .$

Indeed, taking the absolute value of the left-hand side of (7.4.11) gives the left-hand side of (7.4.13). Taking the absolute value of the right-hand side of (7.4.11) and estimating the first two terms as in Corollary 7.2.6 produces an upper bound $C \, (\nabla U, \nabla U)_{L^2(\Omega)}$ for them; the last term can be estimated as $C|\Lambda| \, (U, U)_{L^2(\Omega)}$ (possibly with a different constant C but also depending on $\mathbf{F}$ and the geometry of Ω only). Combining the two bounds and using $|\Lambda| = -\Lambda$, the total bound on the right-hand side becomes

$$C \left((\nabla U, \nabla U)_{L^2(\Omega)} - \Lambda \, (U, U)_{L^2(\Omega)} \right) = C \, (\mathcal{D}_\Lambda u, u)_{L^2(M)} ,$$

thus establishing (7.4.13). The bound (7.4.10) now follows from (7.4.13) by a direct application of Proposition 7.2.8 with $\mathcal{A} = \mathcal{D}_\Lambda$ and $\mathcal{B} = -\Delta_M - \Lambda$, which are both nonnegative for $\Lambda \leq 0$. □

We illustrate Theorem 7.4.11 in Figure 7.13.

Remark 7.4.14. The boundary regularity assumed in the conditions of Theorem 7.4.11 may be relaxed slightly to allow for $C^{1,1}$ boundary; cf. Remark 7.2.11. On the other hand, [**GirKLP22**, Proposition 4.6] shows that for curvilinear polygons the bound (7.4.10) *cannot* hold uniformly over all $k \in \mathbb{N}$ and $\Lambda \leq 0$ for *any* fixed choice of the sequence $\{\nu_k\}$. This observation is based on comparison of the asymptotics of the eigenvalues σ_k^Λ as $\Lambda \to -\infty$ imposed by (7.4.10) with the actual asymptotics for polygons

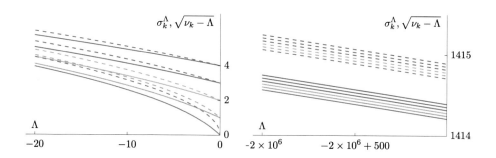

Figure 7.13. Some eigenvalues of the Dirichlet-to-Neumann map $\mathcal{D}_\Lambda$ for the unit disk as functions of Λ (solid curves) and, for comparison, the plots of $\sqrt{\nu_k - \Lambda}$ (dashed curves). In the left figure, $\Lambda \in [-20, 0]$, and k is chosen in the set $\{1, 3, 5, 7, 9\}$. In the right figure, $\Lambda \in [-2 \times 10^6, -2 \times 10^6 + 10^3]$, and k is chosen in the set $\{100, 102, 104, 106, 108\}$.

which can be obtained from the results of [**LevPar08, Kha18, KhaPan18, KhaOuBPan20, Pan20, Pop20**] on the asymptotics of Robin eigenvalues.

■ **Exercise 7.4.15.** The generalised Pohozhaev identity (7.4.12) for the Helmholtz equation has some further applications. Use it first to prove the classical *Rellich's identity* [**Rel40**]: if $\Omega \subset \mathbb{R}^d$ is a domain with a smooth boundary $M = \partial\Omega$ and if Λ, U are an eigenvalue and a corresponding normalised eigenfunction of $-\Delta_\Omega^D$, then

$$(7.4.14) \qquad 2\Lambda = \int_M \langle \mathbf{x}, \mathbf{n} \rangle \, (\partial_n U)^2 \, \mathrm{d}V_M.$$

Then use (7.4.12) and (7.4.14) to prove the following result of A. Hassell and T. Tao [**HasTao02**]: there exist constants $C_1, C_2 > 0$ such that for any eigenvalue Λ and a corresponding normalised eigenfunction U of the Dirichlet Laplacian $-\Delta_\Omega^D$ one has

$$C_1\Lambda \leq \|\partial_n U\|_{L^2(M)}^2 \leq C_2\Lambda.$$

In a similar manner, one can estimate the boundary norm $\|U\|_{L^2(M)}^2$ for a Neumann (or, more generally, Robin) eigenfunction in a domain Ω; see [**RudWigYes21**].

7.4.3. The Dirichlet-to-Neumann map and the eigenvalue counting functions. To make a full circle, we note that the Dirichlet-to-Neumann map is also useful for the study of Laplace eigenvalues. We will sketch Friedlander's original proof of the nonstrict version of the inequality (3.2.9) between the Neumann and Dirichlet eigenvalues of a Euclidean domain Ω which we stated in Theorem 3.2.35. Let $\mathcal{N}^D(\Lambda)$ and $\mathcal{N}^N(\Lambda)$ denote the usual counting functions of the Dirichlet and Neumann Laplacians on Ω, respectively, and let

$$n(\Lambda) := \mathcal{N}^{\mathcal{D}_\Lambda}(0) = \# \left\{ k \in \mathbb{N} : \sigma_k^\Lambda \le 0 \right\}$$

be the number of nonpositive eigenvalues of the Dirichlet-to-Neumann map $\mathcal{D}_\Lambda$. We have already established (Remark 7.4.10) that $n(\Lambda) = 0$ for $\Lambda < 0$. Note also that by the Robin–Dirichlet-to-Neumann duality, zero is an eigenvalue of $\mathcal{D}_\Lambda$ if and only if Λ is an eigenvalue of the Neumann Laplacian on Ω and that the multiplicities of these eigenvalues coincide. Thus, as a varying Λ passes through a Neumann eigenvalue of multiplicity m, exactly m eigenvalue curves σ_k^Λ cross from the upper half-plane to the lower one.

The key result of [**Fri91**] is the formula relating the three counting functions.

Lemma 7.4.16 ([**Fri91**, Lemma 1.2], [**AreMaz12**, Proposition 4]). *Let $\Omega \subset \mathbb{R}^d$ be a Lipschitz domain, and let $\Lambda \in \mathbb{R}$. Then*

$$\mathcal{N}^{\mathrm{N}}(\Lambda) - \mathcal{N}^{\mathrm{D}}(\Lambda) = n(\Lambda).$$

Proof. Since a Robin eigenvalue $\lambda_k^{\mathrm{R},\gamma}$ is strictly monotone increasing in the interval $\left[\lambda_k^{\mathrm{N}}, \lambda_k^{\mathrm{D}}\right)$ as γ increases from zero to $+\infty$, we have

$$I_\Lambda := \left\{ k \in \mathbb{N} : \lambda_k^{\mathrm{N}} \le \Lambda < \lambda_k^{\mathrm{D}} \right\}$$
$$= \left\{ k \in \mathbb{N} : \text{there exists } \gamma \ge 0 \text{ such that } \lambda_k^{\mathrm{R},\gamma} = \Lambda \right\}.$$

By the definition of the eigenvalue counting functions and the first expression for the set I_Λ, we have $\# I_\Lambda = \mathcal{N}^{\mathrm{N}}(\Lambda) - \mathcal{N}^{\mathrm{D}}(\Lambda)$. At the same time, by the Robin–Dirichlet-to-Neumann duality and the second expression for I_Λ, we have $\# I_\Lambda = n(\lambda)$, and the result follows. $\quad\square$

We further have

Lemma 7.4.17 ([**Fri91**, Lemma 1.3], [**AreMaz12**, Lemma 3.2]). *Let $\Omega \subset \mathbb{R}^d$ be a bounded domain with a Lipschitz boundary $M = \partial\Omega$, and let $\Lambda > 0$. Then $n(\Lambda) \ge 1$.*

Proof. Consider, as in the original proof of Theorem 3.2.35, a function $g = \mathrm{e}^{\mathrm{i}\langle \boldsymbol{\omega}, \mathbf{x} \rangle}$, where $\boldsymbol{\omega} \in \mathbb{R}^d$ and $|\boldsymbol{\omega}|^2 = \Lambda$. We have $-\Delta g - \Lambda g = 0$ in Ω and $\mathcal{D}_\Lambda\left(g|_M\right) = \mathrm{i}\, \langle \boldsymbol{\omega}, \mathbf{n} \rangle\, g|_M$. Thus,

$$(7.4.15) \qquad \left(\mathcal{D}_\Lambda\left(g|_M\right), g|_M\right)_{L^2(M)} = \mathrm{i} \int\limits_M \langle \boldsymbol{\omega}, \mathbf{n} \rangle\, \mathrm{d}V_M = 0$$

by the divergence theorem. On the other hand, assuming $n(\Lambda) = 0$ immediately implies $(\mathcal{D}_\Lambda u, u)_{L^2(M)} > 0$ for every $u \in H^{1/2}(M)$, thus contradicting (7.4.15). □

The proof of the nonstrict version $\lambda_{k+1}^{\mathrm{N}} \leq \lambda_k^{\mathrm{D}}$ of (3.2.9) now follows immediately from Lemmas 7.4.16 and 7.4.17: assuming that it is false and choosing any $\Lambda_0 \in \left(\lambda_k^{\mathrm{D}}, \lambda_{k+1}^{\mathrm{N}}\right)$, we have $\mathcal{N}^{\mathrm{N}}(\Lambda_0) = \mathcal{N}^{\mathrm{D}}(\Lambda_0)$, and so $n(\Lambda_0) = 0$ by Lemma 7.4.16, thus contradicting Lemma 7.4.17. The proof of the strict version can be achieved with minor modifications of this argument; see [**AreMaz12**, Theorem 3.3].

Spectral geometry of the Steklov problem and the Dirichlet-to-Neumann map is an actively developing subject, and many interesting questions remain beyond the scope of this chapter. For further reading we refer to survey papers [**GirPol17**], [**ColGGS22**].

A short tutorial on numerical spectral geometry

After a brief overview of the Finite Element Method, we give a hands-on tutorial on solving numerically some of the spectral problems presented in this book using `Mathematica` *and* `FreeFEM`.

Boris
Grigoryevich
Galerkin
(1871–1945)

Ivo
Babuška
(1926–2023)

A.1. Overview

A.1.1. The Finite Element Method.
The aim of this short tutorial is to provide the readers (who may be unfamiliar with numerical analysis or any aspects of computer programming) a direct route to practical calculation of eigenvalues of some of the problems considered in this book. To this end, we neither pretend to give a comprehensive survey of numerical spectral theory nor keep the presentation rigorous, concentrating instead on the practicalities of the *Finite Element Method* (FEM) in its most basic form and in dimension two only and ignoring numerous other available techniques (the finite differences, the method of fundamental solutions, spectral

methods, the boundary element method, to name just a few). For a comprehensive survey of both theoretical and practical foundations of FEM applied to spectral problems see [**SunZho17**].

The Finite Element Method is based on the *Galerkin* (also called the Ritz–Galerkin) method of solving a weak eigenvalue problem (3.1.2); we suppose that all the assumptions made in §3.1.1 about the bilinear form $\mathcal{Q}$ are fulfilled.

Let $V \subset U$ be a finite-dimensional subspace of $U = \mathrm{Dom}\, \mathcal{Q}$. We consider the restriction of (3.1.2) to V; namely, we want to find $\lambda \in \mathbb{R}$ and $u \in V \setminus \{0\}$ such that

$$(\text{A.1.1}) \qquad \mathcal{Q}[u,v] = \lambda \mathcal{B}[u,v] \qquad \text{for all } v \in V,$$

where we set

$$(\text{A.1.2}) \qquad \mathcal{B}[u,v] := (u,v)_{\mathcal{H}}\,.$$

If the subspace V approximates well the span of some eigenfunctions of $\mathcal{Q}$, we expect that the eigenvalues of (A.1.1) will approximate well the corresponding eigenvalues of (3.1.2). One usually studies a family of approximating subspaces V_h depending on a real parameter $h > 0$ in such a way that the projector $U \to V_h$ converges to the identity map as $h \to 0$. Then various estimates of convergence of eigenvalues and eigenfunctions are available. In particular, if λ_k is a simple eigenvalue of (3.1.2) with the corresponding eigenfunction u and if $\lambda_{k,h}$ is the kth eigenvalue of (A.1.1) with $V = V_h$, then with some constant C independent of h we have

$$\lambda_k \le \lambda_{k,h} \le \lambda_k + C \inf_{v \in V_h} \|u - v\|_U^2,$$

where $\|\cdot\|_U$ is the norm induced by (3.1.1); see [**SunZho17**, §1.4.3] and [**BabOsb91**, §8].

Suppose now that $\{v_1, \ldots, v_m\}$ is a basis in V, not necessarily an orthogonal one. Looking for an eigenvector of (A.1.1) in the form $u = \sum_{j=1}^{m} c_j v_j$ with unknown constants c_j, $j = 1, \ldots, m$, and taking $v = v_k$, $k = 1, \ldots, m$, we rewrite (A.1.1) as a generalised matrix eigenvalue problem

$$(\text{A.1.3}) \qquad \mathsf{S}\mathbf{c} = \lambda \mathsf{M}\mathbf{c}, \qquad \mathbf{c} = \begin{pmatrix} c_1 \\ \vdots \\ c_m \end{pmatrix} \in \mathbb{R}^m,$$

where

$$(\text{A.1.4}) \qquad \mathsf{S} := \left(\mathcal{Q}[v_j, v_k] \right)_{k,j=1,\ldots,m}$$

is the so-called *stiffness matrix* and

$$(\text{A.1.5}) \qquad \mathsf{M} := \left(\mathcal{B}[v_j, v_k] \right)_{k,j=1,\ldots,m}$$

is called the *mass matrix*. We now solve the eigenvalue problem (A.1.3) using some numerical linear algebra method.

The Finite Element Method (specifically, in application to spectral problems for the Laplacian in a bounded domain $\Omega \subset \mathbb{R}^2$, and in its simplest form) is usually understood as a particular realisation of the Galerkin method subject to the following conditions:

(a) $\overline{\Omega}$ is represented (or approximated) by a union $\mathcal{T}_h$ of closed triangles, called a *mesh*, where a real parameter h provides an upper bound on the diameter (or some other linear size) of each $T \in \mathcal{T}_h$. The different triangles may only have a common side or a common vertex; see Figure A.1. If Ω is not a polygon, approximating it by a union of triangles obviously introduces some additional errors. There are many alternative choices to triangles, such as quadrilaterals or curvilinear elements, which we do not discuss.

Figure A.1. Examples of automatically constructed meshes for a disk and a domain with a hole.

(b) Let $T \in \mathcal{T}_h$ be a triangle in the chosen mesh, and let $\mathcal{P}_k = \mathcal{P}_k(T)$ be the subspace of all polynomials in two variables of degree at most k. Then $\dim \mathcal{P}_k = \frac{1}{2}(k+1)(k+2) =: s_k$. We choose s_k points $z_1, \ldots, z_{s_k} \in T$, called *nodes*, which lie on $k+1$ straight lines. In particular when $k = 1$ we have $s_1 = 3$ and choose the nodes at the vertices of the triangle, and when $k = 2$ we have $s_2 = 6$ and choose additionally the nodes at the midpoints of the sides.

For a polynomial $p \in \mathcal{P}_k(T)$, the set of functionals $\mathcal{N} := \{\mathcal{N}_j : p \mapsto p(z_j), j = 1, \ldots, s_k\}$ is the set of *degrees of freedom* which is unisolvent; knowing $\mathcal{N}(p) := \{\mathcal{N}_j(p), j = 1, \ldots, s_k\}$ uniquely determines p. In principle, this choice of degrees of freedom is just a specific realisation of the general principle of using any unisolvent set of functionals $\mathcal{N}$.

(c) We can now choose a local basis $\{p_1, \ldots, p_{s_k}\}$ in $\mathcal{P}_k(T)$ by requiring $p_i(z_j) = \mathcal{N}_j(p_i) = \delta_{ij}, \; i, j = 1, \ldots, s_k$. Finally, we set V to be the space of continuous functions on Ω whose restrictions to each $T \in \mathcal{T}_h$ coincides with $\mathcal{P}_k(T)$; thus, the elements of V are piecewise polynomials. Note that continuity is required to ensure $V \subset H^1(\Omega)$ (thus providing the so-called *conforming* finite elements). Such basis functions are called *Lagrangian* finite elements.

Remark A.1.1. The term *finite elements* is variably applied to the whole method, a choice of mesh subdomains (e.g., triangular or quadrilateral finite elements), or a choice of local basis functions (e.g., linear or quadratic conforming finite elements or some other nonconforming finite elements).

A.1.2. Solving spectral problems with `Mathematica`. There are a large number of software packages, either commercial or free to use, which implement the FEM for solving partial differential equations including spectral problems. For an up-to-date review see the corresponding Wikipedia page. In particular, widely available commercial packages `Matlab`[14] (with **PDE Toolbox**[15]) and `Mathematica`[16] (starting from version 10.2) allow one to compute eigenvalues and eigenfunctions of various boundary value problems with relative ease. `Mathematica` is particularly easy to use as it provides two commands, `DEigenvalues`[17] and `NDEigenvalues`[18] for calculating the eigenvalues of a boundary value problem analytically (if possible) and numerically, respectively. The numerical version effectively "hides" all the FEM machinery from the user. The version `NDEigensystem`[19] allows additionally to compute the eigenfunctions.

We do not intend to give any further details of `Mathematica` commands, restricting ourselves to several examples below.

Remark A.1.2. All the scripts listed or discussed in this appendix are available for download; see §A.3.

---

[14] https://www.mathworks.com/products/matlab.html
[15] https://www.mathworks.com/products/pde.html
[16] https://www.wolfram.com/mathematica
[17] https://reference.wolfram.com/language/ref/DEigenvalues.html
[18] https://reference.wolfram.com/language/ref/NDEigenvalues.html
[19] https://reference.wolfram.com/language/ref/NDEigensystem.html

Listing A.1 gives some examples of using **Mathematica** for finding eigenvalues and eigenfunctions *analytically*.[20]

Listing A.1. Finding eigenvalues analytically with **Mathematica**.

```
1 (* Neumann eigenvalues for the unit square *)
2 DEigenvalues[-Laplacian[u[x, y], {x, y}],  u[x, y], {x, y} ∈
     Rectangle[{0, 0}, {1, 1}], 10]
3 (* Dirichlet eigenvalues for the unit disk *)
4 DEigenvalues[{-Laplacian[u[x, y], {x, y}], DirichletCondition[u[x,
     y] == 0, True]},  u[x, y], {x, y} ∈ Disk[], 10]
5 (* Dirichlet eigenvalues for  the isosceles right triangle with
     sides \[Pi] *)
6 DEigenvalues[{-Laplacian[u[x, y], {x, y}],  DirichletCondition[u[x,
     y] == 0, True]},  u[x, y], {x, y} ∈ Triangle[{{0, 0}, {Pi, 0},
     {0, Pi}}], 10]
```

Our main geometric example throughout this tutorial will be the domain

$$\Omega = \Omega' \setminus B,$$

(A.1.6)
$$\Omega' = \left\{ (x,y) : 0 < x < \pi, 0 < y < \pi + x \left(1 - \frac{x}{\pi} \right) \right\},$$
$$B = B_{\left(\frac{\pi}{3}, \frac{\pi}{2} \right), \frac{\pi}{4}};$$

see Figure A.2.

Listing A.2 shows how to compute the first ten Dirichlet and Neumann eigenvalues of Ω with **Mathematica**.

Listing A.2. Computing Dirichlet and Neumann eigenvalues of Ω with **Mathematica**.

```
1 Ω = RegionDifference[ImplicitRegion[0 < x < Pi && 0 < y < Pi + x (
     Pi - x)/Pi, {x, y}], Disk[{Pi/3, Pi/2}, Pi/4]];
2 (* numerical Neumann eigenvalues  for Ω *)
3 NDEigenvalues[-Laplacian[u[x, y], {x, y}], u[x, y], {x, y} ∈ Ω ,
     10]
4 (* numerical Dirichlet eigenvalues for Ω *)
5 NDEigenvalues[{-Laplacian[u[x, y], {x, y}], DirichletCondition[u[x,
     y] == 0, True]}, u[x, y], {x, y} ∈ Ω, 10]
```

---

[20]Listings A.1– A.5 may also be copy-pasted into **Mathematica**.

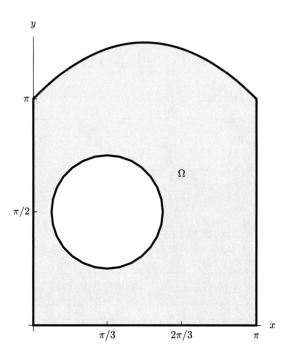

Figure A.2. Domain Ω given by (A.1.6).

Further on, Listings A.3 and A.4 demonstrate how to compute the Robin and Zaremba eigenvalues and eigenfunctions, respectively. The graphical outputs of these scripts are shown in Figures A.3 and A.4. (The actual graphical outputs from these scripts have been slightly edited for presentation purposes.)

Listing A.3. Computing Robin eigenvalues and eigenfunctions of Ω with **Mathematica**.

```
1  Ω = RegionDifference[ImplicitRegion[0 < x < Pi && 0 < y < Pi + x (
       Pi - x)/Pi, {x, y}], Disk[{Pi/3, Pi/2}, Pi/4]];
2  (* numerical Robin (∂ₙu = γu) eigenvalues and contour plots of
       eigenfunctions for Ω *)
3  γ = 2;
4  {eval, efun} = NDEigensystem[-Laplacian[u[x, y], {x, y}] +
       NeumannValue[γ u[x, y], True], u[x, y], {x, y} ∈ Ω , 6];
5  GraphicsGrid[Table[ContourPlot[efun[[3 (i - 1) + j]], {x, 0, Pi}, {
       y, 0, 5 Pi/4}, Contours -> {-0.5, -0.25, 0, 0.25, 0.5},
       RegionFunction -> Function[{x, y, z}, {x, y} ∈ Ω ], Frame ->
       False, AspectRatio -> Automatic, BoundaryStyle -> Thick,
       PlotLabel -> "λ=" <> ToString[eval[[3 (i - 1) + j]]]], {i, 1,
       2}, {j, 1, 3}]]
```

Listing A.4. Computing Zaremba eigenvalues and eigenfunctions of Ω with `Mathematica`

```
1  Ω = RegionDifference[ImplicitRegion[0 < x < Pi && 0 < y < Pi + x (
       Pi - x)/Pi, {x, y}], Disk[{Pi/3, Pi/2}, Pi/4]];
2  (* numerical Zaremba eigenvalues and eigenfunctions for Ω *)
3  (* Dirichlet condition on all sides except Neumann on the curved
       upper side *)
4  {eval, efun} = NDEigensystem[{-Laplacian[u[x, y], {x, y}],
       DirichletCondition[u[x, y] == 0, y <= Pi]}, u[x, y], {x, y} ∈ Ω
       , 6];
5  GraphicsGrid[Table[Plot3D[efun[[3 (i - 1) + j]], {x, 0, Pi}, {y, 0,
       5 Pi/4}, RegionFunction -> Function[{x, y, z}, {x, y} ∈ Ω],
       BoundaryStyle -> Thick, Boxed -> False, Axes -> True,
       AxesOrigin -> {0, 0, 0}, AspectRatio -> Automatic, Ticks ->
       None, PlotLabel -> "λ=" <> ToString[eval[[3 (i - 1) + j]]]], {i
       , 1, 2}, {j, 1, 3}]]
```

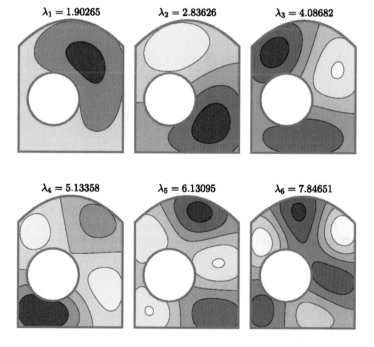

Figure A.3. Contour plots of Robin eigenfunctions ($\gamma = 2$) of Ω given by (A.1.6).

To conclude the `Mathematica` part of our tutorial, we verify in Listing A.5 the Faber–Krahn inequality for regular n-gons P_n for $n = 5, \ldots, 20$.

$\lambda_1 = 1.41388$ $\lambda_2 = 4.1881$ $\lambda_3 = 5.86515$

$\lambda_4 = 7.36363$ $\lambda_5 = 9.33443$ $\lambda_6 = 10.4951$

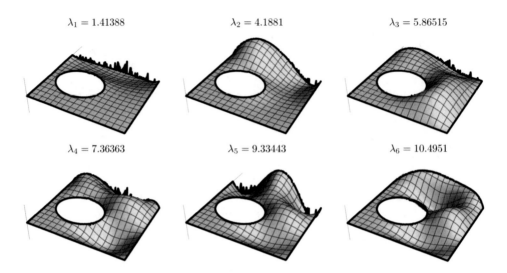

Figure A.4. Height plots of Zaremba eigenfunctions of Ω given by (A.1.6), with the Neumann condition imposed on the curved part of the outer boundary, and the Dirichlet condition elsewhere. Note some spurious oscillations introduced by the numerics.

We additionally compare our numerical result with the asymptotics [**BerGMR21**]

$$(A.1.7) \quad \frac{\lambda_1^{\mathrm{D}}(P_n)}{\lambda_1^{\mathrm{D}}(P_n^*)} = 1 + \frac{4\zeta(3)}{n^3} + \frac{\left(12 - 2j_{0,1}^2\right)\zeta(5)}{n^5} + O(n^{-6}) \qquad \text{as } n \to \infty,$$

where P_n^* is the symmetric rearrangement of P_n and $\zeta(\cdot)$ is the Riemann zeta function. The graphical output from this script (once more, slightly edited for presentation purposes) is shown in Figure A.5.

Listing A.5. Verifying the Faber–Krahn inequality for regular polygons with **Mathematica**

```
1 (* verifying the Faber--Krahn inequality for regular n-gons, n
     =5,...,20 *)
2 evs = Table[NDEigenvalues[{-Laplacian[u[x, y], {x, y}],
     DirichletCondition[u[x, y] == 0, True]}, u[x, y], {x, y} ∈
     RegularPolygon[n], 1][[1]], {n, 5, 20}];
3 asympt = 1 + 4 Zeta[3]/n^3 + (12 - 2 BesselJZero[0, 1]^2) Zeta[5]/n
     ^5;
4 Show[ListPlot[Table[{n, evs[[n - 4]]} /(Pi/Area[RegularPolygon[n]]
     BesselJZero[0, 1]^2)}, {n, 5, 20}], PlotStyle -> {Black,
     PointSize[Large]}, AxesOrigin -> {4, 1}], Plot[asympt, {n, 5,
     20}, PlotRange -> All]]
```

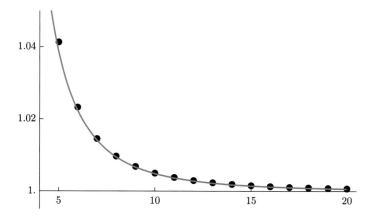

Figure A.5. The ratio $\lambda_1^{\mathrm{D}}(P_n)/\lambda_1^{\mathrm{D}}(P_n^*)$ for regular polygons P_n plotted as a function of n (black dots) and compared to asymptotics (A.1.7).

A.2. Learning FreeFEM by example

A.2.1. The basics. For the rest of this tutorial, we will concentrate on describing the FEM package **FreeFEM**; see [**Hec12**] and the product website https://freefem.org/. As the name suggests, the package is freely available for download. It is powerful enough to cover most of the problems considered in this book, within the usual limitations of the finite element method — for example, one should not expect to perform a reliable computation of sufficiently *large* eigenvalues of any problem. At the same time, it is easy enough to learn very quickly without any prior knowledge of programming or numerical analysis.

Giving a full description of **FreeFEM** is well outside the scope of this tutorial. One should also consult the package documentation for installation instructions and additional details. We will instead show, starting in the next subsection, various examples which should allow the reader to produce their own scripts by mimicking ours.

The general flow of working with **FreeFEM** is somewhat similar to that of LATEX: one creates a **FreeFEM** script (a text file with extension .edp) in an appropriate editing programme; one then executes **FreeFEM** (many editors allow doing so directly from the editing window); one then corrects any script errors reported and then repeats the process until everything works as intended.

A.2.2. The structure of a FreeFEM script: the Neumann problem in a rectangle. The standard structure of a **FreeFEM** script used in spectral problems is more or less the same and roughly complies with the following pattern:

A. Declarations: all user's variables (identifiers) should be declared (and possibly assigned values to) before or during their first appearance.

B. Boundary description.

C. Mesh creation.

D. Choice of FEM basis functions.

E. Description of quadratic forms $\mathcal{Q}$ and $\mathcal{B}$ for (A.1.1).

F. Creation of matrices S and M for (A.1.3).

G. Solving (A.1.3).

H. Results output and/or visualisation.

Listing A.6 shows the script which computes the Neumann eigenvalues for a rectangle $(0, L\pi) \times (0, \pi)$ (with $L = 1$ as shown), and we will go through it in detail to illustrate the realisation of each of the steps A–H in the general scheme.

Listing A.6. The Neumann Laplacian in a rectangle.

```
1  / This FreeFEM script computes eigenvalues of the Neumann Laplacian
       in the rectangle [0.,L pi]x[0, pi].
2  //
3  //---- A. DECLARATIONS ----
4  // IMPORTANT - every command ends with the semicolon!
5  // IMPORTANT - all variables must be declared before(or during)
       first use
6  //
7  real L=1.;      // declare variable L to be real and assign value
       1.0 to it; decimal point indicates it is not integer
8  int npoints=30; // declare variable npoints to be integer and
       assign value 30 to it
9  int N=50; // declare variable N  to be integer and assign value 50
       to it
10 real[int] Evalues(N); // an array of N real numbers parametrised by
       integers; in FreeFem, indices of an array of length N run from
       0 to N-1
11 real t; // real uninitialised variable used later
12 //---- END OF DECLARATIONS so far ----
13 // some more to come later
14 //
15 //---- B. BOUNDARY DEFINITIONS ----
16 // just use parametric curve definitions going in COUNTERCLOCKWISE
       direction; (t=t0, t1) in lines below means that t changes from
       t0 to t1 for this piece
17 //
18 border d1Omega(t=0, L) { x = pi*t; y = 0; label=1; };
19 border d2Omega(t=0, 1) { x = pi*L; y = pi*t; label=1; };
20 border d3Omega(t=L, 0) { x = pi*t; y = pi; label=1; };
21 border d4Omega(t=pi, 0) { x = 0; y = t; label=1; };
```

```
22 // "label=1;" part can be omitted in lines above but will be useful
       in other examples
23 //---- END OF BOUNDARY DEFINITIONS ----
24 //
25 //---- C. MESH CREATION USING buildmesh COMMAND ----
26 // format of the argument: piece_defined_by_border(
       number_of_mesh_points)+...
27 //
28 mesh Th=buildmesh(d1Omega(npoints*L*pi)+d2Omega(npoints*pi)+d3Omega
       (npoints*L*pi)+d4Omega(npoints*pi)); //the number of mesh
       points per side need not be proportional to side length and may
       not be integer (it's rounded down)
29 //---- END OF MESH DEFINITIONS ----
30 //
31 //---- VISUALISE THE MESH, may be commented out
32 //
33 plot(Th,wait=1);
34 //
35 //---- D. DECLARE THE FEM SPACE AND FEM VARIABLES ----
36 //
37 fespace Vh(Th,P2);
38 Vh u,v;
39 //---- END OF FEM SPACE DEFINITIONS ----
40 //
41 //---- E. DEFINE THE QUADRATIC FORMS ----
42 varf q(u,v)=int2d(Th)( dx(u)*dx(v) + dy(u)*dy(v) );
43 varf b(u,v)=int2d(Th)(u*v);
44 //---- END OF QUADRATIC FORMS DEFINITIONS ----
45 //
46 //----   F. CREATE THE MATRICES ----
47 matrix S=q(Vh,Vh);
48 matrix M=b(Vh,Vh);
49 //---- END OF MATRIX CREATION ----
50 //
51 //----   DECLARE THE ARRAY TO HOLD EIGENFUNCTIONS ----
52 Vh[int] Efunctions(N);
53 //
54 //---- G. SOLVE THE PROBLEM ----
55 int k=EigenValue(S,M,sym=true,value=Evalues,vector=Efunctions);
56 //
57 //---- END OF SOLVER ----
58 //---- H. PRINT THE EIGENVALUES ----
59 cout << "We asked for " << N << " eigenvalues and computed " << k
       << " eigenvalues:\n" << Evalues;
60 //
61 //---- PLOT THE 6th EIGENFUNCTION ----
62 plot(Efunctions[5]);
63 // press '?' on the image to see options for graphics
64 //
65 //---- END ----
```

The first eight lines of the script are just the comments; in fact every line starting with the double slash (or any text at the end of a line after a double slash) is ignored by **FreeFEM**and is there just for the ease of reading the script. By the way, empty lines and spaces are also ignored.

Group A of commands, in lines 7–11, contains some declarations. Let us look at them line by line, ignoring the comments.

The line

```
7 real L=1.;
```

declares a variable L to be real and assigns value 1.0 to it. Variable names can be of arbitrary length and consist of upper- and lowercase letters, numbers, and underscore and start with a letter. One can declare several variables at once, not necessarily assigning any values to them; for example one can have

```
real L1, L2=0.5, L3;
```

to define three real variables L1 (unassigned), L2 (with the value 0.5), and L3 (unassigned).

Remark A.2.1. It is very important to remember that *every* individual command should end with the semicolon!

The line

```
8 int npoints=30;
```

declares a variable npoints to be integer and assigns value 30 to it; this variable will be used later.

The line

```
9 int N=50;
```

declares a variable N to be integer and assigns value 50 to it. This variable will denote the number of eigenvalues we want to compute.

The line

```
10 real[int] Evalues(N);
```

declares Evalues to be an array of real numbers of length N indexed by integers from 0 to N−1, which will eventually hold the eigenvalues. Note that interchanging lines 10 and 9 would give an error — we cannot declare an array until we know its size.

The last declaration in line

```
11 real t;
```

declares t as another real variable, left unassigned.

Group B of commands, describing the boundary, is in lines 18–21. The boundary should be defined as a collection of smooth parametrised curves swept in such a way that the domain lies to the *left* of the direction of parametrisation; since in this case we have a simply connected domain, we parametrise in the counterclockwise direction. A definition of a boundary piece usually takes the form

```
border border_name(t=t0,t1) {x=a_function(t); y = another_function(
    t); label=natural_number;}
```

to define a parametric curve. Note that we can have `t1`<`t0` as in lines 20 and 21. Note also that parametrisation parameters are of course a matter of choice; compare lines 19 and 21. The part "`label=...;`" in lines 18–21 is optional — but labels are important if we want to integrate over the boundary, or impose different boundary conditions on different boundary pieces, allowing us to group them together, as we will do later.

The meshing is done in Group C consisting of one line, line 28. Once all the boundary pieces are defined, we create the mesh by executing `buildmesh` command in and assigning the output to variable `Th` declared to be a `mesh`. The general format of `buildmesh` command is

```
buildmesh(boundary1(points1)+...+boundaryX(pointsX));
```

where each `boundary1`, ..., `boundaryX` has been previously defined as a `border` and `points1`, ..., `pointsX` indicate how many mesh points to place on each border, thus determining mesh coarseness. We have used a previously defined variable `npoints` to indicate the number of boundary points per unit length of the boundary, but such a choice is not compulsory, albeit convenient.

We now proceed to describing the finite element space in Group D of commands. The line

```
37 fespace Vh(Th,P2);
```

defines the FEM space `Vh` on the mesh `Th` consisting of Lagrangian quadratic finite elements, as indicated by parameter `P2`. We may have chosen instead Lagrangian linear finite elements (replace `P2` with `P1`) or many other types of finite elements described in the **FreeFEM** manual. Essentially this command introduces the new type `Vh`, and the following command in line 38 declares `u` and `v` to be variables of that type.

Group E, consisting of two commands

```
42 varf q(u,v)=int2d(Th)( dx(u)*dx(v) + dy(u)*dy(v) );
43 varf b(u,v)=int2d(Th)(u*v);
```

defines the quadratic forms (varf)

$$\mathcal{Q}[u,v] = \int_\Omega \left((\partial_x u)(\partial_x v) + (\partial_y u)(\partial_y v) \right) \, \mathrm{d}x\mathrm{d}y,$$

$$\mathcal{B}[u,v] = \int_\Omega uv \, \mathrm{d}x\mathrm{d}y$$

in accordance with (3.1.3) and (A.1.2), where we use built-in two-dimensional integration command int2d and differentiation commands dx and dy.

We now create, in group F of commands in lines 47–48, the matrices S and M associated with the quadratic forms; see (A.1.4) and (A.1.5).

We now proceed to group G of actual computations, first declaring the array Efunctions (of type Vh and length N, parametrised by integers) in line 52 and then solving the problem in line 55. The parameter sym=true to EigenValue indicates that the problem is symmetric (in principle, **FreeFEM** is capable of solving nonselfadjoint problems as well). The output parameter k of EigenValue gives the number of eigenvalues actually computed; in most cases it will coincide with the requested number of eigenvalues N, that is, the length of the output array Evalues declared earlier in line 10. If one is not interested in eigenfunctions but only in eigenvalues, line 52 and the input vector=Efunctions may be omitted.

Finally, group H provides the output: first, the number of eigenvalues computed and the eigenvalues themselves are printed to the standard output (that is, the screen) cout in line 59 (see the **FreeFEM** manual for details on output to a file), and then the contour plot of the sixth eigenfunction is plotted in line 62; press '?' on the plot for help on changing its appearance.

Everything going to plan, one should see, after executing the script, output similar to

```
  --  mesh:  Nb of Triangles =   20768, Nb of Vertices 10573
Real symmetric eigenvalue problem: A*x - B*x*lambda
We asked for 50 eigenvalues and computed 50 eigenvalues:
50
  -7.779660247e-15  1.000000001  1.000000001  2.000000011  4.00000009
  4.000000092  5.000000167  5.00000017   8.000000683  9.000001004
  9.000001013  10.00000134  10.00000137  13.00000296  13.00000304
  16.00000569  16.00000585  17.00000663  17.00000679  18.00000783
  20.00001082  20.00001104  25.00002128  25.00002163  25.00002211
  25.00002219  26.00002377  26.00002461  29.00003341  29.00003427
  32.00004473  34.0000529   34.00005502  36.00006399  36.00006524
  37.00006881  37.00006912  40.00008553  40.00008746  41.00009245
  41.00009489  45.00012209  45.00012669  49.00016273  49.00016362
  50.00016672  50.00016833  50.00017239  52.00019103  52.00019161
times: compile 0.011743s, execution 4.45535s,  mpirank:0
```

```
######## ...
Ok: Normal End
```

One can see that the accuracy of **FreeFEM** is, at least in this case, very reasonable!

🔲 **Numerical Exercise A.2.2.** Experiment with modifying the script from Listing A.6: vary the parameter L, the number of mesh points per unit length of the boundary npoints, the requested number of eigenvalues N, and change the type of finite element from P2 to P1, in various combinations, to see how these modifications affect computational accuracy and time.

A.2.3. Curvilinear boundaries and holes. We now discuss, first, how to modify the **FreeFEM** script from Listing A.6 in order to compute the Neumann eigenvalues of the domain Ω' given by (A.1.6). To do so, we need to change the definition of the boundary piece d3Omega in line 20 to

```
20 border d3Omega(t=pi*L, 0) { x = t; y = pi+t*(pi-t)/pi; label=1; };
```

As the boundary piece d3Omega is now slightly longer, we may additionally increase the number of mesh points on this piece by using ...+d3Omega(1.2* npoints*L*pi)+... in line 28.

Secondly, to incorporate additionally the circular hole and thus consider the domain Ω defined by (A.1.6), we add to group B the command

```
border d5Omega(t=0, 2*pi) { x = pi/3 + (pi/4)*cos(t); y = pi/2 - (
    pi/4)*sin(t); label=1;}
```

Note that in order to keep the domain to the left of this part of the boundary as t changes from 0 to 2π we parametrise the circle *clockwise*. We now change the mesh creation command to

```
mesh Th=buildmesh(d1Omega(npoints*L*pi)+d2Omega(npoints*pi)+d3Omega
    (1.2*npoints*L*pi)+d4Omega(npoints*pi)+d5Omega(2*pi*pi/4*
    npoints));
```

Sample scripts may be downloaded following the links in §A.3.

A.2.4. The Dirichlet, Zaremba, and Robin problems. The Dirichlet conditions are imposed at the stage of defining the quadratic form varf q: if all the boundary pieces have the same label label=1; and then changing the definition of q to

```
varf q(u,v)=int2d(Th)( dx(u)*dx(v) + dy(u)*dy(v)) + on(1,u=0);
```

will impose the Dirichlet conditions on the whole boundary. The general format of the on command is +on(some_label, u=0) or +on(some_label, another_label , ..., last_label, u=0), the latter version imposing Dirichlet conditions on boundaries with all the labels listed. It is important to note (and to re-member) that the Dirichlet boundary conditions are imposed only in the definition of `varf` q and not of `varf` b and only on the first variable of the quadratic form, in our case u.

For Zaremba problem, we use the same approach but we have to change the labels of the boundary pieces where we *do not* want to impose the Dirichlet condition to something else, say `label=2;`, and keep `on(1, u=0)` in the form definition.

The Robin boundary conditions are also imposed by modifying the quadratic form `varf` q according to (3.1.16); to impose this condition with $\gamma = 2$, say, we change the definition of q to

```
real gamma = 2.;
varf q(u,v)=int2d(Th)( dx(u)*dx(v) + dy(u)*dy(v)) + int1d(Th,1)(
    gamma*u*v);
```

It is important to know that the factor γ should appear inside the integral according to **FreeFEM** syntax.

For the sample scripts, see §A.3.

A.2.5. The Laplace–Beltrami operator on manifolds. FreeFEM can additionally handle periodic boundary conditions, which allows us to solve some problems on Riemannian manifolds. All the examples in this sub-section can also be done analytically (thus allowing an easy control on the accuracy of the numerics) but we encourage the reader to modify them fur-ther in order to create more interesting examples; see also [**LevStr21**] for an illustration of the use of **FreeFEM** in computations of eigenvalues and resonances on hyperbolic manifolds.

We start from our basic script for a Neumann problem in a rectangle (Listing A.6) and relabel the sides individually by replacing the lines 18–21 by

```
18  border d1Omega(t=0, L) { x = pi*t; y = 0; label=1; };
19  border d2Omega(t=0, 1) { x = pi*L; y = pi*t; label=2; };
20  border d3Omega(t=L, 0) { x = pi*t; y = pi; label=3; };
21  border d4Omega(t=pi, 0) { x = 0; y = t; label=4; };
```

We now want to identify the sides labelled 2 and 4, thus turning the problem into the one on a flat cylinder. This is achieved at the stage of declaring the FEM space, using **FreeFEM** command `periodic`, by replacing the original line 37 with

```
37  fespace Vh(Th,P2, periodic=[[4,y],[2,y]]);
```

which basically tells **FreeFEM** to identify the value of y on sides 4 and 2.

To solve instead the spectral problem on the flat torus, we additionally have to identify sides 1 and 3 by replacing line 37 with

```
37  fespace Vh(Th,P2, periodic=[[4,y],[2,y],[1,x],[3,x]]);
```

If we instead identify sides 4 and 2 by the mapping $y \mapsto \pi - y$ as in

```
37  fespace Vh(Th,P2, periodic=[[4,y],[2,pi-y]]);
```

we solve the Neumann problem on the Möbius strip.

We note that the boundary pieces identified by `periodic` command need not be either parallel or straight or even have the same length (but must have the same number of boundary mesh points).

The sample scripts are listed in §A.3. We remark that **Mathematica** is also able to handle periodic boundary conditions via `PeriodicBoundaryCondition` [21] command; see a sample script.

We finish this subsection by showing how to compute, in **FreeFEM**, the eigenvalues of the Laplace–Beltrami operator on the sphere $\mathbb{S}^2$. This may be done in several ways; in order to simplify the calculations we will use, first of all, a symmetry trick from §3.2.2 and decompose the spectrum into the union of spectra of the two problems on the hemisphere, one with the Neumann condition imposed on the boundary and another one with the Dirichlet condition. We now use the stereographic projection of the hemisphere onto the unit disk arriving at the Dirichlet and Neumann problems for

(A.2.1) $$-\Delta u = c(x,y)\lambda u,$$

where the conformal factor c is given by

$$c(x,y) = \frac{4}{(1 + x^2 + y^2)^2}$$

and the Laplacian in the left-hand side of (A.2.1) is the usual Cartesian one. Thus, when formulating the corresponding weak problems, we need to replace the quadratic form (A.1.2) with

$$\mathcal{B}[u,v] := \int_{\mathbb{D}} c(x,y)u(x,y)v(x,y)\,\mathrm{d}x\mathrm{d}y.$$

The final trick, in order to solve two problems simultaneously, is to solve (A.2.1) in the disjoint union of two unit disks centred at $(0,\pm2)$, with the Dirichlet condition imposed on one of the circles, and to adjust the conformal factor to

$$c(x,y) = 4\left(1 + x^2 + \left(y - 2\frac{y}{|y|}\right)^2\right)^{-2}.$$

---

[21] https://reference.wolfram.com/language/ref/PeriodicBoundaryCondition.html

The resulting script (this time, uncommented) is shown in Listing A.7.

Listing A.7. The spectrum of the Laplace–Beltrami operator on $\mathbb{S}^2$

```
1  // This FreeFEM script computes eigenvalues of the Laplace-Beltrami
       operator on the unit sphere
2  //
3  //---- A. DECLARATIONS ----
4  int npoints=30;
5  int N=50;
6  real[int] Evalues(N);
7  real t;
8  //---- END OF DECLARATIONS so far ----
9  //
10 //---- B. BOUNDARY DEFINITIONS ----
11 border dOmega1(t=0, 2*pi) { x = cos(t); y = 2+sin(t); label=1; };
12 border dOmega2(t=0, 2*pi) { x = cos(t); y = -2+sin(t); label=2; };
13 //---- END OF BOUNDARY DEFINITIONS ----
14 //
15 //---- C. MESH CREATION USING buildmesh COMMAND ----
16 mesh Th=buildmesh(dOmega1(npoints*2*pi)+dOmega2(npoints*2*pi));
17 //---- END OF MESH DEFINITIONS ----
18 //
19 //---- VISUALISE THE MESH, may be commented out
20 plot(Th,wait=1);
21 //
22 //---- D. DECLARE THE FEM SPACE AND FEM VARIABLES ----
23 fespace Vh(Th,P2);
24 Vh u,v;
25 //
26 //---- E. DECLARE THE QUADRATIC FORMS ----
27 varf q(u,v)=int2d(Th)( dx(u)*dx(v) + dy(u)*dy(v) )+on(1,u=0);
28 varf b(u,v)=int2d(Th)(4/(1+x^2+(y-2*y/abs(y))^2)^2*u*v);
29 //
30 //---- F. CREATE THE MATRICES ----
31 matrix S=q(Vh,Vh);
32 matrix M=b(Vh,Vh);
33 //
34 //---- DECLARE THE ARRAY TO HOLD EIGENFUNCTIONS ----
35 Vh[int] Efunctions(N);
36 //
37 //---- G. SOLVE THE PROBLEM ----
38 int k=EigenValue(S,M,sym=true,value=Evalues,vector=Efunctions);
39 //
40 //---- H. PRINT THE EIGENVALUES ----
41 cout << Evalues;
42 //
```

Executing this script produces an output similar to

```
   -- mesh: Nb of Triangles =  12342, Nb of Vertices 6361
Real symmetric eigenvalue problem: A*x - B*x*lambda
50
   -9.550276753e-15   2.000139729  2.000139735  2.000282213  6.000350531
    6.000350568  6.000701052  6.000707766  6.000707795  12.00061962
   12.00061991  12.00125069  12.0012509   12.0013633   12.00136486
```

```
    12.00150391  20.00095108  20.00095132  20.00191427  20.00191754
    20.00217762  20.00217802  20.00245631  20.00247364  20.0024823
    30.00136367  30.00136728  30.00273512  30.00273578  30.00318084
    30.00319296  30.00368812  30.00369269  30.00378001  30.00380983
    30.00397306  42.00190156  42.00190932  42.00377552  42.00378548
    42.00447397  42.00449093  42.00524994  42.00527772  42.00551986
    42.00553103  42.00587206  42.00590262  42.00600163  56.00263505
    times: compile 0.010895s, execution 3.14007s,  mpirank:0
########  ...
Ok: Normal End
```

This is in good agreement with Theorem 1.2.16 which gives in this case the eigenvalues $k(k+1)$, $k \in \{0\} \cup \mathbb{N}$ of multiplicity $2k+1$.

A.2.6. The Steklov problem, the sloshing problem, and the spectrum of the Dirichlet-to-Neumann map. Our example domain in this subsection is the half-disk

$$D_- = \{(x,y) \in \mathbb{R}^2 : x^2 + y^2 < 1, y < 0\}.$$

To find the eigenvalues of the Steklov problem in D_- we need to recall its weak formulation (7.1.5). Therefore, we need to redefine the form $\mathcal{B}$ in this case as

$$(A.2.2) \qquad \mathcal{B}[u,v] := \int_{\partial D_-} uv\,ds,$$

Otherwise, the treatment is standard; see Listing A.8.

Listing A.8. The spectrum of the Steklov problem in the half-disk.

```
 1 // This FreeFEM script computes eigenvalues of the Steklov problem
      in the half-disk.
 2 //
 3 //---- A. DECLARATIONS ----
 4 int npoints=30;
 5 int N=50;
 6 real[int] Evalues(N);
 7 real t;
 8 //---- END OF DECLARATIONS ----
 9 //
10 //----  B. BOUNDARY DEFINITIONS ----
11 border d1Omega(t=pi, 2*pi) { x = cos(t); y = sin(t); label=1; };
12 border d2Omega(t=1, -1) { x = t; y = 0; label=2; };
13 //---- END OF BOUNDARY DEFINITIONS ----
14 //
15 //---- C. MESH CREATION USING buildmesh COMMAND ----
16 mesh Th=buildmesh(d1Omega(npoints*pi)+d2Omega(npoints*2));
17 //---- END OF MESH DEFINITIONS ----
18 //
19 //---- VISUALISE THE MESH, may be commented out
20 plot(Th,wait=1);
21 //
```

```
22 //---- D. DECLARE THE FEM SPACE AND FEM VARIABLES ----
23 fespace Vh(Th,P2);
24 Vh u,v;
25 //
26 //---- E. DECLARE THE QUADRATIC FORMS ----
27 varf q(u,v)=int2d(Th)( dx(u)*dx(v) + dy(u)*dy(v) );
28 varf b(u,v)=int1d(Th,1,2)(u*v); //note that b changes for Steklov
      and that we integrate over both parts of the boundary
29 //
30 //----  F. CREATE THE MATRICES ----
31 matrix S=q(Vh,Vh);
32 matrix M=b(Vh,Vh);
33 //
34 //---- G. SOLVE THE PROBLEM ----
35 int k=EigenValue(S,M,sym=true,value=Evalues);
36 //
37 //---- H. PRINT THE EIGENVALUES ----
38 cout << Evalues;
39 //
40 //---- END --------
```

To consider the sloshing problem in D_-, with the Steklov condition on the straight part of the boundary and the Neumann condition on the arc, we just need to adjust the definition of the form $\mathcal{B}$ in (A.2.2) in order to integrate over the straight part of the boundary only, therefore replacing line 28 in Listing A.8 by

```
28 varf b(u,v)=int1d(Th,2)(u*v);
```

Finally, to compute the spectrum of the Dirichlet-to-Neumann map $\mathcal{D}_\Lambda$ for a given value of Λ (say, 1.5), we recall the weak statement (7.4.4) and replace line 27 in Listing A.8 by

```
27 real Lambda=1.5; varf q(u,v)=int2d(Th)( dx(u)*dx(v) + dy(u)*dy(v) -
      Lambda*u*v);
```

leaving line 28 unchanged. Note that if Λ is chosen very close to but lower than a Dirichlet eigenvalue of D_-, some low negative eigenvalues of $\mathcal{D}_\Lambda$ may be lost.

A.3. List of downloadable scripts

All scripts mentioned in this appendix are available for download from

https://michaellevitin.net/Book/Scripts

or by clicking directly on the script name in the following table (in the electronic version of the book only). The domains Ω' and Ω are defined by (A.1.6).

Filename	Description	Reference
script1.nb	Mathematica: computing eigenvalues analytically	Listing A.1
script2.nb	Mathematica: Neumann and Dirichlet eigenvalues of Ω	Listing A.2
script3.nb	Mathematica: Robin eigenvalues of Ω	Listing A.3
script4.nb	Mathematica: Zaremba eigenvalues of Ω	Listing A.4
script5.nb	Mathematica: verifying the Faber–Krahn inequality for regular n-gons	Listing A.5
script6.edp	FreeFEM: Neumann eigenvalues of $(0, \pi)^2$	Listing A.6
script7.edp	FreeFEM: Neumann eigenvalues of Ω'	§A.2.3
script8.edp	FreeFEM: Neumann eigenvalues of Ω	§A.2.3
script9.edp	FreeFEM: Dirichlet eigenvalues of Ω	§A.2.4
script10.edp	FreeFEM: Zaremba eigenvalues of Ω	§A.2.4
script11.edp	FreeFEM: Robin eigenvalues of Ω	§A.2.4
script12.edp	FreeFEM: Laplace–Beltrami eigenvalues of a flat cylinder	§A.2.5
script13.edp	FreeFEM: Laplace–Beltrami eigenvalues of a flat torus	§A.2.5
script14.edp	FreeFEM: Laplace–Beltrami eigenvalues of a Möbius strip	§A.2.5
script15.nb	Mathematica: eigenvalues of periodic problems	§A.2.5
script16.edp	FreeFEM: Laplace–Beltrami eigenvalues of $\mathbb{S}^2$	Listing A.7
script17.edp	FreeFEM: Steklov eigenvalues in the half-disk	Listing A.8
script18.edp	FreeFEM: Sloshing eigenvalues in the half-disk	§A.2.6
script19.edp	FreeFEM: Eigenvalues of $\mathcal{D}_\Lambda$ in the half-disk	§A.2.6

Background definitions and notation

We list here some of the standard definitions and notation used throughout the book.

B.1. Sets

We use the standard symbols $\mathbb{N}$, $\mathbb{Z}$, $\mathbb{R}$, $\mathbb{C}$ for the sets of natural, integer, real, and complex numbers, respectively. Our natural numbers do not include zero. We sometimes write

$$\mathbb{N}_0 := \mathbb{N} \cup \{0\}$$

and

$$\mathbb{R}_+ := (0, +\infty).$$

The coordinates of a point $x \in \mathbb{R}^d$ are usually denoted by $(x_1, \ldots, x_d)$. For $x, y \in \mathbb{R}^d$, we write

$$\langle x, y \rangle := \sum_{j=1}^{d} x_j y_j$$

for the usual dot product; we use the same notation in $\mathbb{C}^d$ with the additional complex conjugation over y_j. The length of a vector $y \in \mathbb{R}^d$ is written as $|y| = \sqrt{\langle y, y \rangle}$.

The *complement* of a set $X \subset \mathbb{R}^d$ is denoted by $X^c := \mathbb{R}^d \setminus X$. The *closure* of an open set $U \subset \mathbb{R}^d$ is denoted by $\overline{U}$ and its *boundary* by

$$\partial U = \overline{U} \setminus U.$$

Throughout this book, we say that $\Omega \subset \mathbb{R}^d$ is a *domain* if it is a nonempty connected open set.

We let

$$B_{a,r}^d = B_{a,r} = \left\{ x \in \mathbb{R}^d : |x - a| < r \right\}$$

denote the ball in $\mathbb{R}^d$ with centre a and radius r. We will also write

$$B_r^d := B_{0,r}^d$$

for the ball centred at the origin (or whenever the position of the centre is irrelevant) and

$$\mathbb{B}^d := B_1^d = B_{0,1}^d$$

for the unit ball in $\mathbb{R}^d$. In the planar case, we will also use $\mathbb{D} := \mathbb{B}^2$ for the unit disk.

We denote the volume of the unit ball by

(B.1.1) $$\omega_d = \mathrm{Vol}_d(\mathbb{B}^d) = \frac{\pi^{\frac{d}{2}}}{\Gamma\left(\frac{d}{2} + 1\right)},$$

where Γ is the Gamma function.

Similarly,

$$S_{a,r}^{d-1} = S_{a,r} = \left\{ x \in \mathbb{R}^d : |x - a| = r \right\}$$

denotes the sphere in $\mathbb{R}^d$ with centre a and radius r, and we will also write $S_r^{d-1} = S_r := S_{0,r}$ when it is centred at the origin (or when the position of the centre is irrelevant). We denote the unit sphere in $\mathbb{R}^d$ by

$$\mathbb{S}^{d-1} := S_1^{d-1}$$

and its $(d-1)$-dimensional volume by

(B.1.2) $$\sigma_{d-1} := \mathrm{Vol}_{d-1}(\mathbb{S}^{d-1}) = d\omega_d = \frac{2\pi^{\frac{d}{2}}}{\Gamma\left(\frac{d}{2}\right)}.$$

B.2. Function spaces

If U is an open subset of $\mathbb{R}^d$, we denote by $L^p(U)$, $1 \le p < \infty$, the set of all Lebesgue measurable functions $u : U \to \mathbb{R}$ (or $u : U \to \mathbb{C}$) such that $\int_U |u(x)|^p \, \mathrm{d}x < \infty$. The space $L^p(U)$ equipped with the norm

$$\|u\|_{L^p(U)} := \left(\int_U |u(x)|^p \, \mathrm{d}x \right)^{1/p}$$

is a Banach space, in which we identify elements which coincide almost everywhere. Similarly, $L^\infty(U)$ is the Banach space of all essentially bounded functions u on U with

$$\|u\|_{L^\infty(U)} := \operatorname*{ess\,sup}_{x \in U} u(x) < \infty.$$

We also define

$$L^p_{\mathrm{loc}}(U) := \{u : u|_K \in L^p(K) \text{ for all compact } K \subset U\}.$$

In the special case $p = 2$, $L^2(U)$ equipped with the inner product

$$(u, v)_{L^2(U)} := \int_U u(x)\overline{v(x)}\,\mathrm{d}x$$

is a Hilbert space. Since we are mostly dealing with real-valued functions, we will usually omit the complex conjugation.

For an open set $U \subset \mathbb{R}^d$, let $C(U)$ denote the space of continuous functions on u. We denote the partial derivatives of a function u (if they exist) by

(B.2.1) $$\partial^\alpha u := \frac{\partial^{|\alpha|} u}{\partial x_1^{\alpha_1} \dots \partial x_d^{\alpha_d}},$$

for a multi-index $\alpha = (\alpha_1, \dots, \alpha_d) \in \mathbb{N}_0^d$, where $|\alpha| := \alpha_1 + \dots + \alpha_d$ is the *order* of the derivative. We also write

$$\partial_j u := \frac{\partial u}{\partial x_j}.$$

We denote the space of k-times continuously differentiable functions on U by

$$C^k(U) :=$$

$$\{u : U \to \mathbb{R} : \partial^\alpha u \text{ exists and is continuous in } U \text{ for all } \alpha \text{ with } |\alpha| \le k\},$$

for $k \in \mathbb{N}_0$; obviously, $C^0(U) = C(U)$.

For $u : U \to \mathbb{R}$, $u \in C^1(U)$, we denote its *gradient* by

$$\nabla u := \left(\frac{\partial u}{\partial x_1}, \dots, \frac{\partial u}{\partial x_d} \right),$$

and for a vector-valued function $f : U \to \mathbb{R}^d$, $f = (f_1, \dots, f_d) \subset C^1(U)$, we denote its *divergence* by

$$\operatorname{div} f := \sum_{j=1}^d \frac{\partial f_j}{\partial x_j}.$$

We also set

$$C^\infty(U) := \{u : u \in C^k(U) \text{ for all } k \in \mathbb{N}_0\}$$

and

$$C_0^k(U) := \{u \in C^k(U) : \operatorname{supp} u \Subset U\},$$

$$C_0^\infty(U) := \{u : u \in C_0^k(U) \text{ for all } k \in \mathbb{N}_0\},$$

where $X \Subset Y$ means that X is a compact subset of Y. Further on, we define

$$C^k(\overline{U}) := \{u \in C^k(U) :$$

$\partial^\alpha u$ can be continuously extended to ∂U for all α with $|\alpha| \leq k\}$.

We say that $u \in C(U)$ is *Hölder continuous* with exponent $\beta \in (0,1]$ if there exists a constant $C \geq 0$ such that

$$|u(x) - u(y)| \leq C|x - y|^\beta \qquad \text{for all } x, y \in U.$$

We sometimes use $C^{k,\beta}(U)$ to denote the subspace of functions from $C^k(U)$ whose derivatives of order k are Hölder continuous with exponent β. We say that u is *Lipschitz continuous* (or just *Lipschitz*) if it is Hölder continuous with exponent one; thus the space of all Lipschitz continuous functions on U coincides with $C^{0,1}(U)$.

The *Schwartz space* of rapidly decreasing functions on $\mathbb{R}^d$ is defined as

$$\mathcal{S}(\mathbb{R}^d) :=$$

$$\left\{ u \in C^\infty(\mathbb{R}^d) : \sup_{x \in \mathbb{R}^d} \left| x^\alpha \partial^\beta u \right| < \infty \text{ for all multi-indices } \alpha, \beta \in \mathbb{N}_0^d \right\},$$

where $x^\alpha := x_1^{\alpha_1} \cdots x_d^{\alpha_d}$.

We will use the shorthand notation $L^2(U) = (L^2(U))^d$, $C^k(U) = (C^k(U))^d$, etc., for the spaces of vector-valued functions $f = (f_1, \ldots, f_d) : U \to \mathbb{R}^d$.

B.3. Regularity of the boundary

We follow [**McL00**] and [**ChWGLS12**, Appendix A]. Let $U \subset \mathbb{R}^d$ be a nonempty open set. We say that its boundary ∂U is *Lipschitz* if ∂U is compact and there exist finite families of sets $\{W_i\}$ and $\{U_i\}$ and of functions $\{f_i\}$ of the same cardinality such that

(i) the family $\{W_i\}$, $W_i \subset \mathbb{R}^d$, is a finite open cover of ∂U;

(ii) the family $\{U_i\}$, $U_i \subset \mathbb{R}^d$, is such that $U_i \cap W_i = U \cap W_i$;

(iii) each function $f_i : \mathbb{R}^{d-1} \to \mathbb{R}^d$ is Lipschitz continuous;

(iv) for each i there exists a rigid motion $R_i : \mathbb{R}^d \to \mathbb{R}^d$ such that

$$R_i(U_i) = \{(x', x_d) \in \mathbb{R}^{d-1} \times \mathbb{R} : x_d > f_i(x')\}.$$

We will sometimes say that Ω is a *Lipschitz domain* if it is a domain with a Lipschitz boundary $\partial\Omega$.

In the same manner, we define the C^k or C^∞ boundaries by replacing in part (iii) of the above definition the family of Lipschitz functions $\{f_i\}$ by a family of C^k or C^∞ functions, respectively.

We mention that, for example, all polyhedra have Lipschitz boundary (but not C^1), whereas domains with cusps or slits are not Lipschitz.

Image credits

We are grateful to all organisations listed below for the permission to use their images, and we would like to thank in particular Joanne Foote (The University of Texas at Austin) and Eva Hözl (TUM Archiv) for their assistance. Some images have been cropped for presentation purposes.

Image	Image details: author and source; licence (where known and as given by the source of download, CC = Creative Commons); download address	Page
Babuška	unknown author; included by permission of Oden Institute for Computational Engineering and Sciences at The University of Texas at Austin; downloaded from https://oden.utexas.edu/news-and-events/news/Ivo-Babuska-Mathematician-Finite-Element-Method-Dies-97	265
Beltrami	unknown author; Public Domain; downloaded from https://commons.wikimedia.org/w/index.php?curid=2094478	1
Chladni	by Ludwig Albert von Montmorillon (1794–1854) — Reprinted in U. B. Marvin, *Ernst Florens Friedrich Chladni (1756–1827) and the origins of modern meteorite research*, Meteoritics & Planetary Science, **42**:9 (2007), B3–B68; Public Domain; downloaded from https://commons.wikimedia.org/w/index.php?curid=1424376	101
Chladni's experiment, Figure 4.3	unknown author — from William Henry Stone, *Elementary Lessons on Sound*. Macmillan and Co., London, 1879, p. 26, fig. 12; Public Domain; downloaded from https://commons.wikimedia.org/w/index.php?curid=22901318	102

Image	Image details	Page
Courant	by Konrad Jacobs, Erlangen — Mathematisches Forschungsinstitut Oberwolfach, https://opc.mfo.de/detail?photoID=726; CC BY-SA 2.0 de; downloaded from https://commons.wikimedia.org/w/index.php?curid=3931476	99
Dirichlet	unknown author; Public Domain; downloaded from https://commons.wikimedia.org/w/index.php?curid=90476	25
Faber	unknown author; included by permission of Technical University of Munich, TUM Archiv	137
Fourier	by Julien-Léopold Boilly — derived from https://www.gettyimages.com.au/license/169251384, https://wellcomecollection.org/works/b4qh352u; Public Domain; downloaded from https://commons.wikimedia.org/w/index.php?curid=114366437	3
Galerkin	unknown author — from I. Elishakoff, J. Kaplunov, and E. Kaplunov, *Galerkin's method was not developed by Ritz, contrary to the Timoshenko's statement*, in: *A nonlinear dynamics of discrete and continuous systems*. Advanced Structured Materials **139**. Springer, Cham (2021), p. 66, fig. 5.2; included by permission of Springer Nature, licence number 5564741400002; downloaded from https://doi.org/10.1007/978-3-030-53006-8_5	265
Germain	unknown author — http://www-history.mcs.st-andrews.ac.uk/PictDisplay/Germain.html; Public Domain; downloaded from https://commons.wikimedia.org/wiki/File:Germain.jpeg	99
Hörmander	by By Konrad Jacobs — MFO, https://opc.mfo.de/detail?photoID=1777; CC BY-SA 2.0 de; downloaded from https://commons.wikimedia.org/w/index.php?curid=3902616	213
Kac	by Konrad Jacobs, Erlangen, Copyright is with MFO — Mathematisches Institut Oberwolfach (MFO), https://opc.mfo.de/detail?photoID=1986; CC BY-SA 2.0 de; downloaded from https://commons.wikimedia.org/w/index.php?curid=3897672	183
Krahn	by Heinrich Riedel — TÜR fotokogu; Public Domain; downloaded from https://commons.wikimedia.org/w/index.php?curid=18569489	137
Lamb	by The Royal Society — from Brian Launder, *Horace Lamb and the circumstances of his appointment at Owens College*, Notes Rec. Royal Soc. **67** (2012), 139–158; Public Domain; downloaded from https://commons.wikimedia.org/w/index.php?curid=58935817	219
Laplace	coloured engraving by J. Posselwhite; Public Domain; downloaded from https://commons.wikimedia.org/w/index.php?curid=11128070	1
Minakshisundaram	unknown author — https://mathshistory.st-andrews.ac.uk/Biographies/Minakshisundaram/pictdisplay/; CC BY-SA 4.0; downloaded from https://commons.wikimedia.org/w/index.php?curid=130177799	183

Image	Image details	Page
Neumann	by Photo Deutsches Museum — *Green's Functions with Applications* by Dean G. Duffy, Second Edition; CC BY-SA 4.0; downloaded from https://commons.wikimedia.org/w/index.php?curid=71014721	25
Nirenberg	by Konrad Jacobs — https://opc.mfo.de/detail?photo_id=3062; CC BY-SA 2.0 de; downloaded from https://commons.wikimedia.org/w/index.php?curid=6092917	43
Poisson	by François-Séraphin Delpech — http://www.sil.si.edu/digitalcollections/hst/scientific-identity/CF/display_results.cfm?alpha_sort=P, http://www.sil.si.edu/digitalcollections/hst/scientific-identity/fullsize/SIL14-P005-06a.jpg; Public Domain; downloaded from https://commons.wikimedia.org/w/index.php?curid=378439	194
Pólya	unknown author — ETH-Bibliothek Zürich, Bildarchiv; Public Domain; downloaded from http://doi.org/10.3932/ethz-a-001421808	90
Rayleigh	unknown author — from *Obituary Notices of Fellows Deceased*, Proc. Royal Soc. London. Series A, Containing Papers of a Mathematical and Physical Character, **98**:695 (Mar. 24, 1921); Public Domain; downloaded from https://commons.wikimedia.org/w/index.php?curid=405463	55
Steklov	unknown author – [1] https://www.ras.ru/win/db/show_per.asp?P=.id-52252.ln-en [2] (http://www-history.mcs.st-and.ac.uk/PictDisplay/Steklov.html); Public Domain; downloaded from https://commons.wikimedia.org/w/index.php?curid=19072880	213
Szegő	unknown author – from R. Askey and P. Nevai, *Gabor Szegő: 1895–1985*, The Mathematical Intelligencer **18**:3 (1996), 10–22; included by permission of Springer Nature, licence number 5564411286376; downloaded from https://doi.org/10.1007/BF03024305	159
Weyl	unknown author, ETH-Bibliothek Zürich, Bildarchiv; CC BY-SA 3.0; downloaded from https://commons.wikimedia.org/w/index.php?curid=8098412	55

Bibliography

[AdaFou03]   R. A. Adams and J. J. F. Fournier, *Sobolev spaces*, 2nd ed., Pure and Applied Mathematics (Amsterdam), vol. 140, Elsevier/Academic Press, Amsterdam, 2003. MR2424078 ↑27, 28

[Agm65]   S. Agmon, *Unicité et convexité dans les problèmes différentiels*, Séminaire de Mathématiques Supérieures, No. 13 (Été, vol. 1965), Les Presses de l'Université de Montréal, Montreal, Que., 1966. MR0252808 ↑120, 122

[Agr06]   M. S. Agranovich, *On a mixed Poincaré-Steklov type spectral problem in a Lipschitz domain*, Russ. J. Math. Phys. **13** (2006), no. 3, 239–244, DOI 10.1134/S1061920806030010. MR2262827 ↑230

[Alb71]   J. H. Albert, *Topology of the nodal and critical point sets for eigenfunctions of elliptic operators*, ProQuest LLC, Ann Arbor, MI, 1972. Thesis (Ph.D.)–Massachusetts Institute of Technology. Available online at MIT Libraries, https://dspace.mit.edu/handle/1721.1/45754. MR2940251 ↑132

[Alm00]   F. J. Almgren Jr., *Almgren's big regularity paper, Q-valued functions minimizing Dirichlet's integral and the regularity of area-minimizing rectifiable currents up to codimension 2*, with a preface by J. E. Taylor and V. Scheffer, World Scientific Monograph Series in Mathematics, vol. 1, World Scientific Publishing Co., Inc., River Edge, NJ, 2000, DOI 10.1142/4253. MR1777737 ↑120

[AndClu11]   B. Andrews and J. Clutterbuck, *Proof of the fundamental gap conjecture*, J. Amer. Math. Soc. **24** (2011), no. 3, 899–916, DOI 10.1090/S0894-0347-2011-00699-1. MR2784332 ↑180

[AntFre12]   P. R. S. Antunes and P. Freitas, *Numerical optimization of low eigenvalues of the Dirichlet and Neumann Laplacians*, J. Optim. Theory Appl. **154** (2012), no. 1, 235–257, DOI 10.1007/s10957-011-9983-3. MR2931377 ↑165

[AreCSVV18]   W. Arendt, R. Chill, C. Seifert, H. Vogt, and J. Voigt, *Form Methods for Evolution Equations, and Applications*, 18th Internet Seminar on Evolution Equations, Lecture Notes. Available at seminar's website, https://www.mat.tuhh.de/veranstaltungen/isem18/pdf/LectureNotes.pdf. 38, 62

[AreMaz12]   W. Arendt and R. Mazzeo, *Friedlander's eigenvalue inequalities and the Dirichlet-to-Neumann semigroup*, Commun. Pure Appl. Anal. **11** (2012), no. 6, 2201–2212, DOI 10.3934/cpaa.2012.11.2201. MR2912743 ↑215, 254, 255, 256, 257, 258, 262, 263

[AreMon95]   W. Arendt and S. Monniaux, *Domain perturbation for the first eigenvalue of the Dirichlet Schrödinger operator*, Partial differential operators and mathematical physics (Holzhau, 1994), Oper. Theory Adv. Appl., vol. 78, Birkhäuser, Basel, 1995, pp. 9–19, DOI 10.1007/978-3-0348-9092-2.2. MR1365313 ↑145

[ArmGar01]   D. H. Armitage and S. J. Gardiner, *Classical potential theory*, Springer Monographs in Mathematics, Springer-Verlag London, Ltd., London, 2001, DOI 10.1007/978-1-4471-0233-5. MR1801253 ↑51

[Aro57]   N. Aronszajn, *A unique continuation theorem for solutions of elliptic partial differential equations or inequalities of second order*, J. Math. Pures Appl. (9) **36** (1957), 235–249. MR92067 ↑67, 107, 124

[Ash99]   M. S. Ashbaugh, *Isoperimetric and universal inequalities for eigenvalues*, Spectral theory and geometry (Edinburgh, 1998), London Math. Soc. Lecture Note Ser., vol. 273, Cambridge Univ. Press, Cambridge, 1999, pp. 95–139, DOI 10.1017/CBO9780511566165.007. MR1736867 ↑174

[AshBen89]   M. S. Ashbaugh and R. Benguria, *Optimal lower bound for the gap between the first two eigenvalues of one-dimensional Schrödinger operators with symmetric single-well potentials*, Proc. Amer. Math. Soc. **105** (1989), no. 2, 419–424, DOI 10.2307/2046959. MR942630 ↑180

[AshBen91]   M. S. Ashbaugh and R. D. Benguria, *A sharp bound for the ratio of the first two eigenvalues of Dirichlet Laplacians and extensions*, Ann. of Math. (2) **135** (1992), no. 3, 601–628, DOI 10.2307/2946578. MR1166646 ↑179

[AviDSiKal16] A. Avila, J. De Simoi, and V. Kaloshin, *An integrable deformation of an ellipse of small eccentricity is an ellipse*, Ann. of Math. (2) **184** (2016), no. 2, 527–558, DOI 10.4007/annals.2016.184.2.5. MR3548532 ↑210

[AxlBouWad01] S. Axler, P. Bourdon, and R. Wade, *Harmonic function theory*, Graduate Texts in Mathematics **137**, Springer Science+Business Media, New York, 2001. DOI 10.1007/978-1-4757-8137-3. Also available at author's website https://www.axler.net/HFT.pdf 20, 41, 108

[BabOsb91]   I. Babuška and J. Osborn, *Eigenvalue problems*, Handbook of Numerical Analysis, II, North-Holland, Amsterdam, 1991, pp. 641–787, DOI 10.1016/S1570-8659(05)80042-0. MR1115240 ↑266

[BanParBSh09] R. Band, O. Parzanchevski, and G. Ben-Shach, *The isospectral fruits of representation theory: quantum graphs and drums*, J. Phys. A **42** (2009), no. 17, 175202, 42, DOI 10.1088/1751-8113/42/17/175202. MR2539297 ↑203

[Ban80]   C. Bandle, *Isoperimetric inequalities and applications*, Monographs and Studies in Mathematics, vol. 7, Pitman (Advanced Publishing Program), Boston, Mass.-London, 1980. MR572958 ↑x, 56, 57, 168

[BañCar94]   R. Bañuelos and T. Carroll, *Brownian motion and the fundamental frequency of a drum*, Duke Math. J. **75** (1994), no. 3, 575–602, DOI 10.1215/S0012-7094-94-07517-0. MR1291697 ↑156

[BarBouLeb16] L. Baratchart, L. Bourgeois, and J. Leblond, *Uniqueness results for inverse Robin problems with bounded coefficient*, J. Funct. Anal. **270** (2016), no. 7, 2508–2542, DOI 10.1016/j.jfa.2016.01.011. MR3464049 ↑226

[Beb03]   M. Bebendorf, *A note on the Poincaré inequality for convex domains*, Z. Anal. Anwendungen **22** (2003), no. 4, 751–756, DOI 10.4171/ZAA/1170. MR2036927 ↑180

[BehtEl15]   J. Behrndt and A. F. M. ter Elst, *Dirichlet-to-Neumann maps on bounded Lipschitz domains*, J. Differential Equations **259** (2015), no. 11, 5903–5926, DOI 10.1016/j.jde.2015.07.012. MR3397313 ↑254, 255

[BelBBT18]   F. Belgacem, H. BelHadjAli, A. BenAmor, and A. Thabet, *Robin Laplacian in the large coupling limit: convergence and spectral asymptotic*, Ann. Sc. Norm. Super. Pisa Cl. Sci. (5) **18** (2018), no. 2, 565–591, DOI 10.2422/2036-2145.201601_008. MR3801290 ↑259

[Ben15]   B. Benson, *The Cheeger constant, isoperimetric problems, and hyperbolic surfaces*, arXiv:1509.08993. 154

[Bér86]   P. Bérard, *Spectral geometry: direct and inverse problems*, with appendixes by G. Besson, and by P. Bérard and M. Berger, Lecture Notes in Mathematics, vol. 1207, Springer-Verlag, Berlin, 1986, DOI 10.1007/BFb0076330. MR861271 ↑x, 84

[Bér92]   P. Bérard, *Transplantation et isospectralité. I*, Math. Ann. **292** (1992), no. 3, 547–559, DOI 10.1007/BF01444635. MR1152950 ↑200

[BérBes80]   P. Bérard and G. Besson, *Spectres et groupes cristallographiques. II. Domaines sphériques*, Ann. Inst. Fourier (Grenoble) **30** (1980), no. 3, 237–248, DOI 10.5802/aif.800. MR597025 ↑89

[BérHel14]   P. Bérard and B. Helffer, *Nodal sets of eigenfunctions, Antonie Stern's results revisited*, Sém. de Théorie Spect. et Géom., Grenoble **32** (2014–15), 1–37. DOI 10.5802/tsg.302. 107

[BérMey82]   P. Bérard and D. Meyer, *Inégalités isopérimétriques et applications*, Ann. Sci. École Norm. Sup. (4) **15** (1982), no. 3, 513–541, DOI 10.24033/asens.1435. MR690651 ↑107, 113, 133, 148, 149

[BerNirVar94] H. Berestycki, L. Nirenberg, and S. R. S. Varadhan, *The principal eigenvalue and maximum principle for second-order elliptic operators in general domains*, Comm. Pure Appl. Math. **47** (1994), no. 1, 47–92, DOI 10.1002/cpa.3160470105. MR1258192 ↑113

[Ber72]   F. A. Berezin, *Convex operator functions*, Math. USSR Sb. **17**:2 (1972), 269–277. DOI 10.1070/SM1972v017n02ABEH001504. 94

[vdB83]   M. van den Berg, *On condensation in the free-boson gas and the spectrum of the Laplacian*, J. Statist. Phys. **31** (1983), no. 3, 623–637, DOI 10.1007/BF01019501. MR711491 ↑180

[vdBDryKap14] M. van den Berg, E. B. Dryden, and T. Kappeler, *Isospectrality and heat content*, Bull. Lond. Math. Soc. **46** (2014), no. 4, 793–808, DOI 10.1112/blms/bdu035. MR3239617 ↑203

[vdBSri88]   M. van den Berg and S. Srisatkunarajah, *Heat equation for a region in $\mathbf{R}^2$ with a polygonal boundary*, J. London Math. Soc. (2) **37** (1988), no. 1, 119–127, DOI 10.1112/jlms/s2-37.121.119. MR921750 ↑193

[Ber73]   M. Berger, *Sur les premières valeurs propres des variétés riemanniennes*, Compositio Math. **26** (1973), 129–149. Available at http://www.numdam.org/item/CM_1973__26_2_129_0. MR316913 ↑171

[BerGauMaz71] M. Berger, P. Gauduchon, and E. Mazet, *Le spectre d'une variété riemannienne*, Lecture Notes in Mathematics, Vol. 194, Springer-Verlag, Berlin-New York, 1971, DOI 10.1007/BFb0064643. MR0282313 ↑x, 14, 15, 17, 20, 184, 187, 188, 194

[BerGMR21]   D. Berghaus, B. Georgiev, H. Monien, and D. Radchenko, *On Dirichlet eigenvalues of regular polygons*, arXiv:2103.01057. 272

[Ber17]   G. Berkolaiko, *An elementary introduction to quantum graphs*, Geometric and computational spectral theory, Contemp. Math., vol. 700, Amer. Math. Soc., Providence, RI, 2017, pp. 41–72, DOI 10.1090/conm/700/14182. MR3748521 ↑240

[BerCHS22]   G. Berkolaiko, G. Cox, B. Helffer, and M. P. Sundqvist, *Computing nodal deficiency with a refined Dirichlet-to-Neumann map*, J. Geom. Anal. **32** (2022), no. 10, Paper No. 246, 36, DOI 10.1007/s12220-022-00984-2. MR4456213 ↑148

[BerKuc13]   G. Berkolaiko and P. Kuchment, *Introduction to quantum graphs*, Mathemati-
                 cal Surveys and Monographs, vol. 186, American Mathematical Society, Provi-
                 dence, RI, 2013, DOI 10.1090/surv/186. MR3013208 ↑xviii, 237, 240

[BerKucSmi12]   G. Berkolaiko, P. Kuchment, and U. Smilansky, *Critical partitions and nodal
                 deficiency of billiard eigenfunctions*, Geom. Funct. Anal. **22** (2012), no. 6, 1517–
                 1540, DOI 10.1007/s00039-012-0199-y. MR3000497 ↑148

[BerGetVer04]   N. Berline, E. Getzler, and M. Vergne, *Heat kernels and Dirac operators*, cor-
                 rected reprint of the 1992 original, Grundlehren Text Editions, Springer-Verlag,
                 Berlin, 2004. MR2273508 ↑xviii

[BerDKZ18]   B. C. Berndt, A. Dixit, S. Kim, and A. Zaharescu, *Sums of squares
                 and products of Bessel functions*, Adv. Math. **338** (2018), 305–338, DOI
                 10.1016/j.aim.2018.09.001. MR3861707 ↑20

[Ber55]   L. Bers, *Local behavior of solutions of general linear elliptic equations*, Comm.
                 Pure Appl. Math. **8** (1955), 473–496, DOI 10.1002/cpa.3160080404. MR75416
                 ↑132

[Bes80]   G. Besson, *Sur la multiplicité de la première valeur propre des surfaces rie-
                 manniennes*, Ann. Inst. Fourier (Grenoble) **30** (1980), no. 1, x, 109–128, DOI
                 10.5802/aif.777. MR576075 ↑134

[BhaWei99]   T. Bhattacharya and A. Weitsman, *Estimates for Green's function in terms of
                 asymmetry*, Applied Analysis, J. R. Dorroh, G. R. Goldstein, J. A. Goldstein,
                 and M. M. Tom, eds., AMS Contemporary Math. Series **221**, 31–58. Amer.
                 Math. Soc., Providence, RI, 1999. DOI 10.1090/conm/221. 145

[Bob11]   A. I. Bobenko, *Introduction to compact Riemann surfaces*, Computational ap-
                 proach to Riemann surfaces, Lecture Notes in Math., vol. 2013, Springer, Hei-
                 delberg, 2011, pp. 3–64, DOI 10.1007/978-3-642-17413-1_1. MR2905610 ↑166,
                 169

[Bor16]   D. Borthwick, *Spectral theory of infinite-area hyperbolic surfaces*, 2nd ed.,
                 Progress in Mathematics, vol. 318, Birkhäuser/Springer, [Cham], 2016, DOI
                 10.1007/978-3-319-33877-4. MR3497464 ↑xviii

[BouLev07]   L. Boulton and M. Levitin, *Trends and Tricks in Spectral Theory*. Ediciones
                 IVIC, Caracas, 2007. ix

[Bou15]   J. Bourgain, *On Pleijel's nodal domain theorem*, Int. Math. Res. Not. **2015**
                 (2015), no. 6, 1601–1612. MR3340367 149

[BouWat17]   J. Bourgain and N. Watt, *Mean square of zeta function, circle problem and
                 divisor problem revisited*, `arXiv:1709.04340`. 19

[BraDeP17]   L. Brasco and G. De Philippis, *Spectral inequalities in quantitative form*, Shape
                 optimization and spectral theory, De Gruyter Open, Warsaw, 2017, pp. 201–
                 281, DOI 10.1515/9783110550887-007. MR3681151 ↑164, 225

[BraDePVel15]   L. Brasco, G. De Philippis, and B. Velichkov, *Faber-Krahn inequalities in
                 sharp quantitative form*, Duke Math. J. **164** (2015), no. 9, 1777–1831, DOI
                 10.1215/00127094-3120167. MR3357184 ↑145

[BraPra12]   L. Brasco and A. Pratelli, *Sharp stability of some spectral inequalities*, Geom.
                 Funct. Anal. **22** (2012), no. 1, 107–135, DOI 10.1007/s00039-012-0148-9.
                 MR2899684 ↑164

[Bre11]   H. Brezis, *Functional analysis, Sobolev spaces and partial differential equations*,
                 Universitext, Springer, New York, 2011. MR2759829 ↑28, 43, 48

[BriButPri22]   L. Briani, G. Buttazzo, and F. Prinari, *On a class of Cheeger inequalities*,
                 Ann. Mat. Pura Appl. (4) **202** (2023), no. 2, 657–678, DOI 10.1007/s10231-
                 022-01255-1. MR4552073 ↑158

[Bro01]   F. Brock, *An isoperimetric inequality for eigenvalues of the Stekloff problem*, ZAMM Z. Angew. Math. Mech. **81** (2001), no. 1, 69–71, DOI 10.1002/1521-4001(200101)81:1⟨69::AID-ZAMM69⟩3.0.CO;2-#. MR1808500 ↑225

[Bro88]   R. Brooks, *Constructing isospectral manifolds*, Amer. Math. Monthly **95** (1988), no. 9, 823–839, DOI 10.2307/2322897. MR967343 ↑197

[Bro98]   R. Brooks, *The Sunada method*, Tel Aviv Topology Conference: Rothenberg Festschrift (1998), Contemp. Math., vol. 231, Amer. Math. Soc., Providence, RI, 1999, pp. 25–35, DOI 10.1090/conm/231/03350. MR1705572 ↑197

[Bro93]   R. M. Brown, *The trace of the heat kernel in Lipschitz domains*, Trans. Amer. Math. Soc. **339** (1993), no. 2, 889–900, DOI 10.2307/2154304. MR1134755 ↑210

[Brü78]   J. Brüning, *Über Knoten von Eigenfunktionen des Laplace-Beltrami-Operators*, Math. Z. **158** (1978), no. 1, 15–21, DOI 10.1007/BF01214561. MR478247 ↑117

[BrüGro72] J. Brüning and D. Gromes, *Über die Länge der Knotenlinien schwingender Membranen*, Math. Z. **124** (1972), 79–82, DOI 10.1007/BF01142586. MR287202 ↑115

[BucFNT21] D. Bucur, V. Ferone, C. Nitsch, and C. Trombetti, *Weinstock inequality in higher dimensions*, J. Differential Geom. **118** (2021), no. 1, 1–21, DOI 10.4310/jdg/1620272940. MR4255070 ↑225

[BucFreKen17] D. Bucur, P. Freitas, and J. Kennedy, *The Robin problem*, Shape optimization and spectral theory, De Gruyter Open, Warsaw, 2017, pp. 78–119, DOI 10.1515/9783110550887-004. MR3681148 ↑62, 63, 258

[BucHen19] D. Bucur and A. Henrot, *Maximization of the second non-trivial Neumann eigenvalue*, Acta Math. **222** (2019), no. 2, 337–361, DOI 10.4310/ACTA.2019.v222.n2.a2. MR3974477 ↑164

[BucNah21] D. Bucur and M. Nahon, *Stability and instability issues of the Weinstock inequality*, Trans. Amer. Math. Soc. **374** (2021), no. 3, 2201–2223, DOI 10.1090/tran/8302. MR4216737 ↑225

[Buh16]   L. Buhovsky, *Lecture notes on spectral geometry* (2016), private communication. x, 103

[Bur98]   F. E. Burstall, *Basic Riemannian geometry*, Spectral theory and geometry (Edinburgh, 1998), London Math. Soc. Lecture Note Ser., vol. 273, Cambridge Univ. Press, Cambridge, 1999, pp. 1–29, DOI 10.1017/CBO9780511566165.004. MR1736864 ↑12, 14, 118, 153, 154

[Bus80]   P. Buser, *On Cheeger's inequality $\lambda_1 \geq h^2/4$*, Geometry of the Laplace operator (Proc. Sympos. Pure Math., Univ. Hawaii, Honolulu, Hawaii, 1979), Proc. Sympos. Pure Math., XXXVI, Amer. Math. Soc., Providence, R.I., 1980, pp. 29–77. MR573428 ↑150, 153

[Bus82]   P. Buser, *A note on the isoperimetric constant*, Ann. Sci. École Norm. Sup. (4) **15** (1982), no. 2, 213–230, DOI 10.24033/asens.1426. MR683635 ↑152, 154

[BusCDS94] P. Buser, J. Conway, P. Doyle, and K.-D. Semmler, *Some planar isospectral domains*, Internat. Math. Res. Notices **1994** (1994), no. 9, 391–400. MR1301439 204, 205, 206

[CadFar18] L. Cadeddu and M. A. Farina, *A brief note on the coarea formula*, Abh. Math. Semin. Univ. Hambg. **88** (2018), no. 1, 193–199, DOI 10.1007/s12188-017-0183-4. MR3785792 ↑140, 150

[Can13]   Y. Canzani, *Analysis on manifolds via the Laplacian*, notes of the course given at Harvard University (2013). Available at author's website http://canzani.web.unc.edu/files/2016/08/Laplacian.pdf. x, 17

[Car33]   T. Carleman, *Sur une inégalité différentielle dans la théorie des fonctions analytiques*, C. R. Acad. Sci., Paris **196** (1933), 995–997. 120

[ChWGLS12]   S. N. Chandler-Wilde, I. G. Graham, S. Langdon, and E. A. Spence, *Numerical-asymptotic boundary integral methods in high-frequency acoustic scattering*, Acta Numer. **21** (2012), 89–305, DOI 10.1017/S0962492912000037. MR2916382 ↑29, 30, 52, 53, 214, 230, 232, 290

[Cha95]   S. J. Chapman, *Drums that sound the same*, Amer. Math. Monthly **102** (1995), no. 2, 124–138, DOI 10.2307/2975346. MR1315592 ↑204

[Cha18]   P. Charron, *A Pleijel-type theorem for the quantum harmonic oscillator*, J. Spectr. Theory **8** (2018), no. 2, 715–732, DOI 10.4171/JST/211. MR3812813 ↑149

[ChaHelHoO18] P. Charron, B. Helffer, and T. Hoffmann-Ostenhof, *Pleijel's theorem for Schrödinger operators with radial potentials*, Ann. Math. Qué. **42** (2018), no. 1, 7–29, DOI 10.1007/s40316-017-0078-x. MR3777504 ↑149

[Cha84]   I. Chavel, *Eigenvalues in Riemannian geometry*, including a chapter by B. Randol, with an appendix by J. Dodziuk, Pure and Applied Mathematics, vol. 115, Academic Press, Inc., Orlando, FL, 1984. MR768584 ↑x, 14, 15, 17, 20, 125, 153, 188

[Che71]   J. Cheeger, *A lower bound for the smallest eigenvalue of the Laplacian*, Problems in analysis (Sympos. in honor of Salomon Bochner, Princeton Univ., Princeton, N.J., 1969), Princeton Univ. Press, Princeton, N.J., 1970, pp. 195–199, DOI 10.1515/9781400869312. MR0402831 ↑149, 150

[Che75]   S.-Y. Cheng, *Eigenfunctions and nodal sets*, Comment. Math. Helv. **51** (1976), no. 1, 43–55, DOI 10.1007/BF02568142. MR397805 ↑107, 132, 133, 134

[Chu97]   F. R. K. Chung, *Spectral graph theory*, CBMS Regional Conference Series in Mathematics, vol. 92, Published for the Conference Board of the Mathematical Sciences, Washington, DC; by the American Mathematical Society, Providence, RI, 1997. MR1421568 ↑158

[ChuGriYau96] F. R. K. Chung, A. Grigor'yan, and S.-T. Yau, *Upper bounds for eigenvalues of the discrete and continuous Laplace operators*, Adv. Math. **117** (1996), no. 2, 165–178, DOI 10.1006/aima.1996.0006. MR1371647 ↑180

[CiaFus02]   A. Cianchi and N. Fusco, *Functions of bounded variation and rearrangements*, Arch. Ration. Mech. Anal. **165** (2002), no. 1, 1–40, DOI 10.1007/s00205-002-0214-9. MR1947097 ↑143

[CiaKarMed19] D. Cianci, M. Karpukhin, and V. Medvedev, *On branched minimal immersions of surfaces by first eigenfunctions*, Ann. Global Anal. Geom. **56** (2019), no. 4, 667–690, DOI 10.1007/s10455-019-09683-8. MR4029852 ↑171

[Col17]   B. Colbois, *The spectrum of the Laplacian: a geometric approach*, Geometric and computational spectral theory, Contemp. Math., vol. 700, Amer. Math. Soc., Providence, RI, 2017, pp. 1–40, DOI 10.1090/conm/700/14181. MR3748520 ↑155

[ColDod94]   B. Colbois and J. Dodziuk, *Riemannian metrics with large λ_1*, Proc. Amer. Math. Soc. **122** (1994), no. 3, 905–906, DOI 10.2307/2160770. MR1213857 ↑165, 173

[ColElS03]   B. Colbois and A. El Soufi, *Extremal eigenvalues of the Laplacian in a conformal class of metrics: the 'conformal spectrum'*, Ann. Global Anal. Geom. **24** (2003), no. 4, 337–349, DOI 10.1023/A:1026257431539. MR2015867 ↑173

[ColElSGir11] B. Colbois, A. El Soufi, and A. Girouard, *Isoperimetric control of the Steklov spectrum*, J. Funct. Anal. **261** (2011), no. 5, 1384–1399, DOI 10.1016/j.jfa.2011.05.006. MR2807105 ↑217

[ColGGS22]   B. Colbois, A. Girouard, C. Gordon, and D. Sher, *Some recent developments on the Steklov eigenvalue problem*, arXiv:2212.12528. 263

[ColGirHas18]   B. Colbois, A. Girouard, and A. Hassannezhad, *The Steklov and Laplacian spectra of Riemannian manifolds with boundary*, J. Funct. Anal. **278** (2020), no. 6, 108409, 38, DOI 10.1016/j.jfa.2019.108409. MR4054110 ↑231, 233

[ColMin11]   T. H. Colding and W. P. Minicozzi II, *Lower bounds for nodal sets of eigenfunctions*, Comm. Math. Phys. **306** (2011), no. 3, 777–784, DOI 10.1007/s00220-011-1225-x. MR2825508 ↑114

[CdV73]   Y. Colin de Verdière, *Spectre du laplacien et longueurs des géodésiques périodiques I*, Compos. Math. **27** (1973), no. 1, 80–106. Available at https://eudml.org/doc/89181. 195

[CdV87]   Y. Colin de Verdière, *Construction de laplaciens dont une partie finie du spectre est donnée*, Ann. Sci. École Norm. Sup. (4) **20** (1987), no. 4, 599–615, DOI 10.24033/asens.1546. MR932800 ↑180, 181

[ConSlo92]   J. H. Conway and N. J. A. Sloane, *Four-dimensional lattices with the same theta series*, Internat. Math. Res. Notices **4** (1992), 93–96, DOI 10.1155/S1073792892000102. MR1159450 ↑197

[Cou23]   R. Courant, *Ein allgemeiner Satz zur Theorie der Eigenfunktionen selbstadjungierter Differentialausdrücke*, Nachr. Ges. Göttingen (1923), 81–84. Available at https://eudml.org/doc/59133. 102

[CouHil89]   R. Courant and D. Hilbert, *Methods of mathematical physics*, Volume 1, John Wiley & Sons, 1989. DOI 10.1002/9783527617210. x, 11, 65, 84

[CoxJonMar17]   G. Cox, C. K. R. T. Jones, and J. L. Marzuola, *Manifold decompositions and indices of Schrödinger operators*, Indiana Univ. Math. J. **66**:5 (2017), 1573–1602. Available at https://www.jstor.org/stable/26321061. 148

[CroSha98]   C. B. Croke and V. A. Sharafutdinov, *Spectral rigidity of a compact negatively curved manifold*, Topology **37** (1998), no. 6, 1265–1273, DOI 10.1016/S0040-9383(97)00086-4. MR1632920 ↑211

[Dan11]   D. Daners, *Krahn's proof of the Rayleigh conjecture revisited*, Arch. Math. (Basel) **96** (2011), no. 2, 187–199, DOI 10.1007/s00013-010-0218-x. MR2773220 ↑140, 142, 145

[Dav89]   E. B. Davies, *Heat kernels and spectral theory*, Cambridge Tracts in Mathematics, vol. 92, Cambridge University Press, Cambridge, 1990, DOI 10.1017/CBO9780511566158. MR1103113 ↑x, 50

[Dav95]   E. B. Davies, *Spectral theory and differential operators*, Cambridge Studies in Advanced Mathematics, vol. 42, Cambridge University Press, Cambridge, 1995, DOI 10.1017/CBO9780511623721. MR1349825 ↑x, 38, 57, 58

[DePMon21]   N. De Ponti and A. Mondino, *Sharp Cheeger-Buser type inequalities in $RCD(K, \infty)$ spaces*, J. Geom. Anal. **31** (2021), no. 3, 2416–2438, DOI 10.1007/s12220-020-00358-6. MR4225812 ↑155

[DeSKalWei17]   J. de Simoi, V. Kaloshin, and Q. Wei, *Dynamical spectral rigidity among $\mathbb{Z}_2$-symmetric strictly convex domains close to a circle*, Appendix B coauthored with H. Hezari, Ann. of Math. (2) **186** (2017), no. 1, 277–314, DOI 10.4007/annals.2017.186.1.7. MR3665005 ↑210

[DLMF22]   *NIST Digital Library of Mathematical Functions*. dlmf.nist.gov. Release 1.1.8 of 2022-12-15. F. W. J. Olver, A. B. Olde Daalhuis, D. W. Lozier, B. I. Schneider, R. F. Boisvert, C. W. Clark, B. R. Miller, B. V. Saunders, H. S. Cohl, and M. A. McClain, eds. 10, 11, 153, 158

[Don92]   R.-T. Dong, *Nodal sets of eigenfunctions on Riemann surfaces*, J. Differential Geom. **36** (1992), no. 2, 493–506, DOI 10.4310/jdg/1214448750. MR1180391 ↑118

[DonFef88]   H. Donnelly and C. Fefferman, *Nodal sets of eigenfunctions on Riemannian manifolds*, Invent. Math. **93** (1988), no. 1, 161–183, DOI 10.1007/BF01393691. MR943927 ↑117, 119

[DonFef90]   H. Donnelly and C. Fefferman, *Nodal sets for eigenfunctions of the Laplacian on surfaces*, J. Amer. Math. Soc. **3** (1990), no. 2, 333–353, DOI 10.2307/1990956. MR1035413 ↑118

[DuiGui75]   J. J. Duistermaat and V. W. Guillemin, *The spectrum of positive elliptic operators and periodic bicharacteristics*, Invent. Math. **29** (1975), no. 1, 39–79, DOI 10.1007/BF01405172. MR405514 ↑89

[DyaZwo19]   S. Dyatlov and M. Zworski, *Mathematical theory of scattering resonances*, Graduate Studies in Mathematics, vol. 200, American Mathematical Society, Providence, RI, 2019, DOI 10.1090/gsm/200. MR3969938 ↑xviii

[EdmEva18]   D. E. Edmunds and W. D. Evans, *Spectral theory and differential operators*, Second edition of [MR0929030], Oxford Mathematical Monographs, Oxford University Press, Oxford, 2018, DOI 10.1093/oso/9780198812050.001.0001. MR3823299 ↑28, 36

[Edw93a]   J. Edward, *An inverse spectral result for the Neumann operator on planar domains*, J. Funct. Anal. **111** (1993), no. 2, 312–322, DOI 10.1006/jfan.1993.1015. MR1203456 ↑236

[Edw93b]   J. Edward, *Pre-compactness of isospectral sets for the Neumann operator on planar domains*, Comm. Partial Differential Equations **18** (1993), no. 7-8, 1249–1270, DOI 10.1080/03605309308820973. MR1233194 ↑236

[ElSGiaJaz06] A. El Soufi, H. Giacomini, and M. Jazar, *A unique extremal metric for the least eigenvalue of the Laplacian on the Klein bottle*, Duke Math. J. **135** (2006), no. 1, 181–202, DOI 10.1215/S0012-7094-06-13514-7. MR2259925 ↑171

[ElSIli84]   A. El Soufi and S. Ilias, *Le volume conforme et ses applications d'après Li et Yau*, Séminaire de Théorie Spectrale et Géométrie, Année 1983–1984, Univ. Grenoble I, Saint-Martin-d'Hères, 1984, pp. VII.1–VII.15. Available at http://www.numdam.org/item/TSG_1983-1984__2__A7_0/. MR1046044 ↑170

[Eva10]   L. C. Evans, *Partial differential equations*, 2nd ed., Graduate Studies in Mathematics, vol. 19, American Mathematical Society, Providence, RI, 2010, DOI 10.1090/gsm/019. MR2597943 ↑28, 30, 42, 51, 110, 188

[EvaGar15]   L. C. Evans and R. F. Gariepy, *Measure theory and fine properties of functions*, Revised edition, Textbooks in Mathematics, CRC Press, Boca Raton, FL, 2015. MR3409135 ↑30, 139

[Fab23]   G. Faber, *Beweis, dass unter allen homogenen Membranen von gleicher Fläche und gleicher Spannung die kreisförmige den tiefsten Grundton gibt*, Sitzungberichte der mathematisch-physikalischen Klasse der Bayerischen Akademie der Wissenschaften zu München Jahrgang (1923), 169–172. Available at https://publikationen.badw.de/en/003399311. 138

[Fed14]   H. Federer, *Geometric measure theory*, Classics in Mathematics, Springer, Berlin, Heidelberg, 2014, DOI 10.1007/978-3-642-62010-2. 107, 117

[Fel71]   W. Feller, *An introduction to probability theory and its applications. Vol. II.*, 2nd ed., John Wiley & Sons, Inc., New York-London-Sydney, 1971. MR0270403 ↑191

[Fil04]   N. Filonov, *On an inequality for the eigenvalues of the Dirichlet and Neumann problems for the Laplace operator*, Algebra i Analiz **16** (2004), no. 2, 172–176; English transl., St. Petersburg Math. J. **16** (2005), no. 2, 413–416, DOI 10.1090/S1061-0022-05-00857-5. MR2068346 ↑79

[FilLPS23]   N. Filonov, M. Levitin, I. Polterovich, and D. A. Sher, *Pólya's conjecture for Euclidean balls*, Invent. Math. (2023). DOI 10.1007/s00222-023-01198-1. 93

[Fol95]   G. B. Folland, *Introduction to partial differential equations*, 2nd ed., Princeton University Press, Princeton, NJ, 1995. MR1357411 ↑29, 38, 42, 44

[FoxKut83]   D. W. Fox and J. R. Kuttler, *Sloshing frequencies*, Z. Angew. Math. Phys. **34** (1983), no. 5, 668–696, DOI 10.1007/BF00948809. MR723140 ↑219, 223

[FraLapWei23]   R. L. Frank, A. Laptev, and T. Weidl, *Schrödinger operators: eigenvalues and Lieb-Thirring inequalities*, Cambridge Studies in Advanced Mathematics, vol. 200, Cambridge University Press, Cambridge, 2023, DOI 10.1017/9781009218436. MR4496335 ↑xviii, 97

[FreLagPay21]   P. Freitas, J. Lagacé, and J. Payette, *Optimal unions of scaled copies of domains and Pólya's conjecture*, Ark. Mat. **59** (2021), no. 1, 11–51, DOI 10.4310/arkiv.2021.v59.n1.a2. MR4256006 ↑90

[FreLau20]   P. Freitas and R. S. Laugesen, *From Steklov to Neumann and beyond, via Robin: the Szegő way*, Canad. J. Math. **72** (2020), no. 4, 1024–1043, DOI 10.4153/s0008414x19000154. MR4127919 ↑224

[Fri91]   L. Friedlander, *Some inequalities between Dirichlet and Neumann eigenvalues*, Arch. Rational Mech. Anal. **116** (1991), no. 2, 153–160, DOI 10.1007/BF00375590. MR1143438 ↑79, 259, 262

[Fri21]   L. Friedlander, *On the Weyl asymptotic formula for Euclidean domains of finite volume*, arXiv:2106.07690. 84

[Fri69]   A. Friedman, *Partial differential equations*, Holt, Rinehart and Winston, Inc., New York-Montreal, Que.-London, 1969. MR0445088 ↑41

[Fri00]   T. Friedrich, *Dirac operators in Riemannian geometry*, translated from the 1997 German original by A. Nestke, Graduate Studies in Mathematics, vol. 25, American Mathematical Society, Providence, RI, 2000, DOI 10.1090/gsm/025. MR1777332 ↑xviii

[Fto21]   I. Ftouhi, *On the Cheeger inequality for convex sets*, J. Math. Anal. Appl. **504** (2021), no. 2, Paper No. 125443, 26, DOI 10.1016/j.jmaa.2021.125443. MR4280278 ↑158

[Fus04]   N. Fusco, *The classical isoperimetric theorem*, Rend. Acad. Sci. Fis. Mat. Napoli (4) **71** (2004), 63–107. Available at https://www.docenti.unina.it/webdocenti-be/allegati/materiale-didattico/377226. 145

[Fus08]   N. Fusco, *Geometrical aspects of symmetrization*, Calculus of variations and nonlinear partial differential equations, Lecture Notes in Math., vol. 1927, Springer, Berlin, 2008, pp. 155–181, DOI 10.1007/978-3-540-75914-0_5. MR2408261 ↑144

[Gaf58]   M. P. Gaffney, *Asymptotic distributions associated with the Laplacian for forms*, Comm. Pure Appl. Math. **11** (1958), 535–545, DOI 10.1002/cpa.3160110405. MR99541 ↑184

[GarLin86]   N. Garofalo and F.-H. Lin, *Monotonicity properties of variational integrals, A_p weights and unique continuation*, Indiana Univ. Math. J. **35** (1986), no. 2, 245–268. Available at http://www.jstor.org/stable/24893906. 120

[Gas26]   F. Gassmann, *Bemerkungen zur vorstehenden Arbeit von Hurwitz*, Math. Zeit. **25** (1926), 665–675. DOI 10.1007/BF01283860. 199

[GilTru01]   D. Gilbarg and N. S. Trudinger, *Elliptic partial differential equations of second order*, reprint of the 1998 edition, Classics in Mathematics, Springer-Verlag, Berlin, 2001. MR1814364 ↑42, 43, 50, 51, 111

[Gil04]   P. B. Gilkey, *Asymptotic formulae in spectral geometry*, Studies in Advanced Mathematics, Chapman & Hall/CRC, Boca Raton, FL, 2004. MR2040963 ↑191, 192

[Gin09]   N. Ginoux, *The Dirac spectrum*, Lecture Notes in Mathematics, vol. 1976, Springer-Verlag, Berlin, 2009, DOI 10.1007/978-3-642-01570-0. MR2509837 ↑xviii

[GirKarLag21]   A. Girouard, M. Karpukhin, and J. Lagacé, *Continuity of eigenvalues and shape optimisation for Laplace and Steklov problems*, Geom. Funct. Anal. **31** (2021), no. 3, 513–561, DOI 10.1007/s00039-021-00573-5. MR4311579 ↑224

[GirKLP22]   A. Girouard, M. Karpukhin, M. Levitin, and I. Polterovich, *The Dirichlet-to-Neumann map, the boundary Laplacian, and Hörmander's rediscovered manuscript*, J. Spectr. Theory **12** (2022), no. 1, 195–225, DOI 10.4171/jst/399. MR4404812 ↑230, 233, 235, 236, 259, 260

[GirLPS19]   A. Girouard, J. Lagacé, I. Polterovich, and A. Savo, *The Steklov spectrum of cuboids*, Mathematika **65** (2019), no. 2, 272–310, DOI 10.1112/s0025579318000414. MR3884657 ↑223, 251

[GirNadPol09]   A. Girouard, N. Nadirashvili, and I. Polterovich, *Maximization of the second positive Neumann eigenvalue for planar domains*, J. Differential Geom. **83** (2009), no. 3, 637–661, DOI 10.4310/jdg/1264601037. MR2581359 ↑164

[GirPPS14]   A. Girouard, L. Parnovski, I. Polterovich, and D. A. Sher, *The Steklov spectrum of surfaces: asymptotics and invariants*, Math. Proc. Cambridge Philos. Soc. **157** (2014), no. 3, 379–389, DOI 10.1017/S030500411400036X. MR3286514 ↑236

[GirPol10a]   A. Girouard and I. Polterovich, *Shape optimization for low Neumann and Steklov eigenvalues*, Math. Methods Appl. Sci. **33** (2010), no. 4, 501–516, DOI 10.1002/mma.1222. MR2641628 ↑224

[GirPol10b]   A. Girouard and I. Polterovich, *On the Hersch–Payne–Schiffer inequalities for Steklov eigenvalues*, Funct. Anal. Its Appl. **44** (2010), 106–117. DOI 10.1007/s10688-010-0014-1. 225

[GirPol12]   A. Girouard and I. Polterovich, *Upper bounds for Steklov eigenvalues on surfaces*, Electron. Res. Announc. Math. Sci. **19** (2012), 77–85, DOI 10.3934/era.2012.19.77. MR2970718 ↑229

[GirPol17]   A. Girouard and I. Polterovich, *Spectral geometry of the Steklov problem (survey article)*, J. Spectr. Theory **7** (2017), no. 2, 321–359, DOI 10.4171/JST/164. MR3662010 ↑220, 224, 236, 263

[GorHerWeb21]   C. Gordon, P. Herbrich, and D. Webb, *Steklov and Robin isospectral manifolds*, J. Spectr. Theory **11** (2021), no. 1, 39–61, DOI 10.4171/jst/335. MR4233205 ↑209

[GorWebWol92]   C. Gordon, D. Webb, and S. Wolpert, *Isospectral plane domains and surfaces via Riemannian orbifolds*, Invent. Math. **110** (1992), no. 1, 1–22, DOI 10.1007/BF01231320. MR1181812 ↑200, 204

[GreNgu13]   D. S. Grebenkov and B.-T. Nguyen, *Geometrical structure of Laplacian eigenfunctions*, SIAM Rev. **55** (2013), no. 4, 601–667, DOI 10.1137/120880173. MR3124880 ↑12

[Gre86]   A. G. Greenhill, *Wave Motion in Hydrodynamics*, Amer. J. Math. **9** (1886), no. 1, 62–96, DOI 10.2307/2369499. MR1505437 ↑219

[GréNédPla76]   J. P. Gregoire, J.-C. Nédélec, and J. Planchard, *A method of finding the eigenvalues and eigenfunctions of selfadjoint elliptic operators*, Comput. Methods Appl. Mech. Engrg. **8** (1976), no. 2, 201–214, DOI 10.1016/0045-7825(76)90045-1. MR451767 ↑255

[Gri06]   D. Grieser, *The first eigenvalue of the Laplacian, isoperimetric constants, and the max flow min cut theorem*, Arch. Math. (Basel) **87** (2006), no. 1, 75–85, DOI 10.1007/s00013-005-1623-4. MR2246409 ↑155, 156, 157

[GriNetYau04]   A. Grigor'yan, Y. Netrusov, and S.-T. Yau, *Eigenvalues of elliptic operators and geometric applications*, Surveys in Differential Geometry, vol. 9, Int. Press, Somerville, MA, 2004, pp. 147–217, DOI 10.4310/SDG.2004.v9.n1.a5. MR2195408 ↑172

[Gri11]   P. Grisvard, *Elliptic problems in nonsmooth domains*, reprint of the 1985 original [MR0775683], with a foreword by S. C. Brenner, Classics in Applied Mathematics, vol. 69, Society for Industrial and Applied Mathematics (SIAM), Philadelphia, PA, 2011, DOI 10.1137/1.9781611972030.ch1. MR3396210 ↑29, 30, 62

[GuiKaz80]   V. Guillemin and D. Kazhdan, *Some inverse spectral results for negatively curved 2-manifolds*, Topology **19** (1980), no. 3, 301–312, DOI 10.1016/0040-9383(80)90015-4. MR579579 ↑211

[Gun72]   R. C. Gunning, *Lectures on Riemann surfaces, Jacobi varieties*, Mathematical Notes, No. 12, Princeton University Press, Princeton, N.J.; University of Tokyo Press, Tokyo, 1972. MR0357407 ↑170

[GusAbe98]   K. Gustafson and T. Abe, *The third boundary condition—was it Robin's?*, Math. Intelligencer **20** (1998), no. 1, 63–71, DOI 10.1007/BF03024402. MR1601764 ↑61

[Hal22]   N. Halperin, *Estimates of the Hausdorff measure of nodal sets of Laplace eigenfunctions*, M.Sc. thesis, Hebrew University, 2022. 128

[HarSim89]   R. Hardt and L. Simon, *Nodal sets for solutions of elliptic equations*, J. Differential Geom. **30** (1989), no. 2, 505–522, DOI 10.4310/jdg/1214443599. MR1010169 ↑107, 118, 120

[Har15]   G. H. Hardy, *On the expression of a number as the sum of two squares*, Quart. J. Pure Appl. Math. **46** (1915), 263–283. 19

[HarWri08]   G. H. Hardy and E. M. Wright, *An introduction to the theory of numbers*, Oxford University Press, Oxford, 2008. 7

[HarMic95]   E. M. Harrell II and P. L. Michel, *Commutator bounds for eigenvalues of some differential operators*, Evolution equations (Baton Rouge, LA, 1992), Lecture Notes in Pure and Appl. Math., vol. 168, Dekker, New York, 1995, pp. 235–244. MR1300432 ↑180

[HarStu97]   E. M. Harrell II and J. Stubbe, *On trace identities and universal eigenvalue estimates for some partial differential operators*, Trans. Amer. Math. Soc. **349** (1997), no. 5, 1797–1809, DOI 10.1090/S0002-9947-97-01846-1. MR1401772 ↑174

[Has11]   A. Hassannezhad, *Conformal upper bounds for the eigenvalues of the Laplacian and Steklov problem*, J. Funct. Anal. **261** (2011), no. 12, 3419–3436, DOI 10.1016/j.jfa.2011.08.003. MR2838029 ↑172

[HasKokPol16] A. Hassannezhad, G. Kokarev, and I. Polterovich, *Eigenvalue inequalities on Riemannian manifolds with a lower Ricci curvature bound*, J. Spectr. Theory **6** (2016), no. 4, 807–835, DOI 10.4171/JST/143. MR3584185 ↑155

[HasShe22]   A. Hassannezhad and D. A. Sher, *Nodal count for Dirichlet-to-Neumann operators with potential*, Proc. Amer. Math. Soc. (2022). DOI 10.1090/proc/16207. 256

[HasSif20]   A. Hassannezhad and A. Siffert, *A note on Kuttler-Sigillito's inequalities*, Ann. Math. Qué. **44** (2020), no. 1, 125–147, DOI 10.1007/s40316-019-00113-6. MR4071873 ↑260

[HasTao02]   A. Hassell and T. Tao, *Upper and lower bounds for normal derivatives of Dirichlet eigenfunctions*, Math. Res. Lett. **9** (2002), no. 2-3, 289–305, DOI 10.4310/MRL.2002.v9.n3.a6. MR1909646 ↑261

[Hat01]   A. Hatcher, *Algebraic topology*, Cambridge University Press, Cambridge, 2002. Available at author's website https://pi.math.cornell.edu/~hatcher/AT/AT.pdf. MR1867354 ↑198

[Hay78]   W. K. Hayman, *Some bounds for principal frequency*, Applicable Anal. **7** (1977/78), no. 3, 247–254, DOI 10.1080/00036817808839195. MR492339 ↑156

[Hec12]   F. Hecht, *New development in freefem++*, J. Numer. Math. **20** (2012), no. 3-4, 251–265, DOI 10.1515/jnum-2012-0013. MR3043640 ↑273

[Hed81]   L. I. Hedberg, *Spectral synthesis in Sobolev spaces, and uniqueness of solutions of the Dirichlet problem*, Acta Math. **147** (1981), no. 3-4, 237–264, DOI 10.1007/BF02392874. MR639040 ↑104

[HeiKilMar93] J. Heinonen, T. Kilpeläinen, and O. Martio, *Nonlinear potential theory of degenerate elliptic equations*, Oxford Science Publications, Oxford Mathematical Monographs, The Clarendon Press, Oxford University Press, New York, 1993. MR1207810 ↑51, 104

[Hei30]   W. Heiseinberg, *Über quantentheoretische Umdeutung kinematischer und mechanischer Beziehungen*, Z. Physik **33** (1925), 879–893. DOI 10.1007/BF01328377. 177

[Hel13]   B. Helffer, *Spectral theory and its applications*, Cambridge Studies in Advanced Mathematics, vol. 139, Cambridge University Press, Cambridge, 2013, DOI 10.1007/BF01328377. MR3027462 ↑x, 38

[HelSun16]   B. Helffer and M. Persson Sundqvist, *On nodal domains in Euclidean balls*, Proc. Amer. Math. Soc. **144** (2016), no. 11, 4777–4791, DOI 10.1090/proc/13098. MR3544529 ↑11

[HemSecSim91] R. Hempel, L. A. Seco, and B. Simon, *The essential spectrum of Neumann Laplacians on some bounded singular domains*, J. Funct. Anal. **102** (1991), no. 2, 448–483, DOI 10.1016/0022-1236(91)90130-W. MR1140635 ↑38

[Hen06]   A. Henrot, *Extremum problems for eigenvalues of elliptic operators*, Frontiers in Mathematics, Birkhäuser Verlag, Basel, 2006. MR2251558 ↑x, 146, 147

[Her70]   J. Hersch, *Quatre propriétés isopérimétriques de membranes sphériques homogènes*, C. R. Acad. Sci. Paris Sér. A-B **270** (1970), A1645–A1648. MR292357 ↑165

[HerPaySch75] J. Hersch, L. E. Payne, and M. M. Schiffer, *Some inequalities for Stekloff eigenvalues*, Arch. Rational Mech. Anal. **57** (1975), 99–114, DOI 10.1007/BF00248412. MR387837 ↑225

[HezZel22]   H. Hezari and S. Zelditch, *One can hear the shape of ellipses of small eccentricity*, Ann. of Math. (2) **196** (2022), no. 3, 1083–1134, DOI 10.4007/annals.2022.196.3.4. MR4502596 ↑210

[HilPro80]   G. N. Hile and M. H. Protter, *Inequalities for eigenvalues of the Laplacian*, Indiana Univ. Math. J. **29** (1980), 523–538. Available at https://www.jstor.org/stable/24892878. 174

[HisLut01]   P. D. Hislop and C. V. Lutzer, *Spectral asymptotics of the Dirichlet-to-Neumann map on multiply connected domains in $\mathbb{R}^d$*, Inverse Problems **17** (2001), no. 6, 1717–1741, DOI 10.1088/0266-5611/17/6/313. MR1872919 ↑217

[Hör54]   L. Hörmander, *Uniqueness theorems and estimates for normally hyperbolic partial differential equations of the second order*, Tolfte Skandinaviska Matematikerkongressen, Lund, 1953, Lunds Universitets Matematiska Institution, Lund, 1954, pp. 105–115. MR0065783 ↑230

[Hör18]   L. Hörmander, *Inequalities between normal and tangential derivatives of harmonic functions*, Unpublished manuscripts, 37–41, Springer International Publishing, 2018. DOI 10.1007/978-3-319-69850-2.6 230, 232, 233

[Ivr80]   V. Ya. Ivrii, *Second term of the spectral asymptotic expansion of the Laplace–Beltrami operator on manifolds with boundary*, Funct. Anal. Its Appl. **14** (1980), 98–106. DOI 10.1007/BF01086550. 87

[Ivr19]   V. Ivrii, *Spectral asymptotics for Dirichlet to Neumann operator in the domains with edges*, Microlocal Analysis, Sharp Spectral Asymptotics and Applications V, Springer, Cham, 2019, 513–539. DOI 10.1007/978-3-030-30561-1.31. 251

[JakLNNP05]   D. Jakobson, M. Levitin, N. Nadirashvili, N. Nigam, and I. Polterovich, *How large can the first eigenvalue be on a surface of genus two?*, Int. Math. Res. Not. **2005** (2005), no. 63, 3967–3985. DOI 10.1155/IMRN.2005.3967. ↑71

[JakLNP06]   D. Jakobson, M. Levitin, N. Nadirashvili, and I. Polterovich, *Spectral problems with mixed Dirichlet-Neumann boundary conditions: isospectrality and beyond*, J. Comput. Appl. Math. **194** (2006), no. 1, 141–155, DOI 10.1016/j.cam.2005.06.019. MR2230975 ↑76, 203

[JakNadPol06]   D. Jakobson, N. Nadirashvili, and I. Polterovich, *Extremal metric for the first eigenvalue on a Klein bottle*, Canad. J. Math. **58** (2006), no. 2, 381–400, DOI 10.4153/CJM-2006-016-0. MR2209284 ↑171

[JerKen81]   D. S. Jerison and C. E. Kenig, *The Neumann problem on Lipschitz domains*, Bull. Amer. Math. Soc. (N.S.) **4** (1981), no. 2, 203–207, DOI 10.1090/S0273-0979-1981-14884-9. MR598688 ↑52, 53

[JerKen95]   D. Jerison and C. E. Kenig, *The inhomogeneous Dirichlet problem in Lipschitz domains*, J. Funct. Anal. **130** (1995), no. 1, 161–219, DOI 10.1006/jfan.1995.1067. MR1331981 ↑52

[JimMor92]   S. Jimbo and Y. Morita, *Remarks on the behavior of certain eigenvalues on a singularly perturbed domain with several thin channels*, Comm. Partial Differential Equations **17** (1992), no. 3-4, 523–552, DOI 10.1080/03605309208820852. MR1163435 ↑154

[Joh81]   F. John, *Plane waves and spherical means applied to partial differential equations*, Springer-Verlag, New York-Berlin, 1981. MR614918 ↑41

[JolSha14]   A. Jollivet and V. Sharafutdinov, *On an inverse problem for the Steklov spectrum of a Riemannian surface*, Inverse problems and applications, Contemp. Math., vol. 615, Amer. Math. Soc., Providence, RI, 2014, pp. 165–191, DOI 10.1090/conm/615/12260. MR3221604 ↑236

[JolSha18]   A. Jollivet and V. Sharafutdinov, *Steklov zeta-invariants and a compactness theorem for isospectral families of planar domains*, J. Funct. Anal. **275** (2018), no. 7, 1712–1755, DOI 10.1016/j.jfa.2018.06.019. MR3832007 ↑236

[Kac51]   M. Kac, *On some connections between probability theory and differential and integral equations*, Proceedings of the Second Berkeley Symposium on Mathematical Statistics and Probability, 1950, University of California Press, Berkeley-Los Angeles, Calif., 1951, pp. 189–215. MR0045333 ↑192

[Kac66]   M. Kac, *Can one hear the shape of a drum?*, Amer. Math. Monthly **73** (1966), no. 4, 1–23, DOI 10.2307/2313748. MR201237 ↑190, 209

[KalSor18]   V. Kaloshin and A. Sorrentino, *On the local Birkhoff conjecture for convex billiards*, Ann. of Math. (2) **188** (2018), no. 1, 315–380, DOI 10.4007/annals.2018.188.1.6. MR3815464 ↑210

[Kar16]   M. A. Karpukhin, *Upper bounds for the first eigenvalue of the Laplacian on non-orientable surfaces*, Int. Math. Res. Not. **20** (2016), 6200–6209, DOI 10.1093/imrn/rnv345. MR3579963 ↑170

[Kar18]   M. Karpukhin, *Bounds between Laplace and Steklov eigenvalues on nonnegatively curved manifolds*, Electron. Res. Announc. Math. Sci. **24** (2017), 100–109, DOI 10.3934/era.2017.24.011. MR3699063 ↑229

[Kar19]   M. Karpukhin, *On the Yang-Yau inequality for the first Laplace eigenvalue*, Geom. Funct. Anal. **29** (2019), no. 6, 1864–1885, DOI 10.1007/s00039-019-00518-z. MR4034923 ↑172

[Kar21]   M. Karpukhin, *Index of minimal spheres and isoperimetric eigenvalue inequalities*, Invent. Math. **223** (2021), no. 1, 335–377, DOI 10.1007/s00222-020-00992-5. MR4199444 ↑172

[KarKokPol14] M. Karpukhin, G. Kokarev, and I. Polterovich, *Multiplicity bounds for Steklov eigenvalues on Riemannian surfaces*, Ann. Inst. Fourier (Grenoble) **64** (2014), no. 6, 2481–2502, DOI 10.5802/aif.2918. MR3331172 ↑133, 135

[KarLagPol23] M. Karpukhin, J. Lagacé, and I. Polterovich, *Weyl's law for the Steklov problem on surfaces with rough boundary*, Arch. Rational Mech. Anal. **47** (2023), article no. 77, DOI 10.1007/s00205-023-01912-6. 230

[KarNPP19] M. Karpukhin, N. Nadirashvili, A. V. Penskoi, and I. Polterovich, *Conformally maximal metrics for Laplace eigenvalues on surfaces*, Surveys in Differential Geometry 2019. Differential geometry, Calabi-Yau theory, and general relativity. Part 2, Surv. Differ. Geom., vol. 24, Int. Press, Boston, MA, 2022, pp. 205–256, DOI 10.4310/SDG.2019.v24.n1.a6. MR4479722 ↑169, 172, 173

[KarNPP21] M. Karpukhin, N. Nadirashvili, A. V. Penskoi, and I. Polterovich, *An isoperimetric inequality for Laplace eigenvalues on the sphere*, J. Differential Geom. **118** (2021), no. 2, 313–333, DOI 10.4310/jdg/1622743142. MR4278696 ↑171, 172

[KarNPS21] M. Karpukhin, M. Nahon, I. Polterovich and D. Stern, *Stability of isoperimetric inequalities for Laplace eigenvalues on surfaces*, arXiv:2106.15043. 172

[KarSte20] M. Karpukhin and D. L. Stern, *Min-max harmonic maps and a new characterization of conformal eigenvalues*, arXiv:2004.04086. 172

[KarSte22] M. Karpukhin and D. L. Stern, *Existence of harmonic maps and eigenvalue optimization in higher dimensions*, arXiv:2207.13635. 173

[KarVin22] M. Karpukhin and D. Vinokurov, *The first eigenvalue of the Laplacian on orientable surfaces*, Math. Z. **301** (2022), no. 3, 2733–2746, DOI 10.1007/s00209-022-03009-4. MR4437337 ↑172

[Kaw85] B. Kawohl, *Rearrangements and convexity of level sets in PDE*, Lecture Notes in Mathematics, vol. 1150, Springer-Verlag, Berlin, 1985, DOI 10.1007/BFb0075060. MR810619 ↑144

[Kel66] R. Kellner, *On a theorem of Polya*, Amer. Math. Monthly **73** (1966), 856–858, DOI 10.2307/2314181. MR200623 ↑91

[Kha18] M. Khalile, *Spectral asymptotics for Robin Laplacians on polygonal domains*, J. Math. Anal. Appl. **461** (2018), no. 2, 1498–1543, DOI 10.1016/j.jmaa.2018.01.062. MR3765502 ↑261

[KhaOuBPan20] M. Khalile, T. Ourmières-Bonafos, and K. Pankrashkin, *Effective operators for Robin eigenvalues in domains with corners*, Ann. Inst. Fourier (Grenoble) **70** (2020), no. 5, 2215–2301, DOI 10.5802/aif.3400. MR4245611 ↑261

[KhaPan18] M. Khalile and K. Pankrashkin, *Eigenvalues of Robin Laplacians in infinite sectors*, Math. Nachr. **291** (2018), no. 5-6, 928–965, DOI 10.1002/mana.201600314. MR3795565 ↑261

[Kim22] H. N. Kim, *Maximization of the second Laplacian eigenvalue on the sphere*, Proc. Amer. Math. Soc. **150** (2022), no. 8, 3501–3512, DOI 10.1090/proc/15908. MR4439471 ↑173

[Kin21] J. Kinnunen, *Sobolev spaces*, Lecture notes, Aalto University (2021). Available at author's website https://math.aalto.fi/~jkkinnun/files/sobolev_spaces.pdf. 104

[Kok14] G. Kokarev, *Variational aspects of Laplace eigenvalues on Riemannian surfaces*, Adv. Math. **258** (2014), 191–239, DOI 10.1016/j.aim.2014.03.006. MR3190427 ↑225

[Kor93] N. Korevaar, *Upper bounds for eigenvalues of conformal metrics*, J. Differential Geom. **37** (1993), no. 1, 73–93, DOI 10.4310/jdg/1214453423. MR1198600 ↑172

[KotSmi99]   T. Kottos and U. Smilansky, *Periodic orbit theory and spectral statistics for quantum graphs*, Ann. Physics **274** (1999), no. 1, 76–124, DOI 10.1006/a-phy.1999.5904. MR1694731 ↑240

[Kra25]   E. Krahn, *Über eine von Rayleigh formulierte Minimaleigenschaft des Kreises*, Math. Ann. **94** (1925), no. 1, 97–100, DOI 10.1007/BF01208645. MR1512244 ↑138, 142

[Kra26]   E. Krahn, *Über Minimaleigenschaften der Kugel in drei un mehr Dimensionen*, (German) Acta Commun. Univ. Dorpat. A **9** (1926), 1–44. 147

[Krö92]   P. Kröger, *Upper bounds for the Neumann eigenvalues on a bounded domain in Euclidean space*, J. Funct. Anal. **106** (1992), no. 2, 353–357, DOI 10.1016/0022-1236(92)90052-K. MR1165859 ↑97

[KryLPPS21]   S. Krymski, M. Levitin, L. Parnovski, I. Polterovich, and D. A. Sher, *Inverse Steklov spectral problem for curvilinear polygons*, Int. Math. Res. Not. **2021** (2021), no. 1, 1–37, DOI 10.1093/imrn/rnaa200. MR4198492 251

[Kuh25]   W. Kuhn, *Über die Gesamtstarke der von einem Zustande ausgehenden Absorptionslinien*, Z. Physik **33** (1925), no. 1, 408–412, DOI 10.1007/BF01328322. 177

[KurNow10]   P. Kurasov and M. Nowaczyk, *Geometric properties of quantum graphs and vertex scattering matrices*, Opuscula Math. **30** (2010), no. 3, 295–309, DOI 10.7494/OpMath.2010.30.3.295. MR2669120 ↑240

[KurSuh20]   P. Kurasov and R. Suhr, *Asymptotically isospectral quantum graphs and generalised trigonometric polynomials*, J. Math. Anal. Appl. **488** (2020), no. 1, 124049,15, DOI 10.1016/j.jmaa.2020.124049. MR4079594 ↑252

[KutSig84]   J. R. Kuttler and V. G. Sigillito, *Eigenvalues of the Laplacian in two dimensions*, SIAM Rev. **26** (1984), no. 2, 163–193, DOI 10.1137/1026033. MR738929 ↑12

[KuzKKNPPS14] N. Kuznetsov, T. Kulczycki, M. Kwaśnicki, A. Nazarov, S. Poborchi, I. Polterovich, and B. Siudeja, *The legacy of Vladimir Andreevich Steklov*, Notices Amer. Math. Soc. **61** (2014), no. 1, 9–22, DOI 10.1090/noti1073. MR3137253 ↑214

[LagStA21]   J. Lagacé and S. St-Amant, *Spectral invariants of Dirichlet-to-Neumann operators on surfaces*, J. Spectr. Theory **11** (2021), no. 4, 1627–1667, DOI 10.4171/jst/382. MR4349672 ↑259

[Lam93]   H. Lamb, *Hydrodynamics*, 6th ed., with a foreword by R. A. Caflisch, Cambridge Mathematical Library, Cambridge University Press, Cambridge, 1993. MR1317348 ↑219

[Lam33]   G. Lamé, *Mémoire sur la propagation de la chaleur dans les polyèdres*, J. Éc. Polytech. **22** (1833), 194–251. 76

[Lap91]   M. L. Lapidus, *Fractal drum, inverse spectral problems for elliptic operators and a partial resolution of the Weyl-Berry conjecture*, Trans. Amer. Math. Soc. **325** (1991), no. 2, 465–529, DOI 10.2307/2001638. MR994168 ↑85

[Lap97]   A. Laptev, *Dirichlet and Neumann eigenvalue problems on domains in Euclidean spaces*, J. Funct. Anal. **151** (1997), no. 2, 531–545, DOI 10.1006/jfan.1997.3155. MR1491551 ↑97

[LapSaf96]   A. Laptev and Yu. Safarov, *A generalization of the Berezin-Lieb inequality*, Contemporary mathematical physics, Amer. Math. Soc. Transl. Ser. 2, vol. 175, Amer. Math. Soc., Providence, RI, 1996, pp. 69–79, DOI 10.1090/trans2/175/06. MR1402917 ↑97

[LapWei00]   A. Laptev and T. Weidl, *Sharp Lieb-Thirring inequalities in high dimensions*, Acta Math. **184** (2000), no. 1, 87–111, DOI 10.1007/BF02392782. MR1756570 ↑97

[Lau12]   R. S. Laugesen, *Spectral Theory of Partial Differential Equations — Lecture Notes*, arXiv:1203.2344. x, 58, 67

[Lax02]   P. D. Lax, *Functional Analysis*. John Wiley & Sons Inc., New York, 2002. 33, 35

[Led04]   M. Ledoux, *Spectral gap, logarithmic Sobolev constant, and geometric bounds*, Surveys in Differential Geometry, vol. 9, Int. Press, Somerville, MA, 2004, pp. 219–240, DOI 10.4310/SDG.2004.v9.n1.a6. MR2195409 ↑155

[Lén19]   C. Léna, *Pleijel's nodal domain theorem for Neumann and Robin eigenfunctions*, Ann. Inst. Fourier (Grenoble) **69** (2019), no. 1, 283–301, DOI 10.5802/aif.3243. MR3973450 ↑149

[LevWei86]   H. A. Levine and H. F. Weinberger, *Inequalities between Dirichlet and Neumann eigenvalues*, Arch. Rational Mech. Anal. **94** (1986), no. 3, 193–208, DOI 10.1007/BF00279862. MR846060 ↑82

[LevPar02]   M. Levitin and L. Parnovski, *Commutators, spectral trace identities, and universal estimates for eigenvalues*, J. Funct. Anal. **192** (2002), no. 2, 425–445, DOI 10.1006/jfan.2001.3913. MR1923409 ↑174, 176, 179, 180

[LevPar08]   M. Levitin and L. Parnovski, *On the principal eigenvalue of a Robin problem with a large parameter*, Math. Nachr. **281** (2008), no. 2, 272–281, DOI 10.1002/mana.200510600. MR2387365 ↑261

[LevParPol06] M. Levitin, L. Parnovski, and I. Polterovich, *Isospectral domains with mixed boundary conditions*, J. Phys. A **39** (2006), no. 9, 2073–2082, DOI 10.1088/0305-4470/39/9/006. MR2211977 ↑201, 203

[LevPPS22a]   M. Levitin, L. Parnovski, I. Polterovich, and D. A. Sher, *Sloshing, Steklov and corners: asymptotics of sloshing eigenvalues*, J. Anal. Math. **146** (2022), no. 1, 65–125, DOI 10.1007/s11854-021-0188-x. MR4467995 ↑219, 223, 239, 245, 250, 251

[LevPPS22b]   M. Levitin, L. Parnovski, I. Polterovich, and D. A. Sher, *Sloshing, Steklov and corners: asymptotics of Steklov eigenvalues for curvilinear polygons*, Proc. Lond. Math. Soc. (3) **125** (2022), no. 3, 359–487, DOI 10.1112/plms.12461. MR4480880 ↑237, 239, 240, 246, 251

[LevStr21]   M. Levitin and A. Strohmaier, *Computations of eigenvalues and resonances on perturbed hyperbolic surfaces with cusps*, Int. Math. Res. Not. **2021** (2021), no. 6, 4003–4050. DOI 10.1093/imrn/rnz157. 280

[LevVas96]   M. Levitin and D. Vassiliev, *Spectral asymptotics, renewal theorem, and the Berry conjecture for a class of fractals*, Proc. London Math. Soc. (3) **72** (1996), no. 1, 188–214, DOI 10.1112/plms/s3-72.1.188. MR1357092 ↑85

[LevYag03]   M. Levitin and R. Yagudin, *Range of the first three eigenvalues of the planar Dirichlet Laplacian*, LMS J. Comput. Math. **6** (2003), 1–17, DOI 10.1112/S1461157000000346. MR1971489 ↑179

[Lew46]   H. Lewy, *Water waves on sloping beaches*, Bull. Amer. Math. Soc. **52** (1946), 737–775, DOI 10.1090/S0002-9904-1946-08643-7. MR22134 ↑244

[LiYau82]   P. Li and S.-T. Yau, *A new conformal invariant and its applications to the Willmore conjecture and the first eigenvalue of compact surfaces*, Invent. Math. **69** (1982), no. 2, 269–291, DOI 10.1007/BF01399507. MR674407 ↑170, 171

[LiYau83]   P. Li and S.-T. Yau, *On the Schrödinger equation and the eigenvalue problem*, Comm. Math. Phys. **88** (1983), no. 3, 309–318, DOI 10.1007/BF01213210. MR701919 ↑94

[Lie73]   E. H. Lieb, *The classical limit of quantum spin systems*, Comm. Math. Phys. **31** (1973), 327–340, DOI 10.1007/BF01646493. MR349181 ↑97

[LieLos97]   E. H. Lieb and M. Loss, *Analysis*, Graduate Studies in Mathematics, vol. 14, American Mathematical Society, Providence, RI, 1997, DOI 10.1090/gsm/014. MR1415616 ↑93, 95, 140, 141

[Log18a]   A. Logunov, *Nodal sets of Laplace eigenfunctions: polynomial upper estimates of the Hausdorff measure*, Ann. of Math. (2) **187** (2018), no. 1, 221–239, DOI 10.4007/annals.2018.187.1.4. MR3739231 ↑118, 126, 130

[Log18b]   A. Logunov, *Nodal sets of Laplace eigenfunctions: proof of Nadirashvili's conjecture and of the lower bound in Yau's conjecture*, Ann. of Math. (2) **187** (2018), no. 1, 241–262, DOI 10.4007/annals.2018.187.1.5. MR3739232 ↑118, 126

[LogMal18a]   A. Logunov and E. Malinnikova, *Nodal sets of Laplace eigenfunctions: estimates of the Hausdorff measure in dimensions two and three*, 50 years with Hardy spaces, Oper. Theory Adv. Appl., vol. 261, Birkhäuser/Springer, Cham, 2018, pp. 333–344, DOI 10.1007/978-3-319-59078-3_17. MR3792104 ↑118, 126, 128

[LogMal18b]   A. Logunov and E. Malinnikova, *Review of Yau's conjecture on zero sets of Laplace eigenfunctions*, Current developments in mathematics 2018, Int. Press, Somerville, MA, 2020, pp. 179–212, DOI 10.4310/CDM.2018.v2018.n1.a4. MR4363378 ↑118, 120

[LogMal20]   A. Logunov and E. Malinnikova, *Lecture notes on quantitative unique continuation for solutions of second order elliptic equations*, Harmonic analysis and applications, IAS/Park City Math. Ser., vol. 27, Amer. Math. Soc., [Providence], RI, 2020, pp. 1–33, DOI 10.1090/pcms/027/01. MR4249624 ↑x, 122

[Mak65]   E. Makai, *A lower estimation of the principal frequencies of simply connected membranes*, Acta Math. Acad. Sci. Hungar. **16** (1965), 319–323, DOI 10.1007/BF01904840. MR185263 ↑156

[Mak70]   E. Makai, *Complete orthogonal systems of eigenfunctions of three triangular membranes*, Studia Sci. Math. Hungar. **5** (1970), 51–62. MR272244 ↑76

[MalSha15]   E. G. Mal'kovich and V. A. Sharafutdinov, *Zeta-invariants of the Steklov spectrum of a plane domain*, Sibirsk. Mat. Zh. **56** (2015), no. 4, 853–877, DOI 10.17377/smzh.2015.56.411; English transl., Sib. Math. J. **56** (2015), no. 4, 678–698. MR3492876 ↑236

[Man08]   D. Mangoubi, *On the inner radius of a nodal domain*, Canad. Math. Bull. **51** (2008), no. 2, 249–260, DOI 10.4153/CMB-2008-026-2. MR2414212 ↑115

[Man13]   D. Mangoubi, *The effect of curvature on convexity properties of harmonic functions and eigenfunctions*, J. Lond. Math. Soc. (2) **87** (2013), no. 3, 645–662, DOI 10.1112/jlms/jds067. MR3073669 ↑124

[MaySenStA22] J. Mayrand, C. Senécal, and S. St-Amant, *Asymptotics of sloshing eigenvalues for a triangular prism*, Math. Proc. Cambridge Philos. Soc. **173** (2022), no. 3, 539–571, DOI 10.1017/S0305004121000712. MR4497970 ↑251

[Maz85]   V. G. Maz'ja, *Sobolev spaces*, translated from the Russian by T. O. Shaposhnikova, Springer Series in Soviet Mathematics, Springer-Verlag, Berlin, 1985, DOI 10.1007/978-3-662-09922-3. MR817985 ↑139

[MazShu05]   V. Maz'ya and M. Shubin, *Can one see the fundamental frequency of a drum?*, Lett. Math. Phys. **74** (2005), no. 2, 135–151, DOI 10.1007/s11005-005-0010-1. MR2191951 ↑156

[Maz91]   R. Mazzeo, *Remarks on a paper of L. Friedlander concerning inequalities between Neumann and Dirichlet eigenvalues*, Internat. Math. Res. Notices **1991** (1991), no. 4, 41–48. DOI 10.1155/S1073792891000065. 82

[McC11]   B. J. McCartin, *Laplacian eigenstructure of the equilateral triangle*, Hikari Ltd., Ruse, 2011. Available at publisher's website http://www.m-hikari.com/mccartin-3.pdf. MR2918422 ↑76

[McK70]   H. P. McKean, *An upper bound to the spectrum of* Δ *on a manifold of negative curvature*, J. Differential Geometry **4** (1970), 359–366, DOI 10.4310/jdg/1214429509. MR266100 ↑154

[McL00]   W. McLean, *Strongly elliptic systems and boundary integral equations*, Cambridge University Press, Cambridge, 2000. MR1742312 ↑29, 214, 254, 290

[Mel84]   R. Melrose, *The trace of the wave group*, Microlocal analysis (Boulder, Colo., 1983), Contemp. Math., vol. 27, Amer. Math. Soc., Providence, RI, 1984, pp. 127–167, DOI 10.1090/conm/027/741046. MR741046 ↑87

[Mét77]   G. Métivier, *Valeurs propres de problèmes aux limites elliptiques irrégulières*, Bull. Soc. Math. France Suppl. Mém. **51-52** (1977), 125–219, DOI 10.24033/msmf.235. MR473578 ↑85

[MeySer64] N. G. Meyers and J. Serrin, $H = W$, Proc. Nat. Acad. Sci. U.S.A. **51** (1964), 1055–1056, DOI 10.1073/pnas.51.6.1055. MR164252 ↑27

[Mil64]   J. Milnor, *Eigenvalues of the Laplace operator on certain manifolds*, Proc. Nat. Acad. Sci. U.S.A. **51** (1964), 542, DOI 10.1073/pnas.51.4.542. MR162204 ↑193, 195

[Mil97]   J. W. Milnor, *Topology from the differentiable viewpoint*, based on notes by D. W. Weaver, revised reprint of the 1965 original, Princeton Landmarks in Mathematics, Princeton University Press, Princeton, NJ, 1997. MR1487640 ↑164

[MinPle49] S. Minakshisundaram and Å. Pleijel, *Some properties of the eigenfunctions of the Laplace-operator on Riemannian manifolds*, Canad. J. Math. **1** (1949), 242–256, DOI 10.4153/cjm-1949-021-5. MR31145 ↑189

[MorNir57] C. B. Morrey Jr. and L. Nirenberg, *On the analyticity of the solutions of linear elliptic systems of partial differential equations*, Comm. Pure Appl. Math. **10** (1957), 271–290, DOI 10.1002/cpa.3160100204. MR89334 ↑41

[Mos60]   J. Moser, *A new proof of De Giorgi's theorem concerning the regularity problem for elliptic differential equations*, Comm. Pure Appl. Math. **13** (1960), 457–468, DOI 10.1002/cpa.3160130308. MR170091 ↑50

[Nad87]   N. S. Nadirashvili, *Multiple eigenvalues of the Laplace operator*, Mat. Sb. (N.S.) **133(175)** (1987), no. 2, 223–237, 272; English transl., Math. USSR-Sb. **61** (1988), no. 1, 225–238, DOI 10.1070/SM1988v061n01ABEH003204. MR905007 ↑133, 134, 171, 172

[Nad96]   N. Nadirashvili, *Berger's isoperimetric problem and minimal immersions of surfaces*, Geom. Funct. Anal. **6** (1996), no. 5, 877–897, DOI 10.1007/BF02246788. MR1415764 ↑171

[Nad97]   N. Nadirashvili, *Conformal maps and isoperimetric inequalities for eigenvalues of the Neumann problem*, Proceedings of the Ashkelon Workshop on Complex Function Theory (1996), Israel Math. Conf. Proc., vol. 11, Bar-Ilan Univ., Ramat Gan, 1997, pp. 197–201. MR1476715 ↑145

[Nad02]   N. Nadirashvili, *Isoperimetric inequality for the second eigenvalue of a sphere*, J. Differential Geom. **61** (2002), no. 2, 335–340, DOI 10.4310/jdg/1090351388. MR1972149 ↑172

[NadPen18] N. S. Nadirashvili and A. V. Penskoi, *An isoperimetric inequality for the second non-zero eigenvalue of the Laplacian on the projective plane*, Geom. Funct. Anal. **28** (2018), no. 5, 1368–1393, DOI 10.1007/s00039-018-0458-7. MR3856795 ↑172

[NadSir15] N. Nadirashvili and Y. Sire, *Maximization of higher order eigenvalues and applications*, Mosc. Math. J. **15** (2015), no. 4, 767–775, DOI 10.17323/1609-4514-2015-15-4-767-775. MR3438833 ↑172

[NadSir17]   N. Nadirashvili and Y. Sire, *Isoperimetric inequality for the third eigenvalue of the Laplace-Beltrami operator on* $\mathbb{S}^2$, J. Differential Geom. **107** (2017), no. 3, 561–571, DOI 10.4310/jdg/1508551225. MR3715349 ↑172

[Nam21]   P. T. Nam, *Functional Analysis II*, online lecture notes (2020–21) at LMU Munich. Available at author's website http://www.math.lmu.de/~nam/LectureNotesFA2021.pdf. 93

[NaySho19]   S. Nayatani and T. Shoda, *Metrics on a closed surface of genus two which maximize the first eigenvalue of the Laplacian*, C. R. Math. Acad. Sci. Paris **357** (2019), no. 1, 84–98, DOI 10.1016/j.crma.2018.11.008. MR3907601 ↑171

[NazPolSod05] F. Nazarov, L. Polterovich, and M. Sodin, *Sign and area in nodal geometry of Laplace eigenfunctions*, Amer. J. Math. **127** (2005), no. 4, 879–910, DOI 10.1353/ajm.2005.0030. MR2154374 ↑126

[Nec12]   J. Nečas, *Direct methods in the theory of elliptic equations*, translated from the 1967 French original by G. Tronel and A. Kufner, editorial coordination and preface by Š. Nečasová and a contribution by C. G. Simader, Springer Monographs in Mathematics, Springer, Heidelberg, 2012, DOI 10.1007/978-3-642-10455-8. MR3014461 ↑30

[NetSaf05]   Yu. Netrusov and Yu. Safarov, *Weyl asymptotic formula for the Laplacian on domains with rough boundaries*, Comm. Math. Phys. **253** (2005), no. 2, 481–509, DOI 10.1007/s00220-004-1158-8. MR2140257 ↑84

[Nir59]   L. Nirenberg, *On elliptic partial differential equations*, Ann. Scuola Norm. Sup. Pisa Cl. Sci. (3) **13** (1959), 115–162. Available at http://www.numdam.org/item/?id=ASNSP_1959_3_13_2_115_0. MR109940 ↑42

[NurRowShe19] M. Nursultanov, J. Rowlett, and D. A. Sher, *The heat kernel on curvilinear polygonal domains in surfaces*, `arXiv:1905.00259`. 193, 203

[OsgPhiSar88] B. Osgood, R. Phillips, and P. Sarnak, *Compact isospectral sets of surfaces*, J. Funct. Anal. **80** (1988), no. 1, 212–234, DOI 10.1016/0022-1236(88)90071-7. MR960229 ↑210

[Oss77]   R. Osserman, *A note on Hayman's theorem on the bass note of a drum*, Comment. Math. Helv. **52** (1977), no. 4, 545–555, DOI 10.1007/BF02567388. MR459099 ↑156

[Oss78]   R. Osserman, *The isoperimetric inequality*, Bull. Amer. Math. Soc. **84** (1978), no. 6, 1182–1238, DOI 10.1090/S0002-9904-1978-14553-4. MR500557 ↑153, 156

[OttBro13]   K. A. Ott and R. M. Brown, *The mixed problem for the Laplacian in Lipschitz domains*, Potential Anal. **38** (2013), no. 4, 1333–1364; see also *Correction to: The mixed problem for the Laplacian in Lipschitz domains*, ibid. **54** (2021), 213–217, DOI 10.1007/s11118-012-9317-6. MR3042705 ↑63

[Pan20]   K. Pankrashkin, *An eigenvalue estimate for a Robin p-Laplacian in* C^1 *domains*, Proc. Amer. Math. Soc. **148** (2020), no. 10, 4471–4477, DOI 10.1090/proc/15116. MR4135311 ↑261

[Par17]   E. Parini, *Reverse Cheeger inequality for planar convex sets*, J. Convex Anal. **24** (2017), no. 1, 107–122. Available at https://www.heldermann.de/JCA/JCA24/JCA241/jca24009.htm. MR3619661 ↑158

[Pay55]   L. E. Payne, *Inequalities for eigenvalues of membranes and plates*, J. Rational Mech. Anal. **4** (1955), 517–529, DOI 10.1512/iumj.1955.4.54016. MR70834 ↑79

[PayPólWei56] L. E. Payne, G. Pólya, and H. F. Weinberger, *On the ratio of consecutive eigenvalues*, J. Math. and Phys. **35** (1956), 289–298, DOI 10.1002/sapm1956351289. MR84696 ↑173

[PayWei60]   L. E. Payne and H. F. Weinberger, *An optimal Poincaré inequality for convex domains*, Arch. Rational Mech. Anal. **5** (1960), 286–292 (1960), DOI 10.1007/BF00252910. MR117419 ↑180

[Pet50]   A. S. Peters, *The effect of a floating mat on water waves*, Comm. Pure Appl. Math. **3** (1950), 319–354, DOI 10.1002/cpa.3160030402. MR41593 ↑245

[Pet06]   P. Petersen, *Riemannian geometry*, 2nd ed., Graduate Texts in Mathematics, vol. 171, Springer, New York, 2006. DOI 10.1007/978-0-387-29403-2. MR2243772 ↑14

[Pet14]   R. Petrides, *Maximization of the second conformal eigenvalue of spheres*, Proc. Amer. Math. Soc. **142** (2014), no. 7, 2385–2394, DOI 10.1090/S0002-9939-2014-12095-8. MR3195761 ↑172

[Pet18]   R. Petrides, *On the existence of metrics which maximize Laplace eigenvalue on surfaces*, Int. Math. Res. Not. **2018** (2018), no. 14, 4261–4355, DOI 10.1093/imrn/rnx004. MR3830572 172

[Pet22]   R. Petrides, *Maximizing one Laplace eigenvalue on n-dimensional manifolds*, arXiv:2211.15636. 173

[Pin80]   M. A. Pinsky, *The eigenvalues of an equilateral triangle*, SIAM J. Math. Anal. **11** (1980), no. 5, 819–827, DOI 10.1137/0511073. MR586910 ↑76

[Pin85]   M. A. Pinsky, *Completeness of the eigenfunctions of the equilateral triangle*, SIAM J. Math. Anal. **16** (1985), no. 4, 848–851, DOI 10.1137/0516063. MR793926 ↑76

[Ple56]   Å. Pleijel, *Remarks on Courant's nodal line theorem*, Comm. Pure Appl. Math. **9** (1956), 543–550, DOI 10.1002/cpa.3160090324. MR80861 ↑106, 148

[Poh65]   S. I. Pohožaev, *On the eigenfunctions of the equation $\Delta u + \lambda f(u) = 0$*, Soviet Math. Dokl **6** (1965), 1408–1411. Available at https://www.mathnet.ru/eng/dan/v165/i1/p36. 230

[Pol00]   I. Polterovich, *Heat invariants of Riemannian manifolds*, Israel J. Math. **119** (2000), 239–252, DOI 10.1007/BF02810670. MR1802656 ↑191

[Pol09]   I. Polterovich, *Pleijel's nodal domain theorem for free membranes*, Proc. Amer. Math. Soc. **137** (2009), no. 3, 1021–1024, DOI 10.1090/S0002-9939-08-09596-8. MR2457442 ↑149

[Pól49]   G. Pólya, *Torsional rigidity, principal frequency, electrostatic capacity and symmetrization*, Quart. Appl. Math. **6** (1948), no. 3, 267–277, DOI 10.1090/qam/26817. MR26817. 146

[Pól54]   G. Pólya, *Mathematics and plausible reasoning*, in two volumes, Princeton University Press, Princeton, NJ, 1954. 89

[Pól61]   G. Pólya, *On the eigenvalues of vibrating membranes*, Proc. London Math. Soc. (3) **11** (1961), 419–433, DOI 10.1112/plms/s3-11.1.419. MR129219 ↑91

[PólSze51] G. Pólya and G. Szegö, *Isoperimetric Inequalities in Mathematical Physics*, Annals of Mathematics Studies, No. 27, Princeton University Press, Princeton, N. J., 1951. MR0043486 ↑115, 146

[Pom92]   Ch. Pommerenke, *Boundary behaviour of conformal maps*, Grundlehren der mathematischen Wissenschaften [Fundamental Principles of Mathematical Sciences], vol. 299, Springer-Verlag, Berlin, 1992, DOI 10.1007/978-3-662-02770-7. MR1217706 ↑226

[Pop20]   N. Popoff, *The negative spectrum of the Robin Laplacian*, Spectral theory and mathematical physics, Lat. Amer. Math. Ser., Springer, Cham, 2020, pp. 229–242, DOI 10.1007/978-3-030-55556-6_12. MR4371264 ↑261

[ProStu19] L. Provenzano and J. Stubbe, *Weyl-type bounds for Steklov eigenvalues*, J. Spectr. Theory **9** (2019), no. 1, 349–377, DOI 10.4171/JST/250. MR3900789 ↑233

[RauTay75] J. Rauch and M. Taylor, *Potential and scattering theory on wildly perturbed domains*, J. Functional Analysis **18** (1975), 27–59, DOI 10.1016/0022-1236(75)90028-2. MR377303 ↑72

[Ray77]   Lord Rayleigh [= J. W. Strutt], *The theory of sound*, 1st edition, Macmillan, London, 1877–1878. 138

[ReeSim75]   M. Reed and B. Simon, *Methods of modern mathematical physics*, in four volumes, Academic Press, New York-London, 1975–1979. x, 84

[ReiTho25]   F. Reiche and W. Thomas, *Über die Zahl der Dispersionselektronen, die einem stationären Zustand zugeordnet sind*, Z. Physik **34** (1925), 510–525. DOI 10.1007/bf01328494. 177

[Rel40]   F. Rellich, *Darstellung der Eigenwerte von $\Delta u + \lambda u = 0$ durch ein Randintegral*, Math. Z. **46** (1940), 635–636, DOI 10.1007/BF01181459. MR2456 ↑261

[Ros22a]   A. Ros, *On the first eigenvalue of the Laplacian on compact surfaces of genus three*, J. Math. Soc. Japan **74** (2022), no. 3, 813–828, DOI 10.2969/jmsj/85898589. MR4484231 ↑172

[Ros22b]   A. Ros, *First eigenvalue of the Laplacian on compact surfaces for large genera*, `arXiv:2211.15172`. 172

[Ros97]   S. Rosenberg, *The Laplacian on a Riemannian manifold: An introduction to analysis on manifolds*, London Mathematical Society Student Texts, vol. 31, Cambridge University Press, Cambridge, 1997, DOI 10.1017/CBO9780511623783. MR1462892 ↑x, 13, 15, 184, 190, 200

[RoF15]   G. Roy-Fortin, *Nodal sets and growth exponents of Laplace eigenfunctions on surfaces*, Anal. PDE **8** (2015), no. 1, 223–255, DOI 10.2140/apde.2015.8.223. MR3336925 ↑126

[Roz72]   G. V. Rozenbljum [= G. Rozenblum], *On the eigenvalues of the first boundary value problem in unbounded domains*, Math. USSR-Sb. **18** (1972), no. 2, 235–248. DOI 10.1070/SM1972v018n02ABEH001766 84

[Roz86]   G. V. Rozenblyum [= G. Rozenblum], *On the asymptotics of the eigenvalues of certain two-dimensional spectral problems*, Sel. Math. Sov. **5** (1986), no. 3, 233–244. 236

[Roz23]   G. Rozenblum, *Weyl asymptotics for Poincaré–Steklov eigenvalues in a domain with Lipschitz boundary*, `arXiv:2304.04047`. 230

[RudWigYes21] Z. Rudnick, I. Wigman, and N. Yesha, *Differences between Robin and Neumann eigenvalues*, Comm. Math. Phys. **388** (2021), no. 3, 1603–1635, DOI 10.1007/s00220-021-04248-y. MR4340938 ↑261

[RuzSadSur20] M. Ruzhansky, M. Sadybekov, and D. Suragan, *Spectral geometry of partial differential operators*, Monographs and Research Notes in Mathematics, CRC Press, Boca Raton, FL, 2020, DOI 10.1201/9780429432965. MR4590446 ↑xviii

[SafVas97]   Yu. Safarov and D. Vassiliev, *The asymptotic distribution of eigenvalues of partial differential operators*, translated from the Russian manuscript by the authors, Translations of Mathematical Monographs, vol. 155, American Mathematical Society, Providence, RI, 1997, DOI 10.1090/mmono/155. MR1414899 ↑82, 86, 87, 88, 89

[San55]   L. Sandgren, *A vibration problem*, Medd. Lunds Univ. Mat. Sem. **13** (1955), 1–84. MR72348 ↑230

[Sar90]   P. Sarnak, *Determinants of Laplacians; heights and finiteness*, Analysis, et cetera, Academic Press, Boston, MA, 1990, pp. 601–622, DOI 10.1016/B978-0-12-574249-8.50033-X. MR1039364 ↑210

[Sav01]   A. Savo, *Lower bounds for the nodal length of eigenfunctions of the Laplacian*, Ann. Global Anal. Geom. **19** (2001), no. 2, 133–151, DOI 10.1023/A:1010774905973. MR1826398 ↑117

[Sch90]   A. Schiemann, *Ein Beispiel positiv definiter quadratischer Formen der Dimension 4 mit gleichen Darstellungszahlen*, Arch. Math. (Basel) **54** (1990), no. 4, 372–375, DOI 10.1007/BF01189584. MR1042130 ↑197

[Sch97]   A. Schiemann, *Ternary positive definite quadratic forms are determined by their theta series*, Math. Ann. **308** (1997), no. 3, 507–517, DOI 10.1007/s002080050086. MR1457743 ↑197

[SchYau94]   R. Schoen and S.-T. Yau, *Lectures on differential geometry*, lecture notes prepared by W. Y. Ding, K. C. Chang [G. Q. Zhang], J. Q. Zhong and Y. C. Xu, translated from the Chinese by W. Y. Ding and S. Y. Cheng, with a preface translated from the Chinese by K. Tso, International Press, Cambridge, MA, 1994. MR1333601 ↑x, 150, 165, 174

[Ser73]   J.-P. Serre, *A course in arithmetic*, translated from the French, Graduate Texts in Mathematics, No. 7, Springer-Verlag, New York-Heidelberg, 1973. DOI 10.1007/978-1-4684-9884-4. MR0344216 ↑196

[Sie29]   C. L. Siegel, *Über einige Anwendungen diophantischer Approximationen [reprint of Abhandlungen der Preußischen Akademie der Wissenschaften. Physikalisch-mathematische Klasse 1929, Nr. 1]*, On some applications of Diophantine approximations, Quad./Monogr., vol. 2, Ed. Norm., Pisa, 2014, pp. 81–138, DOI 10.1007/978-88-7642-520-2. MR3330350 ↑11

[Shu01]   M. A. Shubin, *Pseudodifferential operators and spectral theory*, 2nd ed., translated from the 1978 Russian original by S. I. Andersson, Springer-Verlag, Berlin, 2001, DOI 10.1007/978-3-642-56579-3. MR1852334 ↑20, 23, 39, 86

[Shu20]   M. Shubin, *Invitation to partial differential equations*, edited and with a foreword by M. Braverman, R. McOwen and P. Topalov, translated from the 2001 Russian original, Graduate Studies in Mathematics, vol. 205, American Mathematical Society, Providence, RI, 2020, DOI 10.1090/gsm/205. MR4171481 ↑x, 29, 31, 33, 78, 102, 108

[Sog17]   C. D. Sogge, *Fourier integrals in classical analysis*, 2nd ed., Cambridge Tracts in Mathematics, vol. 210, Cambridge University Press, Cambridge, 2017, DOI 10.1017/9781316341186. MR3645429 ↑xvii

[Spi88]   M. Spivak, *A comprehensive introduction to differential geometry*, vol. 4, Publish or Perish, Boston, 1975. MR0394453 16

[Ste14]   S. Steinerberger, *A geometric uncertainty principle with an application to Pleijel's estimate*, Ann. Henri Poincaré **15** (2014), no. 12, 2299–2319, DOI 10.1007/s00023-013-0310-4. MR3272823 ↑149

[Ste70]   W. Stenger, *Nonclassical choices in variational principles for eigen-values*, J. Functional Analysis **6** (1970), 157–164, DOI 10.1016/0022-1236(70)90053-4. MR0259646 ↑57

[Stö07]   H. J. Stöckmann, *Chladni meets Napoleon*, Eur. Phys. J. Spec. Top. **145** (2007), 15–23. DOI 10.1140/epjst/e2007-00144-5. 101

[Str08]   W. A. Strauss, *Partial differential equations: An introduction*, 2nd ed., John Wiley & Sons, Ltd., Chichester, 2008. MR2398759 ↑12

[SunZho17]   J. Sun and A. Zhou, *Finite element methods for eigenvalue problems*, Monographs and Research Notes in Mathematics, CRC Press, Boca Raton, FL, 2017. MR3617375 ↑266

[Sun85]   T. Sunada, *Riemannian coverings and isospectral manifolds*, Ann. of Math. (2) **121** (1985), no. 1, 169–186, DOI 10.2307/1971195. MR782558 ↑193, 199, 200

[Sze54]   G. Szegő, *Inequalities for certain eigenvalues of a membrane of given area*, J. Rational Mech. Anal. **3** (1954), 343–356, DOI 10.1512/iumj.1954.3.53017. MR61749 ↑159, 164

[Tan73]   S. Tanno, *Eigenvalues of the Laplacian of Riemannian manifolds*, Tohoku Math. J. (2) **25** (1973), 391–403, DOI 10.2748/tmj/1178241341. MR0334086 ↑211

[Tay11]   M. E. Taylor, *Partial differential equations I. Basic theory*, 2nd ed., Applied Mathematical Sciences, vol. 115, Springer, New York, 2011, DOI 10.1007/978-1-4419-7055-8. MR2744150 ↑39, 53

[Tay11b]   M. E. Taylor, *Partial differential equations II. Qualitative studies of linear equations*, 2nd ed., Applied Mathematical Sciences, vol. 116, Springer, New York, 2011, DOI 10.1007/978-1-4419-7052-7. MR2743652 ↑166

[Tré82]   F. Tréves, *Introduction to pseudodifferential operators and Fourier integral operators, Vol. 2*, Plenum Press, New York-London, 1980. 86

[Ura17]   H. Urakawa, *Spectral geometry of the Laplacian: Spectral analysis and differential geometry of the Laplacian*, World Scientific Publishing Co. Pte. Ltd., Hackensack, NJ, 2017, DOI 10.1142/10018. MR3701990 ↑174

[Vas86]   D. G. Vasil'ev [= D. Vassiliev], *Two-term asymptotics of the spectrum of a boundary value problem in the case of a piecewise smooth boundary*, Soviet Math. Dokl. **33** (1986), no. 1, 227–230. Available at http://mi.mathnet.ru/dan8733. 85, 87

[Wat95]   G. N. Watson, *A treatise on the theory of Bessel functions*, reprint of the second (1944) edition, Cambridge Mathematical Library, Cambridge University Press, Cambridge, 1995. MR1349110 ↑10, 11, 158

[Wei56]   H. F. Weinberger, *An isoperimetric inequality for the N-dimensional free membrane problem*, J. Rational Mech. Anal. **5** (1956), 633–636, DOI 10.1512/iumj.1956.5.55021. MR79286 ↑159

[Wei54]   R. Weinstock, *Inequalities for a classical eigenvalue problem*, J. Rational Mech. Anal. **3** (1954), 745–753, DOI 10.1512/iumj.1954.3.53036. MR64989 ↑224

[Wel72]   W. Welsh, *Monotonicity for eigenvalues of the Schrödinger operator on unbounded domains*, Arch. Rational Mech. Anal. **49** (1972/73), 129–136, DOI 10.1007/BF00281414. MR336047 ↑65

[Xio18]   C. Xiong, *Comparison of Steklov eigenvalues on a domain and Laplacian eigenvalues on its boundary in Riemannian manifolds*, J. Funct. Anal. **275** (2018), no. 12, 3245–3258, DOI 10.1016/j.jfa.2018.09.012. MR3864501 ↑233

[Yan91]   H. Yang, *An estimate of the difference between consecutive eigenvalues*, preprint IC/91/60 of the Intl. Centre for Theoretical Physics, Trieste, 1991 (revised preprint, Academia Sinica, 1995). Available at https://inis.iaea.org/collection/NCLCollectionStore/_Public/23/015/23015356.pdf. 174

[YanYau80]   P. C. Yang and S.-T. Yau, *Eigenvalues of the Laplacian of compact Riemann surfaces and minimal submanifolds*, Ann. Scuola Norm. Sup. Pisa Cl. Sci. (4) **7** (1980), no. 1, 55–63. Available at http://www.numdam.org/item/ASNSP_1980_4_7_1_55_0/. MR577325 ↑168, 169, 170

[Yau75]   S.-T. Yau, *Isoperimetric constants and the first eigenvalue of a compact Riemannian manifold*, Ann. Sci. École Norm. Sup. (4) **8** (1975), no. 4, 487–507, DOI 10.24033/asens.1299. MR397619 ↑153

[Yau82]   S.-T. Yau, *Problem section*, Seminar on Differential Geometry, Ann. of Math. Stud., vol. 102, Princeton Univ. Press, Princeton, N.J., 1982, pp. 669–706, DOI 10.1515/9781400881918-035. MR645762 ↑117, 172

[Yau86]   S.-T. Yau, *Nonlinear analysis in geometry*, Monographies de L'Enseignement Mathématique, vol. 33, L'Enseignement Mathématique, Geneva, 1986. MR865650 180

[Zar10]   S. Zaremba, *Sur un problème mixte relatif à l'équation de Laplace*, Bull. intern. Acad. des Sciences de Cracovie. Classe des Sciences Math. et Naturelles, Serie A: Sciences math. (1910), 313–344. 63

[Zel09]   S. Zelditch, *Inverse spectral problem for analytic domains. II. $\mathbb{Z}_2$-symmetric domains*, Ann. of Math. (2) **170** (2009), no. 1, 205–269, DOI 10.4007/ annals.2009.170.205. MR2521115 ↑210

[Zwo12]   M. Zworski, *Semiclassical analysis*, Graduate Studies in Mathematics, vol. 138, American Mathematical Society, Providence, RI, 2012, DOI 10.1090/gsm/138. MR2952218 ↑xvii

Index

In most cases, only the first appearance of a term or its definition is listed.

Symbols

$\int$ 108

$\langle \cdot, \cdot \rangle$ 287

$(\cdot, \cdot)_{H^m(\Omega)}$ 27

$\mathcal{A}(\Omega)$ 145

$\boldsymbol{\alpha}$ 237

A^* 34

$A_{u,x}(\cdot)$ 108

$B_{a,r}^d, B_{a,r}, B_r^d, \mathbb{B}^d$ 288

$\beta(\int, Q)$ 126

$\beta(\int, B)$ 118

$C, C^k, C^\infty, C_0^k, C_0^\infty$ 289

$C_{0,\Gamma}^\infty(\Omega)$ 64

$\mathcal{C}_F$ 140

cap 103

card 198

$C_{\mathrm{b},d}$ 87

C_d 83

$\mathbf{Ch}(\boldsymbol{\zeta})$ 240

$\chi(M)$ 134

$\mathbb{D}$ 288

$\mathcal{D}_0$ 214

$\mathcal{D}_\Lambda$ 254

∂_n 8

∂^α 289

Δ

    Euclidean space 1

    Riemannian manifold 15

    spherical coordinates 21

$-\Delta^{\mathrm{D}}$ 36

$-\Delta^{\mathrm{N}}$ 38

$-\Delta^{\mathrm{R},\gamma}$ 61

$-\Delta_\Gamma^{\mathrm{Z}}$ 64

$e(t, x, y)$ 184

$\mathcal{E}_0$ 214

$\mathcal{E}_\Lambda$ 253

$e_M(t)$ 186

η_k 148

$\hat{\int}$ 194

$\mathcal{F}u$ 28

$F_{\boldsymbol{\alpha},\boldsymbol{\ell}}(\tau)$ 240

$\mathcal{H}_0(\Omega)$ 215

$\mathcal{H}_\Lambda(\Omega)$ 254

$H_{0,\Gamma}^1(\Omega)$ 64

$\mathcal{H}^{d-1}(\cdot)$ 117

h_{D} 149

$H^m(\Omega)$ 27

$H_0^m(\Omega)$ 28

$H^m(\partial\Omega)$ 29

$H_{\mathrm{loc}}^m(\Omega)$ 29

h_{N} 151

$\mathcal{H}_m, \widetilde{\mathcal{H}}_m$ 21

$\mathbf{Hes}$ 231

$H_f(\cdot)$ 121

I_m 256

$\mathbf{Jac}$ 231

J_m 10

$j_{m,k}, j_{m,k}'$ 10

$\kappa_{d,m}$ 22

$\boldsymbol{\ell}$ 237

$L(\partial\Omega)$ 224

$\mathcal{L}_F(t)$ 94

$L^p, L^\infty, L^p_{\text{loc}}$ 288

$\overline{\lambda}_1$ 165

Λ_1 170

$\lambda_k(M)$ 60

λ_k^{D} 59

Λ_k 172

λ_k^{N} 60

$\lambda_k^{\text{R},\gamma}$ 62

λ_k^{Z} 64

$\mathcal{M}_{\alpha,\ell}$ 238

$m(\lambda_k)$ 134

μ_k 60

$M_{u,x}(\cdot)$ 108

$\mathbb{N}_0$ 287

$\mathcal{N}(\lambda)$ 17, 78

$\widetilde{\mathcal{N}}(\lambda)$ 78

$\mathcal{N}^{\text{D}}(\lambda)$ 79

$\mathcal{N}^{\text{N}}(\lambda)$ 79

$\mathcal{N}^{\text{S}}(\sigma)$ 230

$N_f(\cdot)$ 121

$N(h, B_r, c)$ 122

$N^b(h, B_r, c)$ 123

ν_k

   eigenvalues of quantum graph ... 239

   Laplace–Beltrami eigenvalues 216

Ω^* 137

$\Omega^\star$ 236

ω_d 288

$\text{ord}_x(f)$ 119

$\mathcal{P}_m, \widetilde{\mathcal{P}}_m$ 21

$\mathcal{P}_{\alpha,\ell}$ 237

$\mathcal{Q}$ 55

$Q_\Omega^{\text{D}}, Q_\Omega^{\text{N}}$ 56

$R[\cdot]$ 57

$R^{\text{S}}[\cdot]$ 217

$\rho_\Omega, \widetilde{\rho}_\Omega$ 155

$\mathbb{R}_+$ 287

R^* 137

Ric 154

$S^{d-1}_{a,r}, S^{d-1}_r, S_r, \mathbb{S}^{d-1}$ 288

$\mathcal{S}(\mathbb{R}^d)$ 290

σ_{d-1} 288

σ_k 215

σ_k^{Λ} 254

τ_m 239

τ_S 74

$T(\Omega)$ 146

$U_F(t)$ 94

$\mathcal{U}_F(t)$ 94

u^* 141

$V_F(t)$ 94

$\mathcal{V}_F(t)$ 94

$\mathcal{Z}_u$ 99

A

angle

   exceptional 241

   special 241

Ashbaugh–Benguria inequality 179

B

bathtub principle 95

Berezin–Li–Yau inequality 94

Bers's theorem 132

Bessel equation 9

Bessel function of the first kind 10

Bolza surface 171

Bourget hypothesis 11

bubbling 172

Buser's inequality 154

C

capacity 103

Cauchy–Riemann operator 45

Cheeger constant

   Dirichlet 149

   Neumann 151

Cheeger's dumbbell 154

Cheeger's inequality

   closed manifolds 152

   Dirichlet 150

   Neumann 152

co-area formula 139

commutator identities 174

conformal

   branched covering 169

   invariance 61

counting function 17, 78

Courant's theorem 103

Courant-sharp eigenvalue 112, 148

D

deck transformation 197

difference quotient 42

diffusion equation 184

directional derivative 12

Dirichlet boundary conditions 3, 7
Dirichlet energy 33
Dirichlet problem 6
   weak 31
Dirichlet–Neumann bracketing 69
Dirichlet-to-Neumann map 214
   for Helmholtz equation 254
divergence 12, 289
domain monotonicity 65
Donnelly–Fefferman bound 119
   local version 123
doubling index
   for balls 118
   for cubes 126

E

eigenfunction 4
eigenvalue 4
elliptic bootstrapping 42
elliptic regularity
   global 46
   local 41
equimeasurable functions 141

F

Faber–Krahn inequality 138
FEM 265
   degrees of freedom 268
   mesh 267
   nodes 267
finite elements
   conforming 268
   Lagrangian 268
   linear 268
   quadratic 268
   triangular 267
Fraenkel asymmetry 145
frequency function 121
Friedlander–Filonov inequality 79
Friedrichs extension 35
fundamental gap 180

G

Galerkin method 266
Gauss's circle problem 18
genus 168
geodesic ball 118

Glazman's lemma 78
gradient 12, 289
Green's formula 30

H

Hardy–Littlewood–Karamata theorem
   191
harmonic
   extension 214
   function 2, 107
Hausdorff measure 117
heat equation 2, 138, 183
   parametrix 187
heat invariants 190
heat kernel 184
heat trace 186
height function 120
Helmholtz extension 253
Hersch's
   lemma 167
   theorem 165
Hersch–Payne–Schiffer inequality ..225
Hile–Protter inequality 173
homogeneous polynomials 21
   harmonic 21
Hörmander's identity 232

I

injectivity radius 125
isospectral
   domains 200
   manifolds 193
   mixed boundary conditions 201
   Steklov 236

K

Klein bottle 171
Korevaar's bound 172
Krahn–Szego inequality 147

L

Laplace equation 2
Laplace operator . *see also* Laplacian, 1
Laplace–Beltrami operator 15
   isothermal coordinates 16

on the sphere 21
Laplacian 1
    Dirichlet 36
    Neumann 38
    p-Laplacian 158
    polar coordinates 9
    Robin 61
    Zaremba 64
lattice 194
layer cake representation 141
length spectrum 210
level set 94
lifting trick 41, 123
Lipschitz boundary 290
Lipschitz function 290

M

mass matrix 267
McKean's inequality 154
mean over a ball 108
method of difference quotients 42
Milnor's example 194
Minakshisundaram–Pleijel expansion
    189
multiplicity 7, 34
    bounds 133

N

Neumann boundary conditions 6, 8
nodal
    deficiency 148
    domain 99
    graph 134
    set 99
nonperiodicity condition 88
normal covering 197
normal derivative 8

O

operator
    adjoint 34
    domain of 33
    extension 33
    positive 35
    selfadjoint 34
    semi-bounded 34
    symmetric 34

P

Payne–Pólya–Weinberger inequality 173
Payne–Weinberger inequality 180
Peters solution 245
Pleijel
    constant 149
    nodal domain theorem 148
Pohozhaev's identity 231
Poincaré's inequality 33
Poisson equation 2
Poisson summation formula 194
Pólya's conjecture 89, 90
Pólya–Szegő
    conjecture 146
    principle 142
principle of not feeling the boundary
    192

Q

quadratic form 55
quantum graph 237
quasi-continuous 104
quasi-eigenvalues 239
quasi-everywhere 104
quasi-mode 246

R

Rayleigh quotient 57
rearrangement
    symmetric decreasing, of a function
        140
    symmetric, of a set 137
reduced inradius 155
Rellich identity 261
Rellich–Kondrachov theorem 28
resolvent 34
Riemannian metric 12
Robin problem 61
Robin–Dirichlet-to-Neumann duality
    256

S

Schrödinger equation 2
Schwartz space 290
sloping beach problem 243
sloshing

problem 219
surface 219
Sobolev
   embedding theorem 28
   space 27
      on a Riemannian manifold 39
   trace theorem 30
spectral invariant 190
spectral prescription 180
spectral theorem
   strong 52
   weak
      Dirichlet 32
      Neumann 38
      on a Riemannian manifold 39
spectrum
   discrete 34
   essential 34
spherical harmonics 21
spherical mean 108
standing wave 4
Steiner symmetrisation 146
Steklov problem 214
stiffness matrix 266
Sturm–Liouville problem 4, 102
subharmonic function 107
sublevel set 94
sum of squares function 16
Sunada
   construction 197
   triple 198
superlevel set 94
Szegő–Weinberger inequality 159

T

Thomas–Reiche–Kuhn sum rule 177
tiling domain 91
torsional rigidity 146
transplantation of eigenfunctions ...201

U

uniformisation theorem 166
universal inequalities 174

V

vanishing order 119
variational principle

Dirichlet Laplacian 59
Dirichlet-to-Neumann map 255
Laplace–Beltrami operator 60
Neumann Laplacian 60
quadratic form 57
Steklov problem 218

W

wave equation 2
   one-dimensional 3
weak derivative 26
weak solution 31, 32, 37, 42
Weinstock inequality 224
Weyl's conjecture 87
Weyl's law 82
   on a Riemannian manifold 191
   Steklov problem 230

Y

Yang's inequalities 174
Yang–Yau bound 169
Yau's conjecture 117

Z

Zaremba problem 63

Selected Published Titles in This Series

237 **Michael Levitin, Dan Mangoubi, and Iosif Polterovich,** Topics in Spectral Geometry, 2023

235 **Bennett Chow,** Ricci Solitons in Low Dimensions, 2023

232 **Harry Dym,** Linear Algebra in Action, Third Edition, 2023

231 **Luís Barreira and Yakov Pesin,** Introduction to Smooth Ergodic Theory, Second Edition, 2023

230 **Barbara Kaltenbacher and William Rundell,** Inverse Problems for Fractional Partial Differential Equations, 2023

229 **Giovanni Leoni,** A First Course in Fractional Sobolev Spaces, 2023

228 **Henk Bruin,** Topological and Ergodic Theory of Symbolic Dynamics, 2022

227 **William M. Goldman,** Geometric Structures on Manifolds, 2022

226 **Milivoje Lukić,** A First Course in Spectral Theory, 2022

225 **Jacob Bedrossian and Vlad Vicol,** The Mathematical Analysis of the Incompressible Euler and Navier-Stokes Equations, 2022

224 **Ben Krause,** Discrete Analogues in Harmonic Analysis, 2022

223 **Volodymyr Nekrashevych,** Groups and Topological Dynamics, 2022

222 **Michael Artin,** Algebraic Geometry, 2022

221 **David Damanik and Jake Fillman,** One-Dimensional Ergodic Schrödinger Operators, 2022

220 **Isaac Goldbring,** Ultrafilters Throughout Mathematics, 2022

219 **Michael Joswig,** Essentials of Tropical Combinatorics, 2021

218 **Riccardo Benedetti,** Lectures on Differential Topology, 2021

217 **Marius Crainic, Rui Loja Fernandes, and Ioan Mărcuţ,** Lectures on Poisson Geometry, 2021

216 **Brian Osserman,** A Concise Introduction to Algebraic Varieties, 2021

215 **Tai-Ping Liu,** Shock Waves, 2021

214 **Ioannis Karatzas and Constantinos Kardaras,** Portfolio Theory and Arbitrage, 2021

213 **Hung Vinh Tran,** Hamilton–Jacobi Equations, 2021

212 **Marcelo Viana and José M. Espinar,** Differential Equations, 2021

211 **Mateusz Michałek and Bernd Sturmfels,** Invitation to Nonlinear Algebra, 2021

210 **Bruce E. Sagan,** Combinatorics: The Art of Counting, 2020

209 **Jessica S. Purcell,** Hyperbolic Knot Theory, 2020

208 **Vicente Muñoz, Ángel González-Prieto, and Juan Ángel Rojo,** Geometry and Topology of Manifolds, 2020

207 **Dmitry N. Kozlov,** Organized Collapse: An Introduction to Discrete Morse Theory, 2020

206 **Ben Andrews, Bennett Chow, Christine Guenther, and Mat Langford,** Extrinsic Geometric Flows, 2020

205 **Mikhail Shubin,** Invitation to Partial Differential Equations, 2020

204 **Sarah J. Witherspoon,** Hochschild Cohomology for Algebras, 2019

203 **Dimitris Koukoulopoulos,** The Distribution of Prime Numbers, 2019

202 **Michael E. Taylor,** Introduction to Complex Analysis, 2019

201 **Dan A. Lee,** Geometric Relativity, 2019

200 **Semyon Dyatlov and Maciej Zworski,** Mathematical Theory of Scattering Resonances, 2019

199 **Weinan E, Tiejun Li, and Eric Vanden-Eijnden,** Applied Stochastic Analysis, 2019

198 **Robert L. Benedetto,** Dynamics in One Non-Archimedean Variable, 2019

For a complete list of titles in this series, visit the
AMS Bookstore at **www.ams.org/bookstore/gsmseries/**.